THE MEANING OF A FORMAT

MP3

Jonathan Sterne

DUKE UNIVERSITY PRESS DURHAM AND LONDON 2012

© 2012 Duke University Press
All rights reserved.
Printed in the United States of
America on acid-free paper ∞
Designed by Amy Ruth Buchanan
Typeset in Minion by Tseng
Information Systems, Inc.
Library of Congress Cataloging-in-
Publication Data appear on the last
printed page of this book.

MP3

SIGN, STORAGE, TRANSMISSION

A series edited by Jonathan Sterne and Lisa Gitelman

FOR CARRIE

The organic has become visible again even within the mechanical complex: some of our most characteristic mechanical instruments—the telephone, the phonograph, the motion picture—have grown out of our interest in the human voice and the human eye and our knowledge of their physiology and anatomy. Can one detect, perhaps, the characteristic properties of this emergent order—its pattern, its planes, its angle of polarization, its color? Can one, in the process of crystallization, remove the turbid residues left behind by our earlier forms of technology? Can one distinguish and define the specific properties of a technics directed toward the service of life: properties that distinguish it morally, socially, politically, esthetically from the cruder forms that preceded it? Let us make the attempt.—**Lewis Mumford**, *Technics and Civilization*

CONTENTS

Acknowledgments ix

Format Theory 1

1. Perceptual Technics 32

2. Nature Builds No Telephones 61

3. Perceptual Coding and the Domestication of Noise 92

4. Making a Standard 128

5. Of MPEG, Measurement, and Men 148

6. Is Music a Thing? 184

The End of MP3 227

Notes 247

List of Interviews 295

Bibliography 299

Index 331

ACKNOWLEDGMENTS

This was supposed to be a short book, a brief comment on a contemporary phenomenon. Instead, I spent the better part of seven years following a major thread in twentieth-century sound history. The book was conceived during an international move. Parts of it have been workshopped in more places than I can count, and it weathered an unplanned temporary detour into departmental administration and an aggressive case of thyroid cancer (in case you are wondering, I recommend the former over the latter). It is impossible to name all the people whose presence or words resonate somewhere in this book. But since this is the acknowledgments section I will give it a try. My apologies for any omissions—they are my mistake.

I must break convention and begin by thanking the love of my life, she of the legendary laugh, the endlessly creative and twisted Carrie Rentschler, to whom this book is dedicated. I have relished our time together over the last twenty-odd years, whether it was dining in, rocking out, or traveling the world. Her company makes everything in life sweeter—even the bitter parts. I am also grateful for her intellectual companionship. No person has read or heard more versions of this book. Her comments have probably touched every page of this book at each stage of its conception. Neither of us needs to say, "now that this book is done, we can finally spend some more time together." But I fulfilled some promises for wine-tasting trips and a lot of time in a saltwater pool between finishing the book and leaving California.

In the same spirit, I would also like to thank my family, all of whom have offered me unconditional support and endless conversation—intellectual and otherwise—as well as needed perspective on what matters in life: my mom and stepdad, Muriel Sterne and Philip Griffin; my brother, sister-in-law, niece, and nephew, David, Lori, Abby, and Adam Sterne; my aunt and uncle, Helen and Mario Avati; and my in-laws, Kay Larson and Judy Anderson and Louis and Dianne Rentschler. Phil, Helen, and Mario didn't live to see the book completed, but I keep reminders of them with me.

In Montreal I have been blessed with a large community that is at once intellectual and personal. The Department of Art History and Communication Studies has been a happy institutional home for me and I owe thanks to all my colleagues. Darin Barney, Jenny Burman, Cecily Hilsdale, Tom Lamarre, Andrew Piper, Marc Raboy, Will Straw, and Angela Vanhaelen all have had conversations with me that directly shaped something in these pages. The rest of my AHCS colleagues, past and present, have provided a wonderful intellectual community, an usually collegial workplace, and good friendship and inspiration as well. I can't possibly measure their good influence on me: Ting Chang, David Crowley, Mary Hunter, Amelia Jones, Becky Lentz, Hajime Nakatani, Charmaine Nelson, Christine Ross, Richard Taws, and Bronwen Wilson. Cornelius Borck, James Delbourgo, and Nick Dew provided a community for me in the History and Philosophy of Science Program/science and technology studies (I am still unsure what it is) and exposed me to countless eras and ideas I wouldn't have otherwise encountered—they are also all wonderful friends and colleagues. I also must thank the AHCS staff, past and present, who not only supported my work from time to time but also became very close colleagues and confidantes in my time as chair. Thanks to Maureen Coote, Matt Dupuis, Maria Gabriel,

Susana Machado, and Jennifer Marleau—I couldn't have made it through my term without them.

Many postdocs, graduate student research assistants, and undergraduate students have helped me out at one stage or another. What is that cliché about learning through teaching? My graduate research assistants were Mitch Akiyama, Mike Baker, Adam Bobbette, Didier Delmas, Daniel Moody-Grigsby, Jeremy Morris, Ariana Moscote Freire, Dylan Mulvin, Matt Noble-Olson, Lilian Radovac, Emily Raine, and Tara Rodgers. Many became collaborators on other projects in the process, and I have learned so much from each of them. Emily and Dylan require special thanks for massive amounts of assistance on editing and checking the manuscript as it moved into publication. Other students (and people whom I first met as students but are now onto other things) helped shape ideas in these pages: Hugh Curnutt, Rob Danisch, Sandra Duric, Saalem Humayun, Erin MacLeod, Heather Mills, Jess Mudry, Zack Furness, Steve Gennaro, Tim Hecker, Randolph Jordan, Gyewon Kim, Paulo Maggauda, Ian Reyes, Danielle Schwartz, Raji Sohal, Davide Tidone, tobias van Veen, and Tom Wilder. I am grateful to my postdocs Damien Charrieras, Carlotta Darò, and John Shiga for seeking me out and for the intellectual enrichment they've provided. Simone Pereira de Sà defied categories and added a lot to the semester she spent with me. I have also been blessed with countless spectacular undergraduates, but a few deserve special thanks for their research assistance and for seeking me out for the opportunity: Stephanie Dixon and David Machinist, who both did research for the book; and especially Agent 99, who taught me binary math and helped me chart an early course through the technical literature.

In addition to my research assistants, I got help from lots of other people. Two librarians in particular—Adam Lauder and Cynthia Lieve—were more like colleagues in the process and tracked down all manner of materials for me. I am grateful to all the people I interviewed about this project for the gift of their time, and I hope I have represented their experiences accurately, if not exactly in the manner they might have imagined. Janice Denegri-Knott sent me a long draft of an essay on the business history of the format out of the blue. And my thinking in the last chapter was shaped in part by my experience co-teaching a master's-level seminar on the future of the music business. I'm grateful to all my colleagues and students over a few years in the course, but especially to David Lametti and

Tina Piper for conversations on intellectual property, and Sandy Pearlman for conversations on the industry and technology.

Ideas in this book have been workshopped in dozens of talks, and I am grateful to my hosts and interlocutors, many new friends I made in the process of writing this book as well as some cherished old friends, teachers, and colleagues with whom I spent time during my travels. Many of their comments and questions were scribbled down on manila folders (or explained to me in e-mails) and have made their way into the book in one form or another. I have no doubt they make me seem more clever than I am. So thanks go out to: Adriana Amaral, Joe Auner, Anne Balsamo, Eric Barry, Nancy Baym, Jody Berland, Wiebe Bijker, Karin Bijsterveld, Georgina Born, Sandra Braman, Jack Bratich, Michael Bull, Jean Burgess, José Cláudio Castanheira, Nicholas Cook, Brady Cranfield, Kate Crawford, Cathy Davidson, Peter Doyle, Catherine Driscoll, Nina Eidsheim, John Erni, Ben Etherington, Bob Fink, Lawson Fletcher, Aaron Fox, Kelly Gates, Tarleton Gillespie, John Michael Gomez Connor, David Goodman, Sumanth Gopinath, Ron Greene, Jocelyne Guilbault, James Hay, Chris Healey, Alison Hearn, John Heijmans, Lisa Henderson, Karen Henson, Michele Hilmes, Brian Horne, Phil Howard, Alison Huber, Myles Jackson, Steve Jones, Jonathan Kahana, Doug Kahn, Anahid Kassabian, Mark Katz, Julia Kursell, Jean-Marc Larrue, Joan Leach, Tim Lenoir, Takura Mizuta Lippit, Justine Lloyd, Alex Mawyer, Meredith McGill, Kembrew McLeod, Tara McPherson, Louise Meintjes, Jairo Moreno, John Mowitt, Lisa Nakamura, Jay Needham, Gina Neff, Dave Novak, Ana Maria Ochoa Gautier, Lisa Parks, Vinicius Pereira, John Durham Peters, Trevor Pinch, Elspeth Probyn, Ron Radano, Jan Radway, Tom Rice, Aimee Rickman, James Riley, Gil Rodman, Christian Sandvig, Kim Sawchuk, Annette Schlichter, Henning Schmidgen, Jeff Sconce, Greg Siegel, Greg Siegworth, Jennifer Daryl Slack, Zoë Sofoulis, Lynn Spigel, Carol Stabile, Jason Stanyek, Charley Stivale, Ted Striphas, Jordan Strom, Peter Szendy, Tim Taylor, John Tebbutt, Catherine Thill, Emily Thompson, Graeme Turner, Greg Wise, Roland Wittje, and ShinJeong Yeo.

Once the book was drafted, it went out for review and I owe many thanks to my two anonymous reviewers. But I also owe thanks to a small group of people who read the book carefully, challenged me on its contents, caught errors, and really helped me improve it. Besides just being a great friend throughout the process, Charles Acland saw ways to extend my arguments about formats and think more broadly about media history. Larry Grossberg challenged me on a number of theoretical points, pushed me on the bigger stakes, and got me to coin concepts instead of using other, more imprecise terms. Lisa Gitelman provided helpful comments on form, format theory, and style (thanks to her, the book now has a sharper, more incisive beginning, and only one of them), and plenty of intellectual inspiration. Fred Turner really pushed me on form and organization of the manuscript, my conceptions of cybernetic and new media history, and my own authorial voice. He convinced me to apply to go to California for sabbatical, and then he and his partner Annie Fischer were incredible hosts when Carrie and I got there, introducing us to everyone they could think of (not to mention Los Charros). I can basically thank him for that entire year of my life, as well as his companionship in it. Mara Mills probably knows more about the history of psychoacoustics and signal processing than anyone else alive. By good fortune we overlapped for six months at Stanford, and I benefited from many hours of conversation with her over the winter and spring (not to mention the field trips). I expect and hope that one day soon our books will sit on the shelf next to one another.

Liz Springate came late to the project as my illustrator, initially in a technical role, redrawing some images for inclusion in the book. But quickly her role became more creative as we batted around ideas for illustrating concepts, or better, creating their visual representations to complement my verbal representations. My only regret is that she wasn't on board sooner and that it wasn't a fuller collaboration. Liz has convinced me of the value of working with visual artists in my academic endeavors, and I hope we can collaborate again soon.

Ken Wissoker always gets thanked for his patience and support, so I must confess to a certain unoriginality there. Ken's DJ skills have somehow transferred into publishing—he's got an ear for scholarship. One of the book's origin stories has the idea first emerging at a dinner we had together with several others, so he's seen it from genesis to completion. I've enjoyed working with him and worrying together about the future of academic publishing. And for the record, I still want to write a short book someday.

My "classmates" at the Center for Advanced Study in the Behavioral Sciences, as well as friends I made (or simply got to enjoy more) while in California, offered both intellectual and social companionship and helped me pass into the liminal state I now occupy. I would especially like to thank Jonathan Abel, Jim Blascovich, Elizabeth and Kurt Borgwardt, Don Brennis, Wendy Brown, Gary Fine, Wynne Furth, Joe Gone, Charles Hirschkind, Allen and Bobbie Isaacman, Lochlann Jain, Heather Love, Steve Macedo, Saba Mahmood, Brenda Major, Tori McGreer, Rhacel Parreñas, Damon Phillips, Robert Proctor, Arvind Rajagopal, Anu Rao, Enrique Rodriguez Allegría, Steven Rubio, Nancy Whittier, Kate Wiegand, and Eric Worby.

A special thank-you goes out to the people with whom I've played music (and in the classic academic fashion of intertwining work and life, *also* talked ideas) over the past seven years: Nick King and Michael Witmore as my musical partners-in-crime; as well as the for-now anonymous members of two bands in California—may we never appear on YouTube.

In addition to people whose roles I can list precisely, there is a group of awesome friends and colleagues not named elsewhere in these acknowledgments who deserve special thanks for being there in one or many ways: Amy Alt, Elaine Arsenault, Steve Bailey, Lisa Barg, Shannon and Craig Bierbaum, David Brackett, Dave Breeden, Kevin Carollo, Nicole Couture, Brian Cowan, James Delbourgo, Manon Derosiers, Jill Didur, Greg Dimitriadis, Deb Dysart-Gale, Jennifer Fischman, Lisa Friedman, Yuriko Furuhata, Loretta Gaffney, George Gale, Kelly Gates, Jayson Harsin, Michelle Hartman, Toby Higbie, Adrienne Hurley, Nan Hyland, Nick King, Sammi King, Laura Kopp, Christine Lamarre, Marie Leger, Peter Lester, Lisa Lynch, Steve Macek, Shoshana Magnet, Bita Mahdaviani, Setrag Manoukian, Dan McGee, Negar Mottahedeh, Dérèq Nystrom, Natalie Oswin, Laila Parsons, Elena Razlogova, Tom and Andrea Robbins, Kellie Robertson, Joseph Rosen, Lorna Roth, Mark Rubel, Jonathan Sachs, Ned Schantz, Wayne Schneider, Bart Simon, Johanne Sloan, Robert Smith-Questionmark, Gretchen Soderlund, Marc Steinberg, Jeremy Stolow, Greg Taylor, Alanna Thain, Andrea Tone, Mrak Unger, Haidee Wasson, Andrew Weintraub, and Robert Wisnovsky.

I would also like to thank by initial a few cancer friends: M, especially, for all the sambar, gossip, and walks over to Zone; also D. S., E. V. J., R. B., D. Z., as well as the whole crew at Thyroid Cancer Canada. I also want to extend thanks to my doctors: Dr. Michael Tamilia, who finally figured out what it was, the Dashing Dr. Ricky Payne™ who oversaw everything, Glenda Falovitch, and Drs. Jacques How, Khalil Sultanem, and Roger Tabbeh as well as all the nurses and hospital staff who cared for me when I was at the Jewish General and Montreal General Hospitals.

Parts of this book have appeared elsewhere in different iterations. A test run of the concept behind the project appeared as "The MP3 as Cultural

Artifact," *New Media and Society* 8, no.5 (November 2006): 825–42, though I've changed my mind on lots of things since then. A more sustained critique of "discontinuity thesis" that humanists apply to digital audio can be found in "The Death and Life of Digital Audio," *Interdisciplinary Science Reviews* 31, no. 4 (December 2006): 338–48. A very compressed version of the Wever-Bray story from chapter 2 appeared as "The Cat Telephone," *The Velvet Light Trap*, no. 64 (fall 2009): 83–84. A condensed version of chapter 5, with some different emphases, appears as "What the Mind's Ear Doesn't Hear," in *Music, Sound and the Transformation of Public and Private Space*, ed. Georgina Born and Tom Rice (Cambridge: Cambridge University Press, 2012, forthcoming as of this writing). And an earlier version of some of the arguments in chapter 6 appears as "How the MP3 Became Ubiquitous," in *The Oxford Handbook of Mobile Music*, ed. Sumanth Gopinath and Jason Stanyek (New York: Oxford University Press, 2012, forthcoming as of this writing). Thanks to my editors and reviewers for their comments and support.

Finally, I want to thank the institutions that generously supported my work on this book and related projects. This book has benefited from financial support provided by McGill University's new faculty startup funds, the Beaverbrook Fund for Media@McGill, the Social Sciences and Humanities Research Council of Canada (SSHRC), and the Fonds québécois de recherche sur la société et la culture (FQRSC). I am also grateful for the generosity of the many institutions that paid for me to talk about my work in progress; and the Annenberg Foundation Fellowship at the Center for Advanced Study in the Behavioral Sciences at Stanford, which provided an exceptional space in which to finish it. It is an incredible privilege to have access to this kind of support, and as a recent immigrant to Quebec and Canada, I have an extra measure of wonder and gratitude for governments and people who put so much behind ideas and culture. Scholars exist because of the institutions that nurture and support independent thought and basic research. Their support is a gift, which we should return in kind by honoring our responsibilities to them, to one another, to our students, and to the people they hope to serve in their best, most ambitious moments.

FORMAT THEORY

At the moment I write this sentence, there are about 10 million people in the world sharing music files over Gnutella, a peer-to-peer file-sharing network. Over the course of a month, that network will move 1 to 1.5 billion music files—and that is only part of its traffic. Add the people who trade MP3 files via BitTorrent, one-click sites like Rapid-Share, and older protocols like Internet Relay Chat and Usenet, and the Gnutella network statistics capture only a small proportion of the overall traffic in MP3s. Measured against traditional CD sales or paid-for downloads, online music sharing alone dwarfs the output of the recording industry. The MP3 is the most common form in which recorded sound is available today. More recordings exist and circulate in MP3 format than in all other audio formats combined. A single file on a single network may be available simultaneously in dozens of countries, without regard for local laws, policies, or licensing agreements.[1] Chances are, if a recording takes a ride on the internet, it will travel in the form of an MP3 file. The traffic in MP3s highlights the distributed character of culture in our age.[2]

The MP3 is a triumph of distribution, but it is also something more.[3] MP3s are so plentiful because they are so small. They use considerably less bandwidth and storage space than the .wav files one finds on standard compact discs. To make an MP3, a program called an encoder takes a .wav file (or some other audio format) and compares it to a mathematical model of the gaps in human hearing. Based on a number of factors—some chosen by the user, some set in the code—it discards the parts of the audio signal that are unlikely to be audible. It then reorganizes repetitive and redundant data in the recording, and produces a much smaller file—often as small as 12 percent of the original file size.[4] The technique of removing redundant data in a file is called *compression*. The technique of using a model of a listener to remove additional data is a special kind of "lossy" compression called *perceptual coding*.[5] Because it uses both kinds of compression, the MP3 carries within it practical and philosophical understandings of what it means to communicate, what it means to listen or speak, how the mind's ear works, and what it means to make music. Encoded in every MP3 are whole worlds of possible and impossible sound and whole histories of sonic practices. Perhaps every sound technology in human history contains within it some model or script for hearing and an imagined, ideal auditor.[6] But MP3 encoders build their files by calculating a moment-to-moment relationship between the changing contents of a recording and the gaps and absences of an imagined listener at the other end. The MP3 encoder works so well because it guesses that its imagined auditor is an imperfect listener, in less-than-ideal conditions. It often guesses right.

Although it is a ubiquitous and banal technology, the MP3 offers an inviting point of entry into the interconnected histories of sound and communication in the twentieth century. To access the format's historical meaning, we need to construct a new genealogy for contemporary digital media culture.[7] Many of the changes that critics mark as particularly salient aspects of contemporary digital or "new" media happened in audio before they surfaced in visual media. As Frances Dyson writes, digital media encapsulate "an accumulation of the auditive technologies of the past." The historical resonance of audio can be extended across the various registers of new media, from their sensual dimensions in both the auditory and visual domains, to their treatment of subjects, to their technical structure and industrial form.[8]

Today, MP3s swim in the waters of the internet and street-level informal economies, but they also carry the traces of other infrastructures. They point to the centrality of telephony to digital history—as an industry, as a set of practices, as an aesthetic field, and as a medium.[9] Telephony is often considered anaesthetic matter in comparison with the usual, more aestheticized subjects of twentieth-century media history such as cinema, television, sound recording, radio, print, and computers. But telephony and the peculiar characteristics of its infrastructure are central to the sound of most audio technologies over the past 130-odd years. The institutional and technical protocols of telephony also helped frame the definitions of communication that we still use, the basic idea of information that subtends the whole swath of "algorithmic culture" from packet switching to DVDs and games, and the protocols and routines of digital technologies we use every day. The centrality of sound history to digital history remains clouded in large swaths of the human sciences, but open any relevant engineering text and this point is immediately clear.[10]

Telephony has also played a crucial role over the last hundred years in shaping our most basic notions of what it means to hear. This has been true both at the level of high science and everyday conversation. The questions, protocols, and findings of much modern hearing research developed out of pressing issues facing the phone system in the early twentieth century. What was the minimum amount of signal that could be sent down the line and still be intelligible as speech? What parts of the audible spectrum were important to hear for intelligibility and which were not? How did the listener's ear react to different sonic conditions, and how did its own processes relate to the economic and technological imperatives of the telephone system? Telephonic transmission drove research into hearing for much of the century and, as a result, shaped both what it means to hear and the notion of the hearing subject that subtends most new audio technologies today. In historical terms, MP3s are usually thought of as an extension of sound recording. Perhaps because of their association with art and music, other kinds of sound recording provide a universe of aesthetic reference for MP3s. But recording owes a tremendous technological and aesthetic debt to telephony. Each major technical iteration of sound recording made use of telephone research: the first phonographs were built in labs funded by telephonic (and telegraphic) research; the first electrical recording and playback technologies were borrowed from innovations in telephone systems in the 1920s; and digital audio recording and playback also used concepts that emerged from AT&T's research wing, Bell Labs, beginning in the 1920s.[11] The telephone system's relationship with the history of hearing is therefore a major theme of this book. If we look into the code of the MP3 for its imagined listening subject, this subject is at least as telephonic as it is phonographic or digital.[12] It owes a debt to radio as well.

In subjecting digital history to the history of sound, we also solve a puzzle that bothers many audiophiles and recording professionals. They have long complained that MP3s have less definition than the CD recordings from which they are made and are therefore said to not sound as good.[13] In an age of ever-increasing bandwidth and processing power, why is there also a proliferation of lower-definition formats? If we have possibilities for greater definition than ever before, why does so much audio appear to be moving in the opposite direction?[14] Presumably, progress should move in one direction. With ever-growing hard-drive sizes, it is now possible to store as many .wav files as one could once accumulate only in MP3 format. Higher-bandwidth systems for moving data are around the corner, or so we are told, where the transmission bottlenecks that necessitate small files will also disappear. By this logic, the MP3 should already be a thing of the past. Some critics describe the format as a regression in terms of sound quality and listener experience. In a recent *Wall Street Journal* article, the political economist Francis Fukuyama waxed nostalgic for the experience of listening to records on a "crazily expensive" Linn Sondek turntable system with the conservative cultural critic Allan Bloom in the 1980s. Of digital audio and photography, he writes, "Don't believe the marketing hype of the techie types who tell you that newer is always better. Sometimes in technology, as in politics, we regress."[15]

There is little to disagree with in Fukuyama's proposition, until you consider his standard of progress. The juxtaposition of decreasing actual definition with increasing available definition is only regressive (or paradoxical) if you assume that the general historical trend of progress in communication technology is toward ever-greater definition and therefore greater verisimilitude. Within that proposition hide a few other common assumptions: (1) that greater definition is the same thing as greater verisimilitude; (2) that increases in definition necessarily enhance end-users' experiences; (3) that increases in bandwidth and storage capacity necessarily lead to higher-definition media for end-users. The account of communication history implied by these propositions outlines a quest for definition, immersion, and richness of experience. Call it the dream of verisimilitude. We are perhaps most familiar with this story because it is also the story used to market new technologies to us as consumers, although it is also quite prevalent in academic and journalistic commentaries on technology. The idea that a new medium is closer to reality and more immersive and interactive than its predecessor is based on this story—whether the new medium in question is the Orthophonic Victrola, compact discs, high-definition TV, virtual reality, or 3D cinema. This idea can be projected backward to the earliest cave paintings and musical instruments, but it generally references the rise of mechanical, electric, and electronic media. Of course, claims about increased definition or the pleasures of immersion are in many cases true, but they are not the whole story.

Aesthetic pleasure, attention, contemplation, immersion, and high definition—these terms have no necessary relationship to one another. They can exist in many different possible configurations. As Michel Chion has argued, definition is not the same thing as correspondence to reality or fidelity: "Current practice dictates that a sound recording should have more treble than would be heard in the real situation (for example when it's the voice of a person at some distance with back turned). No one complains of nonfidelity from too much definition! This proves that it's definition that counts for sound, and its hyperreal effect, which has little to do with the experience of direct audition."[16] Writers have documented how audiences often prefer various kinds of audible distortion in music or speech to an undistorted alternative, and scholars across the human sciences have commented on the affective intensity of low-definition experiences.[17] In recent years, social psychologists and ethnomusicologists have moved from the expectation that their subjects treat music as an end in itself to the presupposition that listening is usually part of a broader sphere of social activity. Radio historians have revised the romantic memory of families in the 1930s crowded around radio sets, insisting that distracted listening—listening while doing other things—was just as prevalent as listening with rapt attention.[18] Even a sound-quality obsessed professional like the mastering engineer Bob Katz has acknowledged that distracted listening is the norm from which careful attention deviates: "There are, of course, specific places where heavy compression is needed: background listening, parties, bar and jukebox playback, car stereos, headphone-wearing joggers, the loudspeakers at record stores, headphone auditioning at the record store kiosk, and so on. . . . I dream of a perfect world where all the MP3 singles are heavily compressed and all the CD albums undamaged."[19]

The history of MP3 belongs to a *general history of compression*.[20] As people and institutions have developed new media and new forms of representation, they have also sought out ways to build additional efficiencies into channels and to economize communication in the service of facilitating greater mobility. These practices often begin close to economic or technical considerations, but over time they take on a cultural life separate from their original, intended use. As with the quest for verisimilitude, compression practices have created new kinds of aesthetic experiences that come to be pleasurable in themselves for some audiences—from the distortion that is a side effect of electrical amplification in radio, phonography, and instrument amplification, to the imagined intimacy of the phone conversation, to the mash-ups that aestheticize the MP3 form and the distribution channels it travels. Histories of contemporary immersive technologies point to a multiplicity of antecedents, going back centuries, even millennia.[21] A general history of compression also connects contemporary practices that are self-consciously understood in terms of compression with a broader history of practices that share the same morphology.

Compression history cuts a wide path through the forest of technological history. It is no accident that so many media technologies are built around spinning mechanisms. A roll of film or tape takes up less room than if it is stretched out from end to end. We could say the same for compact discs and DVDs, the platters of a hard drive, the curves of a record, the spools of paper in teletypes and telegraphs, and the spinning hands of analog clocks. Volume compression figured centrally in radio stations' occupation of slots on the electromagnetic spectrum. Color reduction was central to chromolithography, and to some kinds of painting before it. Telegraphic codes like Morse's were also built around compression in design and in use. The most common letters were the easiest to tap, and telegraphers quickly learned to tap out abbreviations for longer words. Compression history could easily extend back to the invention of the point and the number zero, the codex and the scroll form of the book, the wheel, and perhaps even some kinds of ancient writing and number systems.[22]

Considering the histories of communication and representation in terms of compression also turns our attention from media to formats. It is no accident that a format like the MP3 invites us into compression history. If your goal in designing a technology is to achieve some combination of channel efficiency and aesthetic experience, then the sensual and technical shape of your technology's content is every bit as important as the medium itself. Problems of format have been central to both histories—compression and verisimilitude—and perhaps to others as well. Formats are particularly acute as technical and cultural problems for the compressors, who worry over inefficiencies in the mechanics of transduction, storage, and transmission alongside creation, distribution, and reception. Format problems are obvious material problems in the analog domain, but the same issues obtain in digital media. Software and data have their own materialities, even if their scale seems inhuman. As Matthew Kirschenbaum points out, hard drives are designed to hide their process of magnetic inscription from users' sight, and the result is that invisibility has been conflated with immateriality.[23] Software formats like MP3s exist in the form of configurations of electromagnetic signals. They take up space, which is why MP3s are measured in terms of the bandwidth they require when played back, that is, in kilobits per second (kbps).

The MP3 format may be less than twenty years old, but it orients us toward a much longer history. The MP3 is in some ways a quintessential "new" communication technology with its pedestrian use of computing power that would have been unavailable to individuals a decade before, its nonlinearity, its flexibility, and its careful tuning for limited-bandwidth situations like computer networks, hard drives, and flash memory. But "new" is something of a misnomer, since digital technologies have now been around for over half a century.[24] As much as the MP3 is a creature of a digital world, it is also a creature of the world of electronic (and even electric) transmission. Many of the technological principles behind the MP3 are quite old. The Fourier Transform, originally developed as a mathematical theory of heat, showed that any complex waveform was the sum of more basic sine waves. It is the mathematical basis both for calculating sound frequencies as numbers and a theoretical base line for all forms of digital audio. A simplified function, the Fast Fourier Transform, is also the mathematical basis for some of the important signal-processing innovations in the MP3. Hermann von Helmholtz's dream of unifying the art and science of music through studies of acoustics, sound generation, and perception also shows up in the MP3's algorithm. The ideas of hearing built into MP3s go back decades to Bell Labs. The format became historically legible when its standard was published in 1993, but it is part of much longer historical trajectories.

From Mediality to Formats

If there is such a thing as media theory, there should also be format theory. Writers have too often collapsed discussions of format into their analyses of what is important about a given medium. *Format* denotes a whole range of decisions that affect the look, feel, experience, and workings of a medium. It also names a set of rules according to which a technology can operate. In an analog device, the format is usually a particular utilization of a mechanism. An old record player may play back a variety of formats such as LP, 45, 78, while a tape deck might only take compact cassettes. In a digital device, a format tells the operating system whether a given file is for a word processor, a web browser, a music playback program, or something else. Even though this may seem trivial, it can open out to a broader politics, as an administrative issue across platforms.[25] For instance, digital audio files may appear in formats like .MP3, .wav, .aif, .aac, .ogg, .rm, .aup, and .wma (and many other audio and audiovisual formats); text files may appear as .doc, .ood, .wpd, or even .pdf (which is actually an image file). Programs and devices may play back or display only certain of those file types. Most crucial dimensions of format are codified in some way—sometimes through policy, sometimes through the technology's construction, and sometimes through sedimented habit. They have a contractual and conventional nature. The format is what specifies the protocols by which a medium will operate.[26] This specification operates as a code—whether in software, policy, or instructions for manufacture and use—that conditions the experience of a medium and its processing protocols. Because these kinds of codes are not publicly discussed or even apparent to end-users, they often take on a sheen of ontology when they are more precisely the product of contingency.

Consider entities like "film" and "television," once regarded by scholars as quite stable, but today perhaps less obviously so. "Film" and "television" bundle whole sets of assumptions about the format of sound and image— assumptions that change over time. For instance, since the 1940s, North American analog television has been filmed and broadcast for a screen with a 4:3 horizontal-vertical aspect ratio because at the time that the ratio was set (suggested in 1936 and enshrined in policy in 1941), that was the ratio for screens for Hollywood films. Partly in an attempt to compete with television, Hollywood stepped up ongoing efforts to adopt wider screens. To this day North American analog televisions have a 4:3 aspect ratio, and audiovisual content from other media (such as film, which is often 1.85:1) or formats (such as high-definition, which is 16:9) is reformatted to fit the 4:3 TV screen when we watch it on analog TVs—either through letterboxing or reediting. In either case, the size and shape of the television image is clearly a central part of TV as an aesthetic experience; even casual viewers are made more aware of the different formatting possibilities in television when switching between new HD content and 4:3 reruns.[27] Even more important are things like screen size, sound quality, and color. Similarly, the contexts of exhibition have shaped the size of American film stock and the screens onto which films are projected: 70 mm, 35 mm, and 16 mm all imply different contexts of creation, distribution, and exhibition.[28]

This book endows a nineteen-year-old format with a century-long history in order to highlight some core dimensions of twentieth-century sound history. In the following pages, I consider the development and traffic in MP3s as a massive, collective meditation on the mediality of sound and especially hearing, music, and speech. I use the term *mediality* (and *mediatic* in adjectival form) to evoke *a quality of or pertaining to media* and the complex ways in which communication technologies refer to one another in form or content.[29] In a way, this is an old point. Marshall McLuhan famously wrote that "the 'content' of any medium is always another medium. The content of writing is speech, just as the written word is the content of print, and print is the content of the telegraph"; and Jay Bolter and Richard Grusin have coined the term "remediation" to describe "the representation of one medium in another . . . [to] argue that remediation is a defining characteristic of the new digital media."[30] However, both perspectives tend to focus on the "newness" of media in their definitions, rather than to convey a sense of cross-reference as routine. They also suggest a priori degrees of mediation where media follow one another in a march away from reality. This is perhaps an unfortunate effect of the relationship between the words *media* and *mediation*. But media are not middle terms, intervening in otherwise more primary, fundamental, or organic relationships. Their historical accumulation does not necessarily result in a greater distance from reality or a more refracted consciousness. Mediation is not necessarily intercession, filtering, or representation. Another sense of mediation describes a form of nonlinear, relational causality, a movement from one set of relations to another. As Adorno wrote, "Mediation is in the object itself, not something between the object and that to which it is brought."[31]

Mediality is a mundane term, like *literariness*; it indicates a general web of practice and reference.[32] Expressive forms like literature, poetry, art, cinema, and music refer to themselves in the sense that individual works may refer to others directly, or combine knowledge and practice from their fields (or other allied fields) in new ways. Understanding this web of reference is essential to understanding how they represent, figure, and organize broader realities and relationships. Mediality indicates a similar process in communication. It is a "general condition" in which sonic practices take shape.[33] In the places where MP3s prevail, it is not a question of communication technologies producing sound that is more or less mediated than some other sonic form, but rather that communication technologies are a fundamental part of what it means to speak, to hear, or to do anything with sound.[34] Mediality simply points to a collectively embodied process of cross-reference. It implies no particular historical or ontological priority of communicative forms.

To say that media refer to one another may already be somewhat misleading, since the boundaries of any medium are somewhat fuzzy and change over time.[35] Even the definition of *medium* is itself historically specific. For instance, we may today consider e-mail as a medium. But in 1974, it would likely have been subsumed under "computers" or some other hardware-based definition, despite the fact that mechanical and electronic media have always existed somewhat independently of their technological forms: both sound recording and sound film have existed in several technological forms simultaneously throughout their histories.[36] One of the most interesting aspects of contemporary media theory is that the connotative shadow of hardware looms large over any definition of *media* today, even though media forms, like e-mail, seem ever less attached to any specific form of hardware. You can do your e-mail on a computer, PDA, mobile phone, kiosk, or for that matter print it out and treat it like regular mail—and may in fact do all these things in the same day. Looking back historically, we see that writers tend to associate telephony with telephones, radio with radios, film sound with cameras and movie projectors, sound recording with phonographs, tape recorders, CD players, and portable stereos. Yet the mediality of the medium lies not simply in the hardware, but in its articulation with particular practices, ways of doing things, institutions, and even in some cases belief systems.[37]

Discussions of MP3s have tended to focus on the consumer products associated with them, most notably iPods, smartphones, and other portable MP3 playback devices. They are certainly ubiquitous.[38] Yet the portable MP3 device isn't necessarily the historically significant invention. Whatever epochal questions we can dream up for MP3 players have already been asked of transistor radios and the Walkman. Mainstream magazines of the 1950s pointed to families who would have "music wherever they go" and transistor listeners who "can't stand silence."[39] As Susan Douglas wrote of radio in this period, "Modes of listening were increasingly tied to not just *what* you listened to but to where and how you listened—while falling asleep in your bed, making out on the beach, and especially driving around in the car."[40] Or consider these questions, posed to young Japanese Walkman users in 1981: "whether men with the walkman are human or not; whether they are losing contact with reality; whether the relations between eyes and ears are changing radically; whether they are psychotic or schizophrenic; whether they are worried about the fate of humanity." One of the interviewees replies that the interviewer has missed the point. Soon, "you will have every kind of film on video at home, every kind of classical music on only one tape. This is what gives me pleasure."[41] The dreams associated with MP3 players are old dreams. The anxieties about the human body and the end of public life are old anxieties. While portable devices are certainly significant for all sorts of reasons, we cannot assume their existence or proliferation is the only or central historical trajectory relevant to the development of the MP3. The format is itself a technique for storage and movement of audio.

Format theory would ask us to modulate the scale of our analysis of media somewhat differently. Mediality happens on multiple scales and time frames. Studying formats highlights smaller registers like software, operating standards, and codes, as well as larger registers like infrastructures, international corporate consortia, and whole technical systems. If there were a single imperative of format theory, it would be to focus on the stuff beneath, beyond, and behind the boxes our media come in, whether we are talking about portable MP3 players, film projectors, television sets, parcels, mobile phones, or computers.[42] Format theory also demands greater specificity when we talk in general terms about *media*. As Lisa Gitelman writes, "It is better to specify telephones in 1890 in the rural United States, broadcast telephones in Budapest in the 1920s, or cellular, satellite, corded, and cordless landline telephones in North America at the beginning of the twenty-first century. Specificity is the key."[43]

Just as the concept of mediality refuses an a priori hierarchy of degrees of mediation from reality, format theory refuses an a priori hierarchy of formations of any given medium. Instead, it invites us to ask after the changing formations of media, the contexts of their reception, the conjunctures that shaped their sensual characteristics, and the institutional politics in which they were enmeshed. The 16:9 aspect ratio and 70 mm film are particular historical instantiations, and not peaks toward which progress has been climbing. The same could be said for sound-recording formats. For instance, phonograph preamps in the United States conform to an "RIAA curve," which is a particular frequency response specified by the Recording Industry Association of America, yet many authors will still talk about the sound on a record as having a more direct connection to nature than the sound on a compact disc.[44] The duration of a record is also of historical interest. Reflecting on the long-playing record, Adorno wrote that "looking back, it now seems as if the short-playing record of yesteryear—acoustic daguerreotypes that are always now hard to play in a way that produces a satisfying sound due to the lack of proper apparatuses—unconsciously also corresponded to their epoch: the desire for highbrow diversion, the salon pieces, favorite arias, and the Neapolitan semihits whose image Proust attached in an unforgettable manner to 'O sole mio.' This sphere of music is finished: there is now only music of the highest standards and obvious kitsch, with nothing in between. The LP expresses this historical change rather precisely."[45] Long-playing records could allow listeners to hear entire movements in operas without interruption, opera itself being a musical form designed for long periods of audience attention. The change in format occasions a different relationship between listener and recording, "irrespective of whether, on the one hand, LPs might have been technologically possible from the very start and were only held back by commercial calculations or due to lack of consumer interest, or, on the other, one really only learned so late how to capture extended musical durations without interrupting them."[46] This is an important qualification: we don't have to subscribe to a single model of historical causality or historical change in order to appreciate that a change in format may mark a significant cultural shift.[47]

The stories we tell about formats matter. For instance, most authors will agree that the specifications of a compact disc are somewhat arbitrary. This is not exactly true, but certainly the duration of a compact disc is arbitrary in that it could easily be longer or shorter. There are currently two stories in circulation about the seventy-four-minute length of the original compact disc, one much better known than the other. The standard tale has Norio Ohga, president of Sony, saying that he wanted the CD to be long enough to play Beethoven's Ninth Symphony without interruption; hence CDs had to be made seventy-four minutes long. But this is only one version of why the compact disc came to have the specifications that it did. Executives at Philips told Kees Immink, the lead engineer on the project, that because the compact cassette tape had been such a success, a compact disc should be about the same size so that it would be similarly transportable. Indeed,

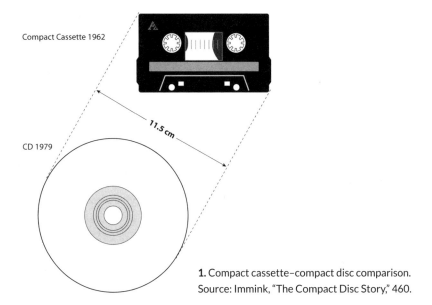

1. Compact cassette–compact disc comparison. Source: Immink, "The Compact Disc Story," 460.

the diameter of a CD is 12 cm, which is just slightly larger than the 11.5 cm diagonal of a cassette tape (see figure 1).[48]

Given the capabilities of equipment in the late 1970s, the physical capacity of the technology and the various industrial compromises surrounding sampling rate and bit depth in digital audio at the time, a CD that was 12 cm in diameter could hold about seventy-four minutes of music. The two other specifications for a CD, a bit depth of 16 bits and sample rate 44.1 kHz (1 Hz = 1 cycle per second; 1 kHz = 1,000 cycles per second), also came from interests in isomorphism with other media. The peculiarities of computer tape in the 1970s meant that 16 bits made more sense for storing audio—8 bits was too few, and 16 bits fit the "byte-oriented nature" of computer data storage. During the 1970s, digital recording techniques were developed that allowed two channels of audio to be stored on a video tape. In Immink's words, "The sampling frequencies of 44.1 and 44.056 kHz were thus the result of a need for compatibility with the NTSC and PAL video formats used for audio storage at the time."[49] The Ohga story appeared repeatedly in the press and persists on the internet, including, amusingly, on Philips's own corporate website, where the company claims that the seventy-four-minute length is the reason why the CD went from 11.5 cm to 12 cm in diameter. Yet this claim seems difficult to substantiate. Immink, who was as much of an insider on the project as one could be, writes that "there were all sorts of stories about it having something to do with the length of Beethoven's 9th Symphony and on, but you should not believe them."[50]

The issue here is not just that one (likely somewhat mythologized) story has been circulating in place of another but that the two stories serve very different ideological functions. The Beethoven's Ninth story authorized the CD as a format fit for high culture, thereby replaying a legitimation tactic used in the service of earlier sound-recording formats—as when Victor sought "legitimate" musicians like Enrico Caruso to record for it. It also, crucially, helps to uphold the myth that the primary referent of sound recording is live music, when in fact recordings—even of supposed live performance—had been referring to other recordings for decades by the time compact discs became feasible.[51] The second story is a tale of mediality, in which one format references extant practices in its design and conception, and in which the design of media involves industrial and cultural politics, as when television tried to mimic cinematic projection with a 4:3 aspect ratio. In the case of Philips, its designers sought a perceived quality of compactness and portability of discrete albums—qualities that are perhaps easy to forget in an age abundantly populated with portable audio media.

Other formats and standards came from operational needs across media. For instance, LP records' 33⅓ rotation-per-minute speed was first developed for synchronized film sound. By 1926 Western Electric was able to produce a disc recording that lasted for the same eleven minutes as a standard reel of film by slowing (then-standard) 10-inch, 78 rpm records down to 33⅓ rpm. (Film reels were eleven minutes because that's how long it took to get through a one-thousand-foot reel of film at twenty-four frames per second.) This alignment of durations of audio and visual media allowed for easy synchronization of soundtrack and image track in projection booths, at least until the record skipped. Later attempts to produce long-playing records retained the 33⅓ speed, but experimented with different materials for the record and needle, as well as the density of the grooves on the disc's surface.[52]

Attention to the mediatic dimension of formats also helps to explain why 128 kbps (kilobits per second) has until very recently been the default bitrate for MP3s made in popular programs like iTunes (though this number is now starting to creep upward). In the 1980s, it was thought that one important means of transmitting digital data would be over phone lines via a protocol named ISDN (ISDN stood for Integriertes Sprach- und Datennetz, though the initials in English generally refer to Integrated Services Digital Network or Isolated Subscriber Digital Network). The ISDN's lines in the 1980s had a capacity of 128 kbps, with an extra 16 kbps for error checking and other network matters. Experiments on digital audio compression in the 1980s were undertaken with the goal of transmitting a continuous, intelligible audio stream within the available bandwidth.[53] Today, even the cheapest DSL connection is wider than 128 kbps for downloads, yet the 128k specification remains a default setting in many programs. Some preliminary tests have even shown that university-aged listeners have a slight preference for 128k files over other rates, no doubt because of familiarity.[54] As suggested by examples like analog television's 4:3 aspect ratio or the diameter of a compact disc—or for that matter, the history of intersections between interior decorating, furniture design, and the shapes of phonographs, radios, and televisions—the issue of reference to other technologies and practices is endemic to the design of mechanical and electronic media, as well as digital media.[55]

The ISDN story points to another set of questions for understanding formats. All formats presuppose particular formations of infrastructure with their own codes, protocols, limits, and affordances. Although those models may not remain constant, aspects of the old infrastructural context may persist in the shape and stylization of the format long after they are needed. Some will be the result of defaults chosen at one moment and left unexamined; others will become core parts of the user's experience and therefore noticeable in their absence. In either case, the sensory and functional shape of the format may emerge from the "objective necessities and constraints of data storage" and transmission at the moment of their development, "but they also accrue phenomenological and aesthetic value" as people actually experience them.[56] The persistence of residual imperatives in custom and sensibility is a well-documented aspect of "embodied history": "a man who raises his hat in greeting is unwittingly reactivating a conventional sign inherited from the Middle Ages, when, as Panofsky reminds us, armed men used to take off their helmets to make clear their peaceful intentions."[57] The same sort of historical process is at work in the shape of any given format or medium. A characteristic that might first appear as the result of numb technological imperatives is actually revealed as something that had an aesthetic and cultural function, even if it is subsequently transformed—as in the case of the 128k standard.[58]

Infrastructures of data networks and later the internet—and the standards upon which they are based—provide the water in which the MP3 fish swim. These technical networks are built on top of one another, like oceanic zones or layers, and depending on the register we consider, the political or cultural articulation of the technology may vary widely. MP3s accumulating on hard drives mean one thing, their presence as a portion of internet traffic means another, their compliance to an ISO standard raises yet a different set of questions, and the psychoacoustic model on which they are based still others.[59] The shape of the infrastructure also matters. It is not simply a question of the size of bandwidth in the network's channel, of how big a data stream can pass through. It is also a question of how the network itself is built: "infrastructures are never transparent for everyone, and their workability as they scale up becomes increasingly complex." Infrastructures tend to disappear for observers, except when they break down.[60]

At first blush, it would appear that the MP3 is perfectly designed for an "end-to-end" network like the internet. The end-to-end ideal is based on a structure where the vast middle of the network does relatively little to its traffic. The network simply passes through that data stream while devices at the ends of the network do the important work. Personal computers are supposed to be "smart" and do most of the processing at either end of internet traffic; routers and switches are supposed to be "dumb" and pass through all traffic without modifying it too much. This is only an ideal; in reality, the internet does not exactly work this way. But the MP3 standard follows an end-to-end logic. The encoding stage has elaborate processing to make the file smaller. The decoding stage has an elaborate and precise set of rules for MP3 playback. In between, the file moves through the internet like any other stream of data. This sort of movement is quite different from other forms of infrastructure.[61]

Whether we are talking about gramophone discs, compact discs, or hard disks, recorded audio has always had a format to it. Changes of format can be at least as significant to consider as changes *of* or *across* media. We are comfortable talking that way in the history of visual art or printing, and we should take equal comfort in thinking this way about communication. Cross-media formats like MP3 operate like catacombs under the conceptual, practical, and institutional edifices of media. Formats do not set us free of constraints or literature from the histories that have already been written. They only offer a different route through the city of mediality. Not all formats are of equal historical or conceptual significance. Many may be analytical dead ends. But if they have enough depth, breadth, and reach, some formats may offer completely different inroads into media history and may well show us subterranean connections among media that we previously thought separate. The study of formats does not mean forgetting what we've learned from the study of media or, more broadly, communication technologies. It is simply to consider the embedded ideas and routines that cut across them.[62]

This book explores the big questions that live inside a diminutive format. It offers a set of concentrically connected histories, winding in from the long-term history of compression and bandwidth management in telephony and twentieth-century sound media to the specific histories of the MP3 format. There is the history of the research that led to the MP3's creation and its social life since being codified in 1992. There is the history of the technology behind the format—perceptual coding of audio—as an idea and a practice, which dates back to changes in ideas about sound and noise in the 1950s, 1960s, and 1970s. Then there is the history of the subject written into the code of the MP3, which first began to appear as AT&T took an interest in hearing research in the 1910s. Each of these three time scales offers insights into the format's meaning that are unavailable through the other two. Although the book presents its history more or less in chronological sequence, my hope is to render these three distinct temporal rhythms as operating simultaneously. The history of MP3 is in this sense polyrhythmic. It exists as part of conjunctures that developed over periods of the last thirty years, the last fifty-five years, and close to the last one hundred years (see figure 2).[63] To explore these conjunctures, I read documents in a wide range of overlapping fields and conducted oral historical interviews to get a sense of the tacit knowledge that lay behind perceptual coding.

Chapter 1 opens as a new approach to hearing research begins to emerge in the 1910s. Perceptual coding gets its hearing science from a field called *psychoacoustics*, which began to take its modern form at this moment. The mathematical model in the MP3 encoder is called a *psychoacoustic model*. Psychoacoustics is the study of auditory perception. More precisely, it uses sound technologies to test and describe the mechanism of human hearing.[64] In many fields, the "discursive rules" of psychoacoustics conditioned the kinds of questions that could be asked about human hearing, how those questions could be investigated, and how they could be applied in research on sound technology.[65] Because of their institutional and cultural situation, the discussions of psychology relevant to the design of the MP3 were

30 years
- engineers start proposing and building perceptual coders;
- MPEG sets standard;
- MP3s get named and come into general use for sharing and distribution of audio.

55 years
- researchers develop predictive theories of masking;
- noise gets domesticated;
- computers become audio media.

100 years
- perceptual technics;
- phone company measures perception as part of its economic program;
- development of psychoacoustics and information theory.

2. Three cycles of MP3 history. Image by Liz Springate. The slices of context get thinner as the time gets longer.

all conditioned by the philosophical, cultural, and political sensibilities of psychoacoustics.

Everything that is known about hearing in its natural state is a result of the interactions between ears and sound technologies. The psychoacoustic construct of *hearing-in-itself* is only accessible through sonic equipment, and through users who are comfortable working with it. This is no accident. Psychoacoustics began the twentieth century as a disorganized field, not fully developed or separated from physiological acoustics. That began to change when American Telephone and Telegraph made basic research into human hearing part of its organizational mission. The phone company realized that by better understanding human hearing, it could tune the phone system to the auditory range of its users. By the 1920s, research into hearing allowed AT&T to quadruple the capacity of parts of its system. By calculating the limits of human hearing, and of the parts of hearing most necessary for understanding speech, AT&T was able to create a system that only reproduced those parts of the signal. With the aid of auditory psychology, AT&T essentially created surplus bandwidth in its phone lines where none had existed before. It was able to monetize the gaps in human hearing and thereby use bandwidth more efficiently to generate more income. I use the term *perceptual technics* to name the application of perceptual research for the purposes of economizing signals. Perceptual technics did for perception what ergonomics did for work. The history of perceptual technics provides insight into a major economic force in twentieth-century communications. It also exerted a great deal of influence on the questions and directions of perceptual science during the same period. Even today, researchers refer back to work first done at Bell Labs; many of the methods, questions, and presuppositions of modern psychoacoustics were developed there.

Today, psychology textbooks routinely describe hearing—and all perception—in terms of information processing. Chapter 2 examines how psychoacoustics intertwined telephony and hearing, and in turn how information theory and cybernetics would later commingle technical and life processes in an analogous fashion. Starting with an experiment in 1929 involving two researchers from Princeton University who wired a series of live cats into a phone system they borrowed from AT&T, I trace a chain of reference in phone research, in which phones and ears were increasingly described in terms of one another. While sound-reproduction technologies had always figured some mechanism of the ear—most notably the tympanum in the diaphragms of phonographs, microphones, and speakers—psychoacousticians gradually established a set of equivalences between telephony and the inner ear and eventually the mind's ear. This model, which began as highly specific to telephony, came to be generalized to animals, and by extension, to people. Telephony and psychoacoustics defined hearing as a problem of information.

At the same time, early theories of signal transmission and redundancy that were developed for the telephone system became the basis not only for psychological theories of the subject, but also for communication in general, culminating in Claude Shannon's mathematical theory of communication. Methodologically, psychoacoustics separated the process of hearing from the meaning of what was heard. It posited that perception and meaning could be disentangled.[66] Information theory separated the transmission and interpretation of messages in an analogous fashion. While information theory's debt to military problems is well documented in the literature, chapter 2 emphasizes its abstraction of acute transmission problems in the telephone system. Shannon built on work done at Bell Labs to formulate his theory of information, developing his general theory from the historical formations of communication technologies in his time, from the residual to the emergent. *The Mathematical Theory of Communication* moves through a variety of communication systems of the time to develop its general, abstract theory: telegraph and teletype, telephone, radio, and television (black and white and color).[67]

Shannon first published his theory in 1948 in a pair of papers in the *Bell System Technical Journal*. It drew wider attention when published as a book a year later with Warren Weaver's introduction, which helped readers see past the complex math. As John Durham Peters has written, Shannon's theory was "many things to many people. It gave scientists a fascinating account of information in terms of the old thermodynamic favorite, entropy, gave AT&T a technical definition of signal redundancy and hence a recipe for 'shaving' frequencies in order to fit more calls on one line, and gave American intellectual life a vocabulary well suited to the country's newly confirmed status as military and political world leader."[68] Information theory also provided a formula for what is now called *lossless compression*, a technique of eliminating redundant data in a transmission with no measurable change to the output. The MP3's encoder combines perceptual coding with Huffman coding, which is a form of lossless compression. Its designer, David Huffman, was a student of Shannon's collaborator Robert Fano.

Perceptual technics and information theory created the intellectual milieu where perceptual coding could happen, but perceptual coding did not emerge in the 1950s. It surfaced on top of currents that developed over the next thirty years. Perceptual coders are built around a specific psychoacoustic theory of hearing; they deal with noise in a particular way, and they make use of considerable computer power. Chapter 3 offers a history of the psychoacoustic concept of *masking*, the domestication of noise, and changing relationships between computers and audio between the 1950s and the 1970s as the basis for the first perceptual coders. Perceptual coding emerged at the intersection of these changes. By the turn of the 1980s, engineers on at least four different continents who were not in touch with one another independently worked out ways to use models of hearing, or more accurately what cannot be heard within the audible spectrum, as the basis for a system of data compression for digital sound transmission. This is possible because they shared a common problem space defined by the intersection of psychoacoustics and information theory, and because of the conjuncture I describe in this chapter. Although there is no exact date on the coinage of the term *perceptual coding*, it appears in published work by 1988 and was probably in common parlance sometime before that. Throughout this book, I will use the term anachronistically to describe those forms of audio coding that use a mathematical model of human hearing to actively remove sound in the audible part of the spectrum under the assumption that it will not be heard. That practice can be dated back to the 1970s, the crucial period under consideration in this chapter.[69]

Early perceptual technics like AT&T's filters put boundaries around the hearable. Perceptual coding sought out the shifting gaps within the field of hearable sounds. Masking and the theory of critical bands are probably the most important psychoacoustic concepts for building a perceptual coder. In the ear, louder sounds mask quieter ones with a similar frequency content.[70] Auditory masking is the process of eliminating similar frequencies, based on the principle that when two sounds of similar frequency content are played together and one is significantly quieter, people will hear only the louder sound. Temporal masking is a similar principle across time. If there are two sounds very close together in time (less than about five milliseconds apart, depending on the material) and one is significantly louder than the other, listeners can only hear the louder sound. The MP3 encoder uses both types of masking to decide which parts of a sound need to be reproduced. The theory of critical bands allowed for the prediction of masking response by positing channels in the ear where sounds will mask one another, and all of the early perceptual coders I have found were built around this theory.

Although masking had been identified as a phenomenon in the nineteenth century, predictive models of masking did not emerge until the middle of the twentieth century. Initially, communication engineers viewed masking as a problem to be overcome. For instance, noise in a telephone line would mask audible speech. Traditionally, the engineering solution had been to minimize the noise. But starting in the 1970s, groups of researchers began to realize that audible sound could mask noise. If the noise of transmission could be moved to the gaps in the audible spectrum, noise could be eliminated as a perceptual phenomenon even though it was still measurably present on the line. This shifting attitude toward noise was reflected in other cultural fields. The idea of using and manipulating noise (rather than eliminating it) had existed as a minor thread since the birth of the historical avant-garde, but it proliferated across fields during the 1960s and 1970s, suggesting methods for everything from the design of cubicle-farm offices to killing pain. That is, while thinkers in a wide range of fields had defined noise as something to be eliminated, dentists, architectural acousticians, artists, and musicians began to explore ways of rendering noise useful. Similar attitudinal shifts toward computers were necessary during this period. Where computers had been used to model perception and as musical instruments, during the 1970s they were increasingly reconceived as sound media in their own right, both in research institutes and in artistic milieus. With a theory of the difference between signal and sound and with a new relationship to noise and to computers, engineers could imagine perceptual coding as a new kind of solution to old problems of perceptual technics and information theory.

We may talk of media as being invented and developed, but the equivalent—and much less spectacular—moment of birth for a format would have to be the moment it becomes a standard. Standards assure that a format that operates on one system will operate on another and govern the protocols for doing so.[71] From the turmoil of standards making, order emerges and a format begins to appear to casual observers as the natural outcome of a contingent and negotiated process.[72] The MP3 is a "standard object," to use Matthew Fuller's term. As he points out, processes become standardized as well as objects. Though we may think of an MP3 as a file, it is also the set of rules governing the process of coding and decoding audio, along with the vast set of processes that at one point or another conform to those rules, that make MP3 what it is. In Fuller's terms, the MP3 is a "metastandard" because it is for audio in general. Just as a shipping container can hold anything smaller than its interior space, an MP3 file can hold any recorded or transmitted sound. Systems built on metastandards require a certain level of self-reference. They must be closed enough to operate independently according to the standards set for them, and they can take on certain self-generating, self-adjusting, or self-perpetuating characteristics. Standard objects like the MP3 bind together "different perspectival scales, technologies, epistemologies, rhythms, and affordances."[73]

Chapter 4 considers the emergence of the MP3 format as a standards-making exercise. It charts the contingencies and compromises that shaped the format.[74] In 1988 the International Organization for Standardization formed the Moving Picture Experts Group to come up with a standard for digital video and audio.[75] For those who believed in standardization, one format was needed that could cut across digital technologies as diverse as compact discs, digital video, high-definition television, teleconferencing, digital broadcast, and satellite communications, to name just a few of the disparate industries interested in a shared standard.[76] To diffuse political differences among participants, the standard MPEG set in 1992 included three different protocols for encoding audio, which they called *layers*. MP3 is *layer 3* of the MPEG-1 audio standard. (There are subsequent MPEG audio standards, but new equipment and software is usually backward-compatible.) It represented a scheme developed by a German Institute called Fraunhofer IIS (Institute for Integrated Circuits—Institut für Integrierte Schaltungen), AT&T, Thomson, and France Télécom. Because no company had cornered the market on perceptual coding in 1988, MPEG was able to serve a regulating function across industries.

The story of MPEG, then, is an instructive if somewhat atypical case of how the standards-making process defines formats. Many technological standards are created in secret and are closely guarded, so MPEG is unusual because some of its history is publicly available. And because of the availability of documentation on the establishment of the MPEG format, its history also illustrates how standards-making manages industrial politics in the sensual design of communication technologies. A settled standard represents a crystallized moment in the negotiation process among different involved actors, whether companies, international standards-making bodies, or governments.[77]

International standards are a gray area in communication policy, but they have played an important and understudied role in what we now call the globalization of media. Without standards, content could not travel as well as it does and could not be as well controlled as it is. Many writers have commented on the centrality of policy to the development of media systems. Rather than viewing technological and industrial change as inexorable forces, they argue, we should turn to analysis of and intervention in governments' policymaking operations.[78] There is much to recommend this approach. But the vast majority of government involvement in standards has been to sort out national broadcasting and telecommunication issues. Standards touch the lives of their users in a much wider range of situations, and audiovisual standards emerge from a vast galaxy of interconnected organizations. Not only are there competing standards in every field, but there are also competing ways of establishing standards in every field. Sometimes they are the result of corporations devising proprietary schemes and leveraging them to dominate the market. Sometimes they emerge from a web of international organizations and consortia. While governmental policymaking can have the veneer of a democratic (or at least bureaucratic) process to which stakeholders are entitled access, much standards-making is rough business, done behind closed doors and conducted under the cloak of secrecy. Yet like government-set communication policy, standards-making is shaped by the interests of the standards-setters.[79] Chapter 4 considers how those interests played out on the ground.

The MP3 represents a confluence of artistic technique, aesthetics, and technology. One of the crucial steps in MPEG's standards-making exercise was a series of listening tests conducted in 1990 and 1991. Chapter 5 tells the story of these tests. MPEG needed a method to judge between different schemes, and to refine the audio processing enough in the standard that it would work for most people in most situations. The consortium considered four different perceptual-coding schemes, each of which was built into a *codec* (a coder-decoder box; today the term used for software that codes or decodes audio; see figure 3). A panel of MPEG-designated expert listeners conducted exhausting tests where a single recording would be run through each of the competing codecs multiple times. After repeated auditions over a period of hours, the listeners had to rate the audio quality without know-

3. The codec as a physical box. This is OCF (Optical Coding in the Frequency Domain), developed in 1989 as a predecessor to the MP3 codec. The hardware box would convert CD-quality audio into perceptually coded audio. Image copyright © Fraunhofer IIS. Used with permission.

ing which sounds came from which box. The statistics from these tests were then used in the MPEG standards exercises.

A whole praxaeology of listening was written into the code of an MP3, where particular kinds of listening subjects and orientations toward listening shaped the format. Through MPEG's listening tests, expert listeners came to represent, in code, an anticipated future listening public. As with the test subjects of psychoacoustics, MPEG's expert listeners weren't supposed to listen to the content of the audio fed through the competing codecs. They were only to listen for its formal characteristics and qualities as sound, as timbre. As the test instructed its listeners to remove their own tastes from the exercise, it offered a synecdoche for the removal of economic interest from the standards-making exercise. The tests were supposed to ensure neutrality among the competing interests within MPEG, even if in matters of aesthetics, there is no possible neutral position.

Any music recorded in a suitable digital format can be coded as an MP3, and in many situations, listeners may not be able to tell the difference. The listening tests reveal how the format negotiated its very particular origins in an attempt to approach a universal model of the listening subject. But the MP3 format itself was unmistakably grounded in a specific cultural milieu at the moment of its formation. Listening tests show the degree to which a professionally defined aesthetic of "good sound" shaped the format as much as more scientific or technical determinations did. MPEG audio followed from the stylistics of mainstream commercial radio and recording in Europe, North America, and Japan. In the end, MP3 audio works so well because it refers directly—in its very conception—to the sounds and practices of other conventions of sound-reproduction practice. Rather than demanding attention, it takes advantage of listeners' distraction from the music. Like a person who slips into a crowd, the MPEG audio was designed to disappear into the global network of communication technologies.

The MP3 format feels like an open standard to end-users. We can download software that encodes or plays back MP3s for free, and we can of course download MP3s for free, whether or not we are supposed to do so. Even developers can download the source code from Fraunhofer (which holds the most patents) for free. The MP3 is a *nonrivalrous* resource because from a user's standpoint, making a copy of an MP3 for someone else doesn't deprive the original user of its use. An MP3 costs almost nothing to make and reproduce—once someone has invested in a computer, software, a relatively reliable supply of electricity, and some kind of internet connection (because of these costs, we cannot say that it is truly free even when it is not directly purchased). An MP3 is a *nonexcludable* resource because it is impossible to prevent people who haven't paid for it from enjoying its benefits.[80] Borrow a CD, rip it, give it back. Now you and the original owner can both listen to the music. Share your MP3 through a peer-to-peer network and millions of others can listen to the same recording at no direct cost to you, or to them (though even this apparently free transaction requires the initial investments listed above).

But MP3 is not open source, and it should not be confused with software like Linux that comes out of the recursive publics of the free-software movement. The MP3 is a proprietary standard that brings in hundreds of millions of euros each year for the companies that hold the rights to it. The website for Thomson, which serves as the licensing representative for Fraunhofer, lists per-unit royalty rates for PC hardware, software, integrated circuits, and digital signal processors; per-title rates for video games; and a straight percentage of broadcasting-related revenue. The list of licensees reads like a who's who of new media distribution. In addition to the companies one would expect to see, like Amazon, Apple, Microsoft, and Motorola, one also sees other kinds of names like Airbus, CNN, Focus on the Family, and Mattel. MP3 has its technical niche in air travel, news, toys, and evangelical Christianity, and for each commercial niche with revenue over US$100,000, there is a fee paid (private uses such as building a personal music library are exempted, so for now you are off the hook).[81]

Chapter 6 explains how this state of affairs came to be. After MPEG established its standard in 1992, the competition that it had sought to manage returned to the surface. Each of MPEG's three layers had patents that went with them, and those patents were only valuable if someone were to start making use of the technology. Initially, the patent holders for layer 2 were much more successful getting their product into lucrative emerging markets, such as video compact discs and satellite radio. But after a series of stalled marketing exercises, Fraunhofer branded layer 3 "MP3" and released software that could encode and decode MP3 files on a computer. Shortly thereafter, an Australian hacker defeated their copy protection and rereleased their codec for free. This turned out to be an incredibly fortunate turn for Fraunhofer. By 1997 MP3 was firmly established as the preferred format for online file sharing. The rest of the story is well known. As CD burners came down in price, unauthorized MP3-filled CDs began to be sold on the streets of major cities. As internet connections proliferated, file-sharing took off among connected users in wealthier countries. File-sharing services like Napster gained international attention, and attempts to defeat them led to even more well-designed schemes to facilitate online traffic, first in music and later in all sorts of audiovisual material and eventually books.

Chapter 6 juxtaposes this wave of file-sharing against earlier waves of unauthorized copying and with ongoing debates over the status of music as a commodity, as a practice, and as a process. Mass file-sharing may amplify a crisis in some quarters of the music industries, but it also affords an opportunity to rethink the social organization of music. Consider that the file-sharing story is usually told in one of two modes. A tragic mode highlights the damage this did to the most powerful players in the recording industry, especially labels that were part of transnational conglomerates. A heroic version of the story holds up file-sharing as part of a social movement which has fought the major-label monopoly over the distribution of music. There is no question that revenue has declined for the largest players in the recording industry, and that the world has not seen unauthorized music exchange on this scale. But ultimately, the current wave of file-sharing has a great deal in common with other episodes of piracy in audio history, such as pirate radio in Britain and cassette piracy in many parts of Asia, Africa, and Latin America. In each of those cases, piracy was itself a market activity. It generated value for someone, and not just the people who managed to sell unauthorized copies.

During the 1980s, blank media companies like BASF and Philips profited greatly from the transnational wave of cassette piracy. The explosion of pirated MP3s is in essence no different. In addition to the MP3's patent holders, purveyors of storage media, playback devices (including a wide range of consumer electronics), and broadband internet have profited handsomely from mass piracy. Unauthorized duplication and circulation have certainly deflated the profitability of recordings, but they are not the antimarket practices we have been led to believe. They fit perfectly well within other kinds of market strategies and with long-standing practices of music consumption, even as they offer opportunities for the transformation of these practices. If we are to reimagine an alternative, better world for musicians and listeners, we will need to look past both the old monopolies over distribution and existing practices of peer-to-peer file-sharing for new models that support a robust musical culture, one not just based on buying and selling.

The conclusion considers the future prospects for MP3s and how MP3 files may fare after the format's fall from prominence. It also situates MP3 technology and MP3 listening against broader contemporary changes in listening practice, consumer electronics, media systems, and sound culture. Compression technologies will continue to develop and proliferate, but today's sonic environment is vastly different from the one that nurtured perceptual technics decades ago. Echoing writers in psychoacoustics, information theory, and philosophy, I end by arguing for a plural ontology and pragmatics of audition. Future work in science, technology, medicine, the arts, humanities, and policy will need to negotiate multiple models of hearing subjects and sonic technologies. The alternative is a sonic monoculture that will be of relevance to an ever-dwindling set of people and contexts.

My history stalls out at the beginning of the twenty-first century, right when MP3s hit mass circulation. It therefore leaves aside commentary on some of the more widely discussed and known aspects of the so-called MP3 revolution. Other writers are already at work on accounts of the role of MP3 players in people's everyday lives, the significance of file-sharing for the recording industry and intellectual property battles, the differing roles of MP3s outside of wealthy nations, the format's future or lack thereof, the interweaving logics of musical collections and databases, MP3-specific music like mashups, and the relationship of the internet to musical production and distribution.[82] Though those issues will arise in the following pages, this book steps back from those well-established contemporary concerns to consider the MP3 format in some of the longer flows of history in sound, communication, and compression.

It may seem strange that the history of a recent invention covers the better part of a century and spends the bulk of its first two chapters on events before 1950. I certainly didn't plan to write it this way. My initial readings in perceptual coding and psychoacoustics led me to ideas first hatched in the 1960s and 1970s. I expected the book to begin somewhere in what is now chapter 3, where I would explain the immediate origins and "prehistory" of perceptual-coding technology. But the limits of that periodization were driven home for me when I interviewed JJ Johnston, an engineer who worked at AT&T Bell Labs and developed one of the first working audio perceptual coders (he also may have coined the phrase *perceptual coding* in its current usage). During our interview at his home, Johnston took me downstairs to his bookshelf and we talked about where his various ideas came from. Johnston showed me a copy of Harvey Fletcher's 1929 edition of *Speech and Hearing* (his may have been the 1953 version), and we discussed how ideas and methods in that book shaped his work in the 1980s. In researching this book as a history of the possibility of the technology as well as the brand, it became clear that much of what we think of as native to "new" media was developed close to a century ago, and that to get at the historical meaning of the MP3 format, I would have to rethink the place of digital communication in the larger universe of twentieth-century communication technologies. In doing research, we challenge the presuppositions we bring to problems.

The work of cultural studies is to redescribe context, to analyze conjunctures, to attend to the relations of people, power, and practices built into any phenomenon or problem. "Any event can only be understood relationally, as a condensation of multiple determinations and effects."[83] As I suggested in the first two sections of this introduction, MP3s invite us to consider a not-very-well-documented history of compression; they also push us toward other strands of context that are not well documented in the available histories. Although psychoacoustics has played a major role in the design of almost every sound technology in the twentieth century (as has psychophysics more broadly in other sensory technologies), there are few critical assessments of its import or effects. In addition to my own reading and reflection, I have profited from the work of Mara Mills, who has followed something of a parallel path in her history of telephone research and hardness-of-hearing; Georgina Born, who has studied the role of psychoacoustics in avant-garde composition; and Emily Thompson, who places the phone company at the center of the history of electroacoustics. But there is still much conceptual work to be done. Meanwhile, information theory has seen quite a renaissance in communication historiography, but apart from Paul Edwards, David Mindell, and a few others, its deep connection to psychoacoustics and phone research has not been a major theme. This points to a larger argument in the book. While much of the existing history of information theory focuses on its connection with military problems, I highlight the connections between information theory and corporate capitalism. This book is a story of how a set of problems in capitalism taught us to think about hearing and communication more broadly. The principles first articulated as part of Bell's program of perceptual technics in the 1920s are essential to the workings and profitability of MP3s, and also to the entire economy of new media systems.

The MP3 format has seen many competitors, and with good reason. It is a proprietary format, meaning that to create software for it you must pay a licensing fee. The format itself was the result of political compromise. It is possible to improve on the psychoacoustic assumptions and signal-processing protocols built into the MP3 format, and it is possible to get better sound quality with lower bitrates. The format itself has continued to develop. Technically, the MPEG standard is a standard for decoding audio; it also allows for innovation. Newer encoders have improved on the early ones in many aspects of processing, including the psychoacoustic model. But the past few years have seen a proliferation of alternative formats to MP3, for instance AAC, which improves on MP3s' signal processing, handles more channels of audio, and encodes at lower bitrates (and avoids some of the industrial compromises of the original MPEG format), and Ogg Vorbis, which is an open-source format, meaning that one does not need to pay a fee to write software for it. Increased technical capacity is just as likely to occasion a new proliferation of compressed formats, though the 128k standard is starting to give way to higher bitrates for newer files. For now, MP3 remains the world's preeminent audio format. One day, other formats will eclipse the MP3 if they are more efficient, more accessible, or otherwise more convivial. But the conjuncture that gave rise to MP3s will retain a certain significance—as the moment when the dream of verisimilitude was publicly troubled, when perceptual processing washed over the digital landscape, and when an uneasy truce over the materiality of sound and music was spectacularly disturbed.

1. PERCEPTUAL TECHNICS

In 1938 the unfortunately named Professor Edwin G. Boring wrote that "really not so very much happened in the sixty years after" Hermann von Helmholtz published *On the Sensations of Tone* in 1863.[1] By this, he meant that Helmholtz inaugurated a new paradigm for understanding hearing—one grounded in the methods and ideas of physiology and physics—that other researchers played out for six decades. The judgment may have been a bit skewed by Boring's interests. But something did change in hearing research—especially American hearing research—in the 1910s and 1920s. Researchers and money poured into the field, and hearing studies began to shift their touchstones from physiological and physical models to a particular set of psychological models. During this period, psychologists and engineers devised ways of thinking about sound, hearing, and technology that shaped media for the rest of the twentieth century.

The shadow of a listener imprinted in every MP3 has its origins in the history of psychoacoustics as an academic subject and the history of a particular kind of imagined human subject within that field. Starting about one hundred years ago, the needs and interests of telephone research increasingly conditioned the problems, materials, and methods of hearing research, transforming the field and providing the foundations for modern psychoacoustics (as well as speech therapy and several other related fields). The models of the hearing subject, and the ideas of information that make possible the MP3 format and digital audio, were rooted in specific problems faced by AT&T as it sought to increase its profits as an industrial monopoly. This chapter focuses on the origins of modern psychoacoustics, while the next chapter considers the development of information theory. The Bell System and a burgeoning electrical components industry provided both fields with plenty of sunshine and fertilizer to nurture their growth.

This chapter and the next explore how the nature and tenor of the connections and analogies between ears and sound technologies shifted in the early twentieth century. In part, this was a result of the change from the dominance of physiological acoustics to psychoacoustics in hearing research. Not only did each field construct a different model of the ear, but the two fields also posed hearing as different kinds of problems and in different kinds of spaces. In the history of aural-electrical thought, the 1910s and 1920s marked the beginning of a shift from the middle ear as the site of inquiry toward the inner ear and the mind. Similarly, conceptions of what was transmitted by ears and media changed. Telegraphic electricity signified a message (as did the messages in the nervous system), but telephonic electricity was transduced from sound and back into sound. Between these moments, sound would come to exist as information. An electrics of a resonant middle ear gave way to an electronics and informatics of the inner ear. Bell's drive to conserve bandwidth led the company to participate in the development of a new way of thinking about perception, its limits, and the limits of communication technologies that still echoes in the design of new media a hundred years later. Perceptual coding descends from Bell's initial quest to squeeze more profit out of its infrastructure, as do many of the electronic and digital communications technologies that surround us. Riffing on Friedrich Nietzsche's discussion of aesthetics in *Nietzsche Contra Wagner*, John Durham Peters writes that early sound media were forms of "applied physiology."[2] As research into the telephone and research into hearing collided, the converse also came to be true. Psychoacoustics and information theory were theoretical extrapolations of communication technologies.

Audiometry and the Emergence of Modern Psychoacoustics

Standard stories about the history of hearing research mark two major shifts in the twentieth century: an orientational shift from physiological acoustics to psychoacoustics, and a series of major technological changes that transformed the field. Boring's history of hearing research listed specific technologies as turning points in ideas about the ear. Technology also plays a starring role in Audrey B. Davis's and Uta Merzbach's history of hearing research. They note that American research generally followed European work before the First World War in substance, method, and equipment.[3] The authors lament that early twentieth-century university-based psychological researchers could not follow "the path of technology" unless they aligned themselves with engineers in industrial laboratories and clinicians in state laboratories: "These were no longer the days when one could rely on the munificence of a King Maximillian of Bavaria to underwrite an apparatus for vowel construction, as had Helmholtz, or make history as a Bell or an Edison on the basis of invention and entrepreneurship alone. In the industrial society of the twentieth century, scientific advances based on technological breakthroughs have been made with the support of either government or industry."[4] Though their view of Bell and Edison is somewhat romantic (in fact, both benefited greatly from an association with Western Union and private investors), Davis and Merzbach rightly identified a new set of principles guiding scientific research into hearing in the early twentieth century.

Davis and Merzbach write that in the approximate half-century between the publication of the final edition of Hermann Helmholtz's *On the Sensations of Tone* and 1930, "audition in psychology passed through a well-defined phase." By this they mean that psychological studies of hearing began the period as a mélange of laboratory psychology, theoretical exploration, and physiological experimentation but ended with the dominance of electrophysiology in the study of hearing.[5] The employment of hearing history by Boring, Davis, and Merzbach echoed an important paper published by the Bell Labs researchers Harvey Fletcher and R. L. Wegel in 1922 titled "The Frequency-Sensitivity of Normal Ears." They list prior attempts to determine absolute thresholds of human hearing. In the process, they directly connect knowledge of hearing to available sound technologies. Consider the technologies used and dates in the list:

1870: organ pipe
1877: whistle
1883: tuning fork
1889: telephone receiver
1904: modified "phone"
1905: telephone receiver
1921: thermal receiver

The list notes a shift from musical-mechanical technologies to electrical apparatus connected with the telephone and suggests a story of progress that ends with the authors: "The development of the vacuum tube amplifier and oscillator, condenser transmitter and thermal receiver has given us precision apparatus which has made it possible to make accurate measurements of ear sensitivity."[6] Georg von Békésy's and Walter Rosenblith's history of hearing research, published in 1948, also marks a break between a fourth phase characterized by microscopic observation in physiological research and a fifth phase characterized by experiments with living animals and electrical effects. For them, the shift was technological.[7]

The psychoacoustician Hallowell Davis's reflections on his career, published in 1977, place the period from about 1910 to about 1930 as the key moment (and especially around 1930) when a device called the audiometer came into general use, displacing "tuning forks, ticking watches, and the whispered voice as a clinical tool for diagnosis of hearing loss." Davis approvingly cites Fletcher's preference for phonograph records, sound film, condenser microphones, vacuum tube amplifiers, and high-speed mirror oscillographs for the study of human speech and hearing. In making those choices, Fletcher privileged the trappings of modern sound media over earlier tools of acoustic research.[8]

The audiometer encapsulates this shift. An audiometer is a meter that electrically generates a continuous tone that can vary in pitch and loudness. It is important because it allows the standardization of responses and measurements in experiments, and thus enables researchers to create and compare aggregate data and to reproduce experiments and results. Psychoacoustics emerged at a moment when psychology as a whole aspired to the condition (and social status) of a physical science, when psychology labs were first being developed and experimental methods were crucial to the field's self-understanding and to the knowledge it aimed to create. The problem was that psychology dealt with the intensely private, interior dimensions of human experience, an issue that had a longer history in empirical thought. According to Kurt Danziger, "In ordinary scientific observation the experiences that are described are private too—no two scientists see exactly the same phenomenon—but this problem is overcome by the use of special media of communication that make agreement on crucial aspects of the experience possible. These media are partly physical . . . and partly linguistic."[9] The audiometer represented just such a special medium—one that combined a physical instrument with a new, more precise and more measurable way of talking about people's perceptions of sound. It took a central place in psychoacousticians' own accounts of their field's history because it represented the moment when the psychology of hearing found a possible solution to the problem of how to walk and talk like a physical science.

The earliest published references to audiometers appeared in 1879, shortly after the invention of the telephone, and telephone technology would be important to their development. A biography of Alexander Graham Bell attributes the term to him, although at least two other researchers were calling their devices for measuring hearing "audiometers" by 1879. Audiometers appear to have first been used to test for partial hearing in the Deaf. Bell, for instance, reported in his experiments that many children originally classified as totally deaf actually had some partial hearing. For Bell, even a little hearing could be useful in teaching Deaf children to speak.[10] But despite these early reports, the technology itself was not in wide circulation for decades. As late as 1896, we can still see psychological researchers struggling with how to measure the perceived intensity of sound and how to talk about it. A *Science* article from 10 April of that year describes the two then-reigning methods for determining loudness: (1) "dropping a ball successively from two different heights and recording the minimum difference in height necessary to enable the observer to determine which fall gives rise to the louder sound" and (2) "moving an object producing a constant sound, such as a ticking watch, or a tuning fork, uniformly towards or away from the ear."[11] From the point of view of the emerging scientific mind-set in psychology, neither approach was sufficient, since they lacked reproducibility, standardization, and calibration, making both methods impressionistic at best. The author, Joseph Jastrow (founder of the University of Wisconsin's psychology department), makes reference to an electrical audiometer as a possible solution but deems it unfit for experimental use. Instead, Jastrow constructed a gas-powered appa-

4. Seashore audiometer (1899). Source: Seashore, "An Audiometer."

ratus to continuously vary the volume of a steady pitch. With it, he was able to mark the points at which a sound became just audible or just inaudible to an experimental subject. Jastrow's device was, however, limited in that it was difficult to standardize and reproduce.[12]

The first widely used audiometer in North America was developed at another Midwestern land-grant university by Carl Seashore (see figure 4). Seashore was an experimental psychologist and a student of Edward Wheeler Scripture who had trained in Germany with Wilhelm Wundt and was thus aware of the emerging German psychophysical tradition. There is a possibly apocryphal story that Yale's psychology laboratory was founded the very day that Seashore enrolled as a student. Seashore's interests were steered toward the then-new field of applied psychology, and especially apparatuses for testing subjects. As an assistant professor at the University of Iowa in 1898, Seashore developed an audiometer in collaboration with a physics instructor named Charles Bowman. Their unit used a telephone receiver attached to a battery-powered electrical circuit. The circuit produced forty gradations of a clicking sound, which the operator could vary to measure the points at which a test subject could hear differences. The Stoelting Company sold the Seashore audiometer commercially for decades, but it did not initially catch on as a device for psychoacousticians. In part, its reliance on the electrical and telephonic technology of the late nineteenth century limited its applications for research, and in part, its potential user base was quite small.[13]

The device that did catch on, and that in some ways still defines audiometry, was developed by Edmund Fowler, Harvey Fletcher, and Robert Wegel. It was a vacuum tube–based audiometer that amplified pitched tones generated by a wave filter. The amplifying vacuum tube (or audion, as the inventor Lee de Forest called it) allowed for the amplification of small electrical signals. Until transistors were developed in the late 1940s, vacuum tubes were the best way to amplify electrical signals. Tubes made it possible for experimental psychologists and physiologists to fine-tune their measurements, to generate sounds for study, and to measure tiny electrical signals in the brain and elsewhere. For AT&T, tubes would also be essential for long-distance service and other amplification applications.[14]

As a social field and as an intellectual practice, the psychology of hearing underwent many changes that made the tube-based audiometer possible and its use meaningful—and these changes are registered in the device itself (see figures 5 and 6). Western Electric's audiometers differed from Seashore's model in more than just their tubes: they combined many strains of emergent audio technology, from telephony, sound recording, radio, and other fields in order to make precise measurements of hearers' sensitivities to different frequencies and different volumes. M. D. Fagen's history of the Bell System describes it thus: "The new tools making these more accurate measurements of audition possible were, of course, the vacuum tube oscillator and amplifier, the condenser transmitter, and the thermal receiver, together with a special attenuator whose range of variation in current output was more than three-millionfold."[15] Fagen's "of course" comes from the fact that the components in this new audiometer all emerged from other strands of Western Electric's research. During the First World War, Western Electric engaged in research on a string galvanometer for amplification of small signals on telephone lines, extremely sensitive microphones for the study of sounds emitted by large guns, portable transmitters with vacuum tubes that could withstand battle conditions, and secret-signaling methods.[16] Immediately before the war, they worked on telephony, radio, and public-address systems as well as other devices useful for building and testing such technologies. Western Electric hired a physicist named Harold Arnold in 1911, and by 1913 Arnold had convinced them that the best way

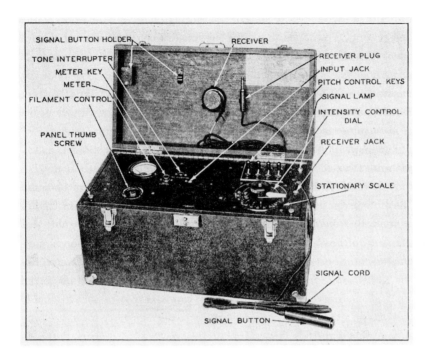

5. Western Electric–Bell Labs 2A-audiometer with Harvey Fletcher's legend. Source: Fletcher, *Speech and Hearing*, 216.

6. Face of the 2-A audiometer. Source: Smithsonian Institution photo 72–11334; USNMHT 306749.12.

to improve the telephone system was through basic research into sound and hearing. They hoped to outline the smallest perceptible thresholds for changes in sounds. With this knowledge, it would be possible to design a telephone system that transmitted sound that could be heard well but was maximally efficient in its use of current and bandwidth.[17] Thus, Western Electric's audiometer combined a new organizational ethos with the fruits of its ongoing research programs for military and corporate interests.

The institutional shift to corporate and military applications is as important as the technological shift to tubes and condenser microphones, because in many senses the former is a condition of possibility for the latter. The motives for psychoacoustic research and its driving questions change as the sound media industry provides both the tools and the rubric for the study of hearing. The change is already apparent in Seashore's audiometer, but the process is more complete and thorough in the Western Electric model, where each component came out of other military or industrial applications. Harvey Fletcher and Georg von Békésy were both originally motivated in their acoustical research by "a responsibility for developing and improving the telephone." Hallowell Davis recalls a conversation with Békésy, who said, "I was put in charge of the government telephone laboratory. I wanted to find out first of all whether the human ear was enough better acoustically than a telephone to justify further efforts to improve the telephone."[18] It wasn't just hearing research; work in noise abatement from this period also owes much of its sensibility and instrumentation to basic research in radio and telephony. Paradigmatic shifts in acoustical research were certainly technological, but they were also conceptual and institutional.[19] All of this was well known to people working in the field at the time. Fletcher's and Wegel's article, published in 1924, explicitly states, "It is important for the proper engineering of the telephone plant to know in absolute terms the sensitiveness of the ears of the average telephone user."[20]

Psychoacoustics as Theoretical Telephony

What does it mean for research into the faculty of hearing to be so closely tied to the interests of media corporations? People working at Bell had a very clear idea. Fletcher wrote on the first page of the author's preface to *Speech and Hearing*, a canonical work of psychoacoustics, that "if we could accurately describe every part of the system from the voice through the telephone instruments to and including the ear, we could engineer the parts at our disposal with greater intelligence."[21] In imagining the faculty of hearing as encased within a technological system of human creation, Fletcher casts the ear and the fact of perception as partial objects, elements of something bigger. He groups ears and voices with vacuum tubes, nests of wires, switching systems, carbon and condenser microphones, and a host of other devices. The business of sound reproduction fostered a growing interest in the economic utility of hearing itself.

Corporations like AT&T joined a growing list of modern institutions that sought to bring "life and its mechanisms into the realm of explicit calculations" and make "knowledge-power an agent of transformation of human life," to use Michel Foucault's turn of phrase.[22] The more AT&T knew about human hearing, the more income it could extract from its infrastructure. By conducting hearing research, AT&T planned to incorporate its users' auditory abilities (and inabilities) into the design of its telephone system. At the moment that Bell sought to become a legal monopoly, a utility that would provide "universal service" in an environment where prices were regulated, AT&T began a program to systematically incorporate its users' bodies into its infrastructure. If the price for a given commodity was relatively fixed, a corporation like AT&T had to look for other ways to increase its profits.

Psychoacoustic research became the means through which the phone system could administer the human capacity to hear for its own benefit. Arnold is clear on this matter as he winds down his own introduction to Fletcher's 1929 edition of *Speech and Hearing*:

> Economically the most important outcome of the work has been the increase of exact knowledge as to the requirements and limitations to be placed upon the transmission of speech in the telephone system. As time goes on there must be an evolution toward even greater perfection in those particular elements which are most important to intelligibility. The system is so large that the cost of such an evolution is immense and changes undertaken without an accurate knowledge of their value might lead to burdensome expenditures for disproportionate results; but, with the facts established by this investigation in hand, we can weigh any contemplated change and judge whether it is the one that offers most improvement at the moment and what its ultimate effect will be in its joint operation with other elements of the system.[23]

In this passage, Arnold suggests that from the standpoint of perception (and more specifically, intelligibility), the phone system produces a form of surplus value. It produces more sound than its users' ears need to hear. If AT&T could find the most effective frequencies for its users to understand speech, it could eliminate other less important frequencies. It would get more value out of its existing infrastructure by only transmitting the "necessary" bits of speech and cramming more phone calls into the existing bandwidth on the lines.[24] Although AT&T was interested in surplus value, in the first instance its enterprise had to assign different kinds of economic values to different frequencies.

AT&T's interest in basic perceptual research was part of its move to corporate research and development (R&D), for which historians still cite it as an innovator. Good business sense changed in the second and third decades of the twentieth century because the nature of the modern corporation changed during that period. Stagnating price competition was at the heart of the new corporate economy. The pattern in a monopoly market is simple. All other things being equal, a corporate monopolist will cut its price to the point at which selling *more* makes up for the profit lost in revenue from the higher price. But for a variety of reasons, groups of corporations in relatively stable markets (oligopolies) began behaving in a similar fashion. The lack of price competition enabled more regimented administration and planning within corporations, and also forced them to look for other ways to increase their value and profitability—either through finding cost-saving measures in the production process, through other innovations, by expanding their operations, or by acquiring other corporations. More than ever before, corporate capitalism became a matter of administration, and administration itself became one of the central dimensions of both economy and culture in twentieth-century life.[25]

While there is more than one story to be told of American corporate history, AT&T stands out as a representative case because of its aspiration to monopoly status and regulated pricing.[26] American Bell Telephone began its life as a classic nineteenth-century corporation, owned by a small group of capitalists in the Boston area who sought to expand their fortunes. By 1899 when it became capitalized as AT&T (with the Bell System as a subsidiary), it had been doing its best to overcome competition from independent telephone operations through expansion of its own. But this expansion came with increasing debt, and AT&T required outside financing to continue. The eventual result was a partial buyout by J. P. Morgan and

Company (along with a variety of its associates) in the winter of 1907. The Bell System was then effectively under the control of one of the Gilded Age capitalists. Indeed, Morgan had originally hoped to consolidate a national monopoly in telephony and telegraphy in exactly the same way he had built a monopoly in steel. By May 1907 the Morgan group installed Theodore N. Vail as the company's president.[27]

Almost immediately, Vail brought together AT&T's research and development activities under the umbrella of Western Electric. Previously, the AT&T engineering department in Boston had handled standards for equipment and service provision, as well as specifications for offices, while the Western Electric office in New York City carried out development work. The resulting establishment was among the first modern corporate R&D operations in the world. The use of R&D itself has a longer history, including nineteenth-century American corporations hiring individual scientists to solve problems and Thomas Edison's laboratory, which was essentially a freelance R&D firm. Modern R&D is defined by its collectivism and directedness: "industrial laboratories set apart from production facilities, staffed by people trained in science and advanced engineering who work toward deeper understandings of corporate-related science and technology, and who are organized and administered to keep them somewhat insulated from immediate demands yet responsive to long-term company needs." American companies imported this model from Germany in the first decade of the twentieth century, and AT&T, like DuPont, General Electric, and others, saw research and development as a way to better manage its market and increase its profits.[28]

AT&T's interest in maximizing the capabilities of its infrastructure was not simply a technical matter of improving service or equipment—it was directly tied to its status as an aspiring monopoly. As AT&T's annual report, published in 1908, put it, research was central to its "Bell System" concept: "The Engineering Department takes all new ideas, suggestions and inventions, and studies, develops, and passes upon them. It has under continuous observation and study all traffic methods and troubles, improving and remedying them. . . . When it is considered that some of these questions involve the permanency, duration and usefulness of a telephone plant costing millions of dollars, and changes costing hundreds of thousands, some idea of its importance can be formed."[29]

Because of AT&T's affiliation with J. P. Morgan, they were easily able to acquire smaller independent telephone companies. At the first sign that a company was in trouble, Morgan's banking trust would cut off its credit, at which point AT&T could move in and make the purchase. A far more ambitious version of this move came with AT&T's acquisition of Western Union in 1909. AT&T merged telephone and telegraph service for the first time, and its increased size gave it ever-greater leverage against independents. But regulatory attitudes were changing toward monopolistic business practices in the United States at this time. By 1912 AT&T became one of the targets of the attorney general and the Interstate Commerce Commission. When J. P. Morgan died in March 1913, the possibility for some kind of compromise in the pursuit of monopoly appeared. By December of that year, a vice president at the company wrote a memorandum explaining that AT&T would sell off Western Union in such a way that AT&T would no longer control it; it would stop purchasing independent phone companies; and, for the first time, it would allow independent telephone companies to connect with its infrastructure. This came to be known as the "Kingsbury Commitment," and it relieved AT&T of dealing with antitrust investigations. While regulation was slow and uneven at a national level, and despite a brief government takeover of telephone operations during the military participation of the United States in the First World War, AT&T was able to effectively operate as a monopoly from this point forward. It only helped that technical innovations in long-distance telephony enabled AT&T to firmly establish telephony as a national medium.[30]

By 1915 Vail was actively promoting the regulated monopoly idea. As Vail put it, regulation should "encourage the highest possible standard in plant, the utmost extension of facilities, the highest efficiency in service, and to that end should allow rates that will warrant the highest wages for the best service, some reward for high efficiency in administration, and such certainty of return on investment as will induce investors . . . to supply all the capital needed to meet the demands of the public."[31] Vail's argument puts forward exactly the needs and values of the modern corporation specified above: limited price competition, opportunities for innovation and expansion, and predictable profits achieved through careful administration.

It is in this context that we must understand AT&T's move toward basic research in general, and specifically basic research in hearing. The corporation had an interest in maximizing profits in a context of regulated price, increasing the importance of infrastructural and technical innovation. It also, perhaps not for the first time but certainly for the first time on such a grand scale, pointed AT&T toward new research and greater efficiency in its operations as a way to overcome limits on other kinds of capital accumulation. By the first decades of the twentieth century, it was old news that capitalists used machinery to extract the greatest possible value from their workers—either by subjecting the worker's body to the timing and logic of the machine or through union-busting tactics that made machinery profitable even when it was less efficient than the workers it replaced.[32] But the move to hearing research took this one step further, for it brought the user's body into the process of increasing corporate profit margins: "The giant corporation withdraws from the sphere of the market large segments of economic activity and subjects them to scientifically designed administration."[33] AT&T encountered hearing as an economic problem once its options for extracting additional profit through price were limited. Among other strategies, it sought to learn which frequencies could be excluded from the market for telephone signals. It aimed to render the user's ear an object of its own administration, and thus avoid "burdensome expenditures for disproportionate results," to use Harold Arnold's phrase.

AT&T's establishment of a general equivalence among ears, frequencies, and value—where the ear could be measured in terms of the others—was key to this endeavor. Based on AT&T's estimation of the limits of human hearing, some frequencies were essential for the reproduction of sound. Call them *necessary* frequencies. Other frequencies were unnecessary for the intelligible transmission of speech. Call them *surplus* frequencies. These surplus frequencies could be repurposed and sold if (and only if) they could be separated from the necessary frequencies. For this purpose, AT&T developed some new modulation and filtering techniques, and borrowed others from radio engineering. By measuring the minimum bandwidth needed for intelligible speech and then building filters to limit calls to that bandwidth, the company was able to effectively quadruple the capacity of phone lines by 1924 (see figures 7, 8, and 9). Where a phone line once transmitted one call (and sometimes also a telegraph message), it could now transmit four, each filtered into its own band.[34] Where AT&T could once bill for one call, it could now bill for four—with minimal modifications of infrastructure and no price increase. The phone company invented a kind of surplus value and then invested into its production process as, to use a metaphor, a lubricant for the machinery.

What AT&T did in a specific case became a general feature of communication engineering in the twentieth century. By using engineering to make use of the limits of human perception, systems could be made more effi-

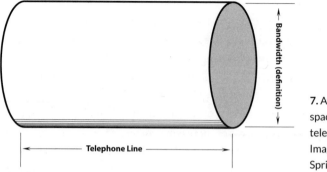

7. Available space inside a telephone line. Image by Liz Springate.

cient and thereby generate more profit for their owners, or simply carry more signal, more content, for more users. Following my discussion in the introduction, I use the term *definition* to describe the available bandwidth or storage capacity of a medium in terms of how much of its content can be presented to an end-user at any given moment. I will use the term *surplus definition* to describe the difference between the definition that a media system can theoretically produce and what its users need to perceive in order for it to work, *so long as that difference is put to use*.

Two caveats are immediately in order:

1. Surplus definition only exists relationally. If the managers of a media system do not make use of its higher definition capacities to squeeze more signal into the channel, then it is by definition not surplus definition, but some other kind of excess. For instance, if a driver on a freeway blasts an expensive, high-definition car stereo with the windows down on an open freeway, the system will produce all sorts of definition that he or she will never hear. Those frequencies are not surplus definition, however, because they are not being folded back into the process of technological reproduction. They simply wander off into the world, in search of other ears that won't hear them.

2. The term *need* in "need to perceive" is a social need and not simply shaped by biological, psychological, or physiological limits. It is initially determined by the managers of a communication system (or the designers of technologies) on the basis of their imagination of their user population. But actual use and the history of use may in fact change the relationship between necessary and surplus frequencies. How much definition is *enough* or *necessary* is a contextual question. For instance, decades ago Mark Twain and Franz Kafka built short stories around the difficulty of hearing some-

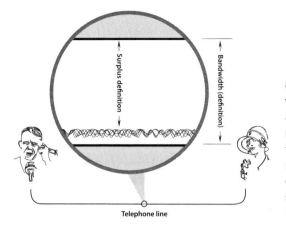

8. Bell analyzes the bandwidth of intelligible speech and determines the minimal bandwidth it needs to transmit two sides of a phone call. The rest of the channel is now available as surplus definition. Image by Liz Springate.

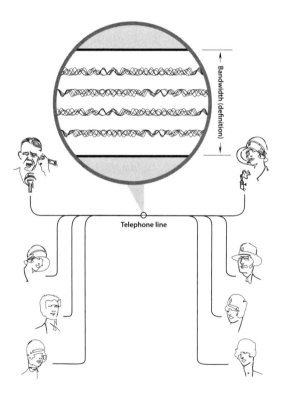

9. Bell designs a system of filters to render the surplus definition useful. The company can now fit four phone calls in a single line, instead of just one. Image by Liz Springate.

one on the other end of an analog phone line. Yet by the 1920s, those landlines were already of a higher definition than some mobile phone connections today. Now, analog landlines are the object of nostalgia. The *New York Times* media critic Virginia Heffernan writes of the analog telephone that "your phone voice was distinctive; your phone manner was distinctive. You thought a great deal about people who rhythmically and mysteriously inhaled and exhaled cigarette smoke while they talked or left long silences or didn't hang up immediately after saying goodbye." Today, "sound signals, so unfaithful to the original they hardly seem to count as reproductions, come through shallow. You can hardly recognize voices. Fragile, fleeting connections shatter in the wind. You don't know when to talk and when to pause; voices overlap unpleasantly. You no longer have the luxury to listen for over- and undertones; you listen only for content."[35]

When a communication system harnesses surplus definition and puts it to use, it generates *perceptual capital*. Perceptual capital is the accumulated value generated by a surplus definition. It can be reinvested in a system to generate additional value, as in the case of AT&T quadrupling its infrastructural capacity. It can also be invested into systems that would not otherwise work or work as well. Just as automation can speed up mass production (or in some cases make it viable in the first place), perceptual capital can be used to intensify communication technologies' processes of reproduction. In these terms, we could say that MPEG utilized perceptual capital in order to make systems like ISDN lines work for audio and video, or to cram both video and audio onto a formerly audio-only medium like the compact disc. Many of the twentieth century's most important electronic media operated according to a logic by which they utilized some form of perceptual capital. Telephony, radio, color television, satellite image transmission, computers, and computer networks all make use of this principle in one way or another. Surplus definition exists around the horizon of perceptibility, and so we might also think of perceptual capital as imperceptual capital, since it is generated from the accumulation of media material that users do not perceive.[36]

Perceptual capital is thus a special kind of capital, with the following characteristics:

1. Like other kinds of capital, perceptual capital can be transformed into cash profits, or it can become part of an investment, a form of fixed capital, that allows a bigger technological system to perform. In this respect it is a fairly ordinary form of fixed capital.[37] There is no such thing as *perceptual*

capitalism. The method of generating surplus value from surplus definition can work in corporate-liberal and neoliberal economies alike, and probably in others as well.

2. Perceptual capital exists only in the aggregate and does not inhere in individuals. It is the product of a relationship between users and communication technologies. As a form of fixed capital, it exists on the production side, and not in the distributed, contested, or metaphorized forms of symbolic capital that users may employ in the social worlds they inhabit.[38]

3. Media and consumer electronics companies' desires to accumulate and use perceptual capital have served as an important motivating force to shape perceptual research in the twentieth and twenty-first centuries. Sometimes this may occur through direct agenda-setting, as at Bell Labs. At other times it may occur through the inheritance of methods, approaches, or presuppositions of this research in other, unrelated work, as when engineers working on perceptual coding import ideas from psychoacoustics. At times this motivating force competes with (or exists alongside) other motivations, but it is an important element of the history of perceptual science. Perceptual science is probably no different from most other sciences, but because it is a science of the subject, we must maintain a certain mindfulness of its conditioning when considering its conclusions.

4. Perceptual capital is not directly generated from labor, because it makes use of users' abilities and inabilities to perceive economic or other instrumental ends. Of course, perceptual capital could not exist without the general intellect: "To the degree that large-scale industry develops, the creation of real wealth comes to depend less on labour time and on the amount of labour employed than on the power of the agencies set in motion during labour time, whose 'powerful effectiveness' is itself in turn out of all proportion to the direct labour time spent on their production, but depends rather on the general state of science and on the progress of technology, or the application of this science to production. (The development of this science, especially natural science, and all others with the latter, is itself in turn related to the development of material production. . . . Labour no longer appears so much to be included within the production process; rather, the human being comes to relate more as watchman and regulator to the production process itself."[39] As Mara Mills writes, Bell's research in this period is part of a historical process of "molding and regulation of technology according to human norms."[40] But what does it mean to say that surplus definition and perceptual capital are not directly produced by labor? Of course,

there is labor going on when we turn to the researchers who design AT&T's hearing studies and apply that knowledge to its infrastructure. But there is something else happening, beyond the labor of research and development, when we consider the relationship of the medium to its users. Several important strands of political economy have made their analytical mark precisely by expanding the concept of labor to better describe the economic implications of communication practices. In addition to that project, there are other processes of value-generation in communication, and it is worth following their threads as we spin our social critiques. Alongside an expanded concept of labor, we must also look to circulation as another mode through which media systems generate and manage value.

A brief detour through Marx's critique of liberal political economists like Adam Smith and David Ricardo (who first proposed the labor theory of value) and its application in the study of media will clarify my point. Marx argued that profits came from the difference between the labor for which workers were compensated—the work they needed to do in order to "reproduce themselves," which he called necessary labor—and the work for which they were uncompensated, which he called surplus labor. Surplus labor generated surplus value, which explained the difference between the cost of wages and the prices of products workers produced: profit. A capitalist could then pocket profit or reinvest it in the means of production, for instance, through the purchase of technology (which Marx considered "fixed capital" since it was money bound up in machinery).[41] Although I have lifted some terminology and morphology in my description of necessary and surplus definition, the parallel doesn't quite fit in the AT&T case. AT&T generated surplus frequencies (a kind of surplus value) from its estimation of the capacities and incapacities of its users. But its users are not laboring for the company in any usual or meaningful sense of the term.

The usual response of political economists in this situation would be to expand the notion of labor. This has been a useful approach to media audiences and communities of users, as well as to the kinds of economic relations media corporations expect. For instance, Dallas Smythe presciently wrote that for broadcast audiences "all non-sleeping time is work time" because even if audiences weren't being paid, they were generating profits for media providers by paying attention to content, being measured, and then being sold to advertisers.[42] So Smythe expanded the concept of labor to include audience attention. Smythe's formulation works for advertiser-supported broadcast media, as well as other advertiser-supported media, from newspapers and magazines to web pages and search engines. Though he would not necessarily characterize his analysis in this fashion, Henry Jenkins, in his studies of fan participation in a variety of media, clearly demonstrates that leisure activities generate value for the franchises in which they participate. Again, the concept of labor is expanded. Building off Maurizio Lazzarato's concept of immaterial labor and his distinction between work and labor, Tiziana Terranova argues that even though some activities are free and unpaid, they can generate value for others as *free labor*, and thereby may in fact be exploitative in a manner analogous to waged labor. Free labor is unwaged and generally occurs under the rubric of leisure—posting reviews on Amazon.com, moderating internet fora, and "liking" products or services on Facebook are all current examples. For Terranova free labor is "a fundamental moment in the creation of value in the digital economies," where "knowledgeable consumption of culture is translated into productive activities that are pleasurably embraced and at the same time often shamelessly exploited."[43]

While this expansion of the concept of labor has been fruitful in the examples above, it does not offer the same level of insight in our case. Even if we subsume categories that used to be called consumption—like talking on the phone—under the broader rubric of "immaterial labor," we are still left with a problem: the telephone's surplus frequencies come from people *not* hearing. One could argue that surplus frequencies are harvested from users in the sense that AT&T based them on a calculation of users' abilities and inabilities to hear. But whatever you want to call not hearing (or not seeing), it dilutes the term *labor* to meaninglessness if you include this as labor. A different process of value-generation is happening here. A capitalist enterprise (the phone system) expands by finding new areas to endow with value (in this case the relationship between frequencies and the limits of hearing), but in the process it reduces these things (frequencies, the threshold of hearing) to their exchange value. That is why the separation of the process of hearing and what is heard is so essential to psychoacoustics' theory and method. The quantifiable parts (audible and inaudible frequencies) are economically relevant; the unquantifiable parts (the content of speech, the meaning of what is said) are not.[44]

Thus, I use the term *perceptual technics* to describe the process of creating surplus definition and transforming it into perceptual capital. Through perceptual technics, a company can economize a channel or storage medium in relation to perception. In the phone system, it is the process of coding the flows of frequencies in terms of thresholds of hearing and endowing those flows with value where they had none. With perceptual coding and the MP3, perceptual technics is the process of coding the flows of audio from the compact disc (or, more accurately, the red-book standard on which the CD is based) in terms of a dynamically changing relationship between the signal content and the ear's masking behavior. To use a term always fashionable in business, perceptual technics *monetizes* the flow of media in terms of a measured, estimated, and modeled perceptual capacity.[45] In the phone example, it sets up a relationship between the phone line and a model of an ear and monetizes it for the purposes of making money in telephony. In the MP3 example, it sets up a relationship between red-book audio and a model ear, and monetizes it in terms of transmission through a range of other possible channels, from digital audio broadcasting to the internet.

As a concept, *perceptual technics* cuts a path through several related histories. It is derived from the term *psychotechnics*, popularized in English by Hugo Münsterberg in his book *Psychology and Industrial Efficiency* (1913) (though Münsterberg built on earlier work in Germany). Psychotechnics originally referred to the application of psychology to industry, business, and occupational choice.[46] Friedrich Kittler draws the explicit connection, calling media "psychotechnologies" based on the insight that

> everyday reality itself, from the workplace to leisure time, has long been a lab in its own right. Since the motor and sensory activities of so-called Man (hearing, speaking, reading, writing) have been measured under all conceivable extreme conditions, their ergonomic revolution is only a matter of course. The second industrial revolution enters the knowledge base. Psychotechnology relays psychology and media technology under the pretext that each psychic apparatus is also a technological one, and vice-versa. Münsterberg made history with studies on assembly-line work, office data management, combat training. . . . Film technology itself (as with phonography in Guyau's case) became a model of the soul—initially as philosophy and, eventually, as psychotechnology.[47]

But psychotechnics was a wide-ranging field of industrial and applied psychology. In this period, psychotechnics thus covered the whole span of industrial psychology, and while AT&T's research into perception might have been a subset, psychotechnics includes a much broader range of problems than perceptual technics—from the behavior of workers, to optimal pedagogical methods, to psychological approaches to advertising. Thus, I prefer to use the term *perceptual technics* to refer to the specific economization of definition through the study of perception in the pursuit of surplus value. I also mean for it to resonate with Lewis Mumford's use of the term *technics* to describe a mode of civilization; not just technology but "a will-to-order . . . a reorientation of wishes, habits, ideas, goals."[48]

Perceptual technics is "lodged between mechanics and biology," to borrow a turn of phrase from Bernard Stiegler. It is a perceptual relative of ergonomics, which takes its name from the economization of work. The "ergo" in *ergonomics* implies work, motion, and activity. While ergonomics is about capacity, fit, and interaction, perceptual technics is about playing the threshold of perception. Today, ergonomics is considered part of a larger field called *human factors ergonomics* (HFE), which suffuses the design of media and a wide range of physical spaces. The field of HFE defines itself in terms of studying and optimizing the productivity and comfort of people as they interact with machines, which requires systematic knowledge of human-machine interaction. Retrospectively, it might be possible to argue that perceptual technics is a subset of human factors ergonomics, although the field's own advocates do not do so. In his history of the field, David Meister characterizes the period before the Second World War as "premodern," and the period between the two world wars as a "dull period" in the development of the field. Yet this is exactly the period when AT&T actualized knowledge of its users' limits in developing the phone system. Although clearly Meister's exclusion could be read as a simple oversight in retrospect, psychology enters human factors through cognition and disposition. For instance, aviation psychology developed as a field during the First World War, when it aimed to determine which people were best suited to become pilots on the basis of emotional stability, tilt perception, and mental alertness.[49] In other words, it tried to figure out who would be crazy enough to fly a plane, but sane and stable enough not to crash it.

Like psychotechnics, ergonomics, and human factors, perceptual technics owes a political and philosophical debt to Taylorism. It also owes a debt to corporate liberalism, although the ubiquity of perceptual coding today suggests it is equally at home in neoliberal and global economies. Perceptual technics involves corporations, engineers, and others managing technologies in relation to statistical aggregates of people. AT&T's innovation was thus to focus its application of psychology to industry at the threshold of perception, to give that threshold an economic value, and to build a series of research problems, research methods, and industrial techniques around the pursuit of that value. In the world of perceptual technics, *hearing by itself* was an ironic concept, because there was no sophisticated knowledge of hearing available beyond that acquired through instrumental inquiry. Normal hearing was deduced for the purposes of telephony. For the remainder of the century, a pattern would repeat itself in perceptual research, even in its most therapeutic guises. The normal case would be defined by its instrumental use. This is why hearing aids and cochlear implants are still largely designed to privilege speech over other kinds of sounds (because the way of thinking is derived from telephone research). It is also why when hard-of-hearing mobile phone users began to complain of feedback from mobile phones, their manufacturers replied that the problem must be with the hearing aids, not the phones.[50] Perceptual technics conjoined normal and abnormal through the application of instrumental cases tied to communication technologies. To get to hearing-in-itself, twentieth-century researchers first had to pass through media.

Although the Bell Labs researchers' interests in normality were driven by phone interests, this period also coincides with the growth of the New York League for the Hard of Hearing (founded in 1909), and the creation of hardness-of-hearing as an identity category, at least in the New York social milieus where Bell Labs researchers traveled. Mara Mills has argued that the hard-of-hearing became essential to understandings of normal hearing during this period. If the responses of hearing people to the Deaf shaped early telephony, disability again played a crucial role in the early twentieth-century definition of hearing, telephony, and ultimately communication. As Mills writes of the league, "Members undertook a number of projects during this time, some of which would now be classified as medicalizing, others as building disability culture: they arranged lip-reading competitions, organized dance classes and afternoon teas, assisted members with employment, offered an on-site 'hand-work shop,' and wired theaters for the new electric hearing aids." By 1919 the league had begun to focus its efforts on the prevention of childhood hearing loss, and collaborations with doctors like Edmund Fowler led to testing. By 1925 the league and Bell Labs collaborated on testing over four thousand students in New York City public schools. In its quest to establish clear boundaries between pathological and normal hearing (and because the earlier tests were not properly standardized), AT&T would continue these tests for well over a decade, culminating in the phone company's presence at the world's fairs in New York and San Francisco, where it tested more than 750,000 people.[51]

While one normally thinks of having a normal range of human hearing from which pathologies and Deafness might deviate, in fact, the same body of research established what could be called a normal range for human hearing and related zones and criteria of pathology. As Georges Canguilhem wrote, "It is not paradoxical to say that the abnormal, while logically second, is existentially first."[52] In the case of hearing research, it seems that the abnormal is both existentially and logically primary. Alexander Graham Bell first applied the audiometer to find vestiges of hearing in the Deaf, and Fletcher and the American Federation of Organizations for the Hard of Hearing sought to find thresholds of deafness in the hearing. This is a common historical pattern even outside scientifically oriented fields. For instance, eighteenth-century ideas of deafness shaped the then-emergent practice of silent reading (though reading out loud would continue to be an important practice well into the nineteenth century). "Normalism," as disability scholars call it, suffuses the history of ideas with what it means to be human and to communicate in the most basic forms by listening or reading. Even more, it animates scientific attempts to quantify, measure, and define aspects of humanity.[53]

It would be easy to deduce grim implications from the scenario I've set out thus far. There is a dark tale to be told of big corporations alienating perceptual capacities from people in order to extract more profit from them. But like ergonomics, perceptual technics is an ambivalent innovation. One may be tempted to feel nostalgic for a prior moment in history when the human body itself was unalienated and was not subjected to politics or economy, but that would amount to nostalgia for the Middle Ages. As hearing research moved toward living subjects, even to objectify them, there was a new level of social engagement and exchange. The definition of normal and pathological hearing was a classic case of an able-bodied group (doctors, communication engineers) medicalizing a population, but it was also the result of a disabled group's own advocacy project. In addition to being economistic and instrumental, AT&T's interest in a better coupling between ear and receiver was expansive, inclusive, and lively.

Psychoacoustic Subjects

AT&T's was not the first encounter between psychology and economy, especially around the question of population. Already in the middle of the eighteenth century, texts on government identified the organization and management of population—to the maximum economic benefit of the state and the household—as one of their expressed goals. Nineteenth-century branches of psychology invented categories like degeneracy in order to make arguments about who should and should not have children, and where people should and should not move.[54] But the psychology of the 1910s was different in both register and attitude. Apart from its collaboration with the New York League for the Hard of Hearing, AT&T offered economic rather than moral justifications for its mission. Its psychological investigations into the population were for its own good first, and for the convenience of its users second. Moreover, AT&T's investigations were sub-individual. In its effort to stretch its infrastructure and maximize profits, it sought to understand thresholds and tendencies within any body, as well as across the population. As with earlier psychologists, it would eventually develop means for classifying listeners as normal or not normal, but its goal was to make use of the normal field. In the nineteenth century, auditory physiologists often experimented with mute or dead bodies and body parts to make them speak, and earlier auditory psychologists often dealt with theoretical subjects but not real people.[55] The members of Harvey Fletcher's group at Western Electric hooked up their machinery to live people. As the history of audiometers shows, they were not the first to do so, but by the time Western Electric began its work, experimental methods and live subjects were becoming more of a norm for acoustic research. So at the exact same time that the forebears of psychoacoustics amplify the tradition of objectifying the faculties of speech and hearing, large numbers of living people start appearing on the scene again in the practice of experimentation.

The shift from physiologists of the dead like Hermann Helmholtz to engineers of the living like Fletcher did not happen overnight or in a single break (the dead continued to exert an important force in psychoacoustics). There is a long line of connections between physiological and psychological investigation; the former was a major influence on the latter. Wilhelm Wundt, one of the pioneering figures of modern experimental psychology, dedicated two-thirds of his first textbook to the physiology of the nervous system. In the introduction to part 1, titled "The Bodily Substrate of Mental Life," Wundt wrote that "psychology is called upon to trace out the relations that obtain between conscious processes and certain phenomena of the physical life." In borrowing physiology's experimental method (and moving away from philosophical psychology), Wundt had in mind addressing problems of life itself—"both the vital processes of the physical organism and the processes of consciousness," as well as the relationship between them.⁵⁶ In this way, Wundt positioned psychological experience as directly tied to the living body, and he positioned psychology as one of the disciplines through which life processes themselves could be managed.

Medicine had seen a similar shift a few decades before. The word *subject* was originally used in medicine to describe a corpse intended for anatomical dissection, but by the 1820s it was used to describe a "person who presents himself for or undergoes medical or surgical treatment." In psychology, the term *subject* crept in through studies of hypnosis during the 1880s, and by the end of the century came to signify a "normal human adult data source."⁵⁷ As a new paradigm emerged in psychology, these subjects were arranged into anonymous aggregates, statistical groups meant to represent general tendencies but who were actually artifacts of the psychologist's own creation. As Kurt Danziger writes, "In order to make universalistic knowledge claims, psychologists took to presenting their data as the attributes of collective rather than individual subjects. Very frequently, these collectivities were constructed by psychologists for this specific purpose, and in constructing them they postulated the existence of a collective organism that already exhibited the assumed general characteristics on which their knowledge claims depended. Thus, to demonstrate the effects of supposedly stable characteristics, they constructed experimental groups defined by such assumed characteristics, and to demonstrate the modifying effects of experimental intervention they constructed groups entirely defined by exposure to such intervention."⁵⁸ In other words, psychologists grouped individuals together for the purposes of experiments and then wrote about the individuals as instances of a collective organism. Rooted in this approach were several other assumptions. Psychologists studied "normal" subjects so that they might better generalize about populations. As they moved toward greater generalization, psychologists became less and less likely to identify the actual subjects in their studies by name—a practice common before the First World War. They also became increasingly interested in statistics and populations, rather than individuals. At the extreme was a psychologist like L. A. Quetelet, who took regular events in a population (such as commission of crimes) and transformed them from acts of individuals within given populations to propensities for the action within individual members of that population. In other words, if one in ten people committed a crime, then each person had a 10 percent chance of becoming a criminal. In general, this is now treated as a kind of racist logic—for instance, when used to assign propensities for crime or intelligence quotients to members of different racial or ethnic groups.[59] But a logic that appears racist when applied to one set of measurements of people appears bureaucratic or economic in another setting. For instance, if Fletcher's group could determine the sounds that only 10 percent of the population was likely to hear anyway, those frequencies would be less essential for the development of telephone technology. Analogous reasoning accompanied psychology's entrance both into the politics of race and into the conduct of everyday business.

Herein lies the connection between the academic subject of psychoacoustics and the human subjects that psychoacousticians would imagine. Like their predecessors, psychoacousticians conducted their experiments as if they were investigating hearing *as such*. But since their experiments were guided by the need to measure not only ears but also technological systems for (and of) ears, psychoacoustic experiments involving people had irreducible dimensions of intersubjectivity and culture, however superficial they might have seemed. Even as they reached for universality, they were situated in the politics of the lab, in their engagement with specific measurement technologies, and in their larger institutional settings. For instance, in a discussion of the measurement of hearing loss, Fletcher points out that the average decibel level of a city office is around 30 db, and therefore, "a speech test made in such an office would not differentiate between ears that are normal and those having a 30 db loss in hearing. The importance of making hearing tests in soundproof booths is thus evident."[60] Fletcher also points to the importance of compensating for individual differences through the use of a volume control, thereby allowing for the testing of "all degrees of hearing from normal hearing to total deafness."[61] It can also be used to modulate the voice of the researcher as he speaks directly to the patient through the earphone. Thus, for each test, there is a careful interaction between researcher, research subject, and machinery. The hearing test itself thus has some basic level of intersubjectivity, since the subject must indicate when she or he can hear the tones being played by the audiometer, and the researcher must base her or his calibration of the equipment on the subject's responses. Though these are simple things, they do show the degree to which the subjective experience of hearing—and speaking—is central to a technological system by which hearing is measured and objectified.

Harvey Fletcher offers no formal discussion of Bell Labs' process for acquiring research subjects, but the general outlines are known to us. In the section on hearing loss, for instance, Fletcher mentions a test performed over three days on a thousand schoolchildren in New York City to highlight the efficiency of his apparatus.[62] In the section on research into the perception of musical tones, Fletcher mentions that "the judgments of pitch and quality of the musical tones were made by three persons familiar with music."[63] In both cases we actually know very little about the subjects (apart from the likely guess that the schoolchildren were compelled to participate while the music lovers were not), but it does point to a whole series of interactions in advance of the experiments. In the case of the schoolchildren, it was a collaboration between Bell Labs and the New York League for the Hard of Hearing. By 1927 a quarter of a million children had been tested for hearing loss by the American Federation of Organizations for the Hard of Hearing, a project made possible by the "cooperative efforts" of Bell Labs. By 1939 Bell Labs would have run over 1.5 million tests with audiometers.[64]

A version of Quetelet's aggregate subject lives on in auditory pathology. There were really two sympathetic projects at work in this series of audiometer tests: Bell Labs' desire to map the boundaries and topography of a hearing subject so that it could better calibrate its equipment for profit, and the medical-political desire to identify and remedy (or even prevent) deficiencies in hearing. One sought to establish normality in order to use it for financial gain. The other sought to establish pathology in order to ameliorate it. But normalism only tells part of the story. There is also a less systematic approach evidenced in Fletcher's use of "music lovers" in his research.

We only know that Fletcher implicitly equated "knowledge of music" with enhanced ability to discriminate among tones, but I can imagine the researchers trolling around the company for colleagues who had big record collections and professed great aesthetic knowledge. Thus the social practices of hearing music were also posited prior to investigations into hearing itself. As with the children in schools, there is a blurring between hearing *as such* and the experiences of actual people in actual historical events, especially as those people are made anonymous through the procedure of the experiment and the rhetoric of the article that follows it. Beyond the experiments on schoolchildren (and later fair-goers), Fletcher's group was typical in its choice of subjects. Students and other psychologists made up the vast majority of psychological research subjects in this period.[65] Similarly, in a classic study of masking published in 1924, R. L. Wegel and C. E. Lane offered no discussion of the acquisition of experimental subjects. Instead they concentrated on the apparatuses used for creating masking cts in subjects' ears. Given that Wegel played a major role in Bell Labs' ovation of the audiometer, a focus on apparatus only made sense.⁶⁶ But e complete absence of a discussion of the acquisition of research subjects ɔints to the degree in which the anonymous, aggregated psychological subject of Quetelet and others was fully in play at Bell Labs.

For Fletcher and for Wegel and Lane, a theory of hearing-as-such depended on the constant interactions among subjects, technologies, and researchers in the lab. More than once, telephone receivers stand in for ears in Fletcher's *Speech and Hearing*. This was not just analogical reasoning. Alexander Graham Bell and Clarence Blake had placed a tympanic membrane (eardrum) in a telephone receiver in 1878. Since then, a more electrical understanding of hearing suggested that the ear *as a system* had a great deal in common with the telephone.⁶⁷ Indeed, one of the available theories of hearing at the time was named *the telephone theory* for the obvious reason that it conceived of the auditory nerve as a kind of electrical telephone line to the brain. Variations in pitch and intensity in sound would be reproduced in the inner ear, which would then cause more or fewer nerve fibers to fire.⁶⁸ If the ear was like a telephone system, then the two systems might operate according to the same logics, and a more direct mode of translation between them could be found. Perhaps, via electrophysiological research, living ears could be literally wired into technological systems for the purposes of study and analysis, and even for the purpose of telephony. But not human ears, at least not yet.

2. NATURE BUILDS NO TELEPHONES

In 1929, two psychologists at Princeton University built a cat telephone. Following a procedure developed by physiologists, Ernest Glen Wever and Charles W. Bray removed part of a cat's skull and most of its brain in order to attach an electrode—in the form of a small wire hook—to the animal's right auditory nerve, and a second electrode to another area on the cat's body. Those electrodes were then hooked up to a vacuum tube amplifier by sixty feet of shielded cable located in a soundproof room (separate from the lab that held the cat). After amplification, the signals were sent to a telephone receiver. One researcher made sounds into the cat's ear, while the other listened at the receiver.[1] What they found astonished them. The signals picked up off the auditory nerve came through the telephone receiver as sound: "Speech was transmitted with great fidelity. Simple commands, counting, and the like were easily received. Indeed, under good conditions the system was employed as a means of communication between operating and sound-proof rooms."[2] Given Helmholtz's prevailing theory of hearing and the established accounts of nerve sensations, the researchers had not expected the nerve itself to transmit pulses that could be transduced back into sound. But it appeared to do just that. Wever and Bray checked for all other possible explanations for the transmission of sound down the wire. They even killed the cat, to make sure that there was no mechanical transmission of the sounds apart from the cat's nerve: "After the death of the animal the response first diminished in intensity, and then ceased."[3]

Though their findings were later overturned, their experiments marked an important moment in the history of psychoacoustic research, and their approach was paradigmatic. Through perceptual technics, AT&T sought a way to incorporate the capacities of its users' hearing into its infrastructure. Wever and Bray actually incorporated a being's hearing into a telephone circuit. With the correct transformer in place, the ear was a component of the phone system, not its object. Wever's and Bray's experiment approached the logical limit of the psychoacoustic subject, a subject literally wired into the system to pass electrical signals between brain and machine, a subject whose brain signals could be calibrated and correlated to vibrations in other sound technologies or out in the world. Persistent and popular analogies between the nervous system and electrical telegraphy were common in the nineteenth century. Laura Otis quotes the German physiologist Emil DuBois-Reymond identifying "an agreement not merely of the effects, but also perhaps of the causes" in the nervous system and the electrical telegraph.[4] Otis also recounts delicious stories of telegraphers applying circuits to their fingers and tongues. In the 1800s, Vorselmann de Heer proposed an electrical telegraph that would send electrical signals directly into the fingers of a receiving subject, but the practicalities of varying the voltage for each individual proved too much. In the 1860s, the physicist William Thomson claimed he could taste the differences among signals.[5] The articulation of telephones, electricity, and ears gestures back to this longer history and extends it forward.

In theory, anyway, sound-reproduction technologies have always had a degree of interchangeability with aspects of human hearing. Even the earliest sound-reproduction technologies were shaped from a model of the tympanic membrane of the middle ear (to the point of actually using cadavers' tympanic membranes in experimental apparatuses). Like that membrane, the membranes in microphones and speakers vibrate, to transduce sounds to electrical signals (or the vibrations of a stylus), or vice versa. So even the earliest telephones, phonographs, and radios were essentially chains of diaphragms vibrating in sympathy with one another. Their juxtaposition of ear and machine becomes even more important in early accounts of sound-reproduction technologies as social processes, but even there, the machine only made it to the middle ear.[6]

In Wever's and Bray's experiment, the interchange between ears and machines extends from the middle ear to the inner ear. This shift is paradigmatic, because it literally places sound-reproduction technology *inside* the mind's ear. To put a Zen tone to it, the telephone existed both inside and outside Wever's and Bray's cat. By extension, the phone system existed both inside and outside people. The cat's ear was like a condenser microphone on the front end of their system. It needed power (in the form of life) to operate. Invented during the First World War, the condenser microphone was hailed as the "most nearly perfect electro-acoustical instrument in existence" and the origin point of modern acoustics. It used two charged diaphragms separated by a small air gap to transduce sound into electricity.[7] Unlike previous microphones, it required a power supply. When the power was cut, the sound would fade away. So as the sound faded from their cat microphone, it demonstrated in the animal's death that life itself could power a phone, or any other electro-acoustic system.

While Wever and Bray thought they were measuring a single set of signals coming off the auditory nerve, they were in fact conflating two sets of signals. The auditory nerve either fires or does not fire, and therefore doesn't have a directly mimetic relationship to sound outside it—there is no continuous variation in frequency or intensity, as you would have with sound in air. A series of experiments in 1932 revealed that the mimetic signals were coming from the cochlea itself. Called *cochlear microphonics*, these signals were responsible for the sounds coming out of Wever's and Bray's speaker in the soundproof room. Hallowell Davis wrote in a paper published in 1934 that when wired into a phone system and properly isolated, nerves could not reproduce high-frequency sounds well on their own. But the cochlea could do so, "with the accuracy of a microphone."[8]

Davis and his collaborators' work on cochlear transmissions paved the way for a wide range of subsequent research, and cochlear microphonics are still important today. While they did challenge Wever's and Bray's conclusions, Davis and his collaborators continued down the same epistemological path where ears and media were interchangeable; in fact, one was best explained in terms of the other. One of the most widely acknowledged and controversial achievements of this work has been the development of cochlear implants. Previous treatments for hardness-of-hearing or deafness involved interventions in the middle ear; cochlear implants resulted from the project of intervening in the inner ear, a practice that was only possible because of the line of research begun by Wever and Bray. Meanwhile, the brain's work of translation—from firing neurons into the perception of sound—became a major preoccupation of psychoacousticians as well and remains an open question down to the present day.[9]

This chapter offers a history of information that places Wever's and Bray's collapse of life and phone against the conceptual collapse of life and machine later effected by cybernetics and information theory. In the process, I argue for the centrality of the phone company and its construction of hearing to the ways that communication was conceived more generally in communication engineering—culminating in the development of modern information theory in Claude Shannon's *The Mathematical Theory of Communication*. The history will appear in this chapter as largely institutional and intellectual, but it is also meant to be directly technological. Alongside its use of perceptual coding, an MP3 encoder uses Huffman coding to eliminate redundancy in its code. Huffman was a student of Robert Fano at MIT, and he improved a coding algorithm developed by Shannon and Fano.[10] Thus, the origins of information theory are in a very direct way the origins of a central part of the MP3's code. Because they so often stand in for people in auditory research, our story begins with cats, who will make frequent appearances in the history of information theory and cybernetics.[11]

Communicating with Cats: A Short History of the "With"

It becomes even clearer how important Wever and Bray were if we consider their relationship to the cats in their experiments and contrast it with the longer history of auditory research conducted on cats. Brain researchers interested in hearing have long used cats as research subjects. The cat brain is structurally similar enough to the human brain to allow for useful comparison, and cats were relatively inexpensive both to acquire and maintain in the lab (in comparison with, for instance, monkeys). Meanwhile, cat-telephone interactions appear as early as 1877, when Eli W. Blake Jr. wrote to Alexander Graham Bell that "the family cat was brought to the telephone and 'kitty kitty' called through the instrument at the other end. The poor animal was prodigiously excited and sprang away from the instrument, nor could she be induced to come near it again. I wish I had a dog to experiment on!"[12] What sets Wever and Bray apart is that they do not address the cat. They use its ear as a microphone, but the cat's consciousness is never in question. The cat itself could not have been "conscious" in any meaningful sense, since its entire cerebral cortex and most of its midbrain had been removed.

Despite the absence of anything like cat consciousness in these decerebrated animals, earlier researchers treated their decerebrated cats as, well, cats. This sets apart Wever's and Bray's approach. For instance, an article published in 1914 by Alexander Forbes and Charles S. Sherrington

> reported that "sharp hissing three feet from the right ear caused . . . a raising of the head, the head being carried to the right." One common response to sound in the decerebrate animal was a lashing of the tail. Even more striking responses have been recorded by Bazett & Penfield who studied animals that had been decerebrated some days beforehand. These workers stated that "the most effective sound was a small scratching noise, and an animal would often react promptly by raising its head if a piece of paper was crumpled up at a distance of one or two yards. These sounds are similar to those made by a mouse. . . . Animals would react to these slight noises even though there was a constant loud noise from the motors and stirrers to which no reaction was shown except when it was started after an interval of rest."[13]

In all of these examples (drawn from a survey in a physiology textbook published in 1964), the cats are not enjoying a fully intersubjective relationship with the researchers. The cats weren't enjoying anything at all. At no time did the researchers imagine anything approaching a relationship of equality with the live animals they were cutting open. But the researchers did still think of their cats as, pardon the phrase, "cat subjects." The researchers compared their decerebrated cats' responses and behaviors to those expected of regular cats. Henry Cuthbert Bazett's and Wilder Penfield's invocation of the similarity of their scratching and crumpled paper sounds to "those made by a mouse" (as opposed to a squirrel, or some other animal) is a perfect example. Here the researchers imagined their research cat as they would any other cat. Bazett and Penfield followed a path set out by Forbes and Sherrington in 1913:

> The reaction of lashing of the tail-tip was in our experience one of the most persistent and characteristic of the responses. This mimetic move-

ment is in the normal cat indicative of affective reaction. Darwin describing the behavior of a cat "when threatened by a dog" wrote "the whole tail or the tip alone is lashed or curled from side to side." This lashing of the tail-tip was in our Experiment VI seen to be accompanied by bristling of the hairs of the tail, a phenomenon which supports the inference that the mimetic reflex was such as in the intact animal would have been attributed to emotion. The mimesis thus displayed may be taken as signifying aggression. . . . The character of the reflex mimesis in our experiments is further suggested by the efficiency of sounds resembling the barking of a dog, yowling of a cat, or the whistling of birds. . . . As exhibited in our experiments the mimesis was doubtless purely reflex.[14]

They explicitly compared the decerebrated cat to the normal, healthy cat. The entire study was based on the animals' reactions. Whereas Wever and Bray used cats as a means of communication, earlier researchers used decerebrated cats as the object of their communications. Forbes and Sherrington sought to get to essential truths of hearing by treating their decerebrated cats *as cats*.

Scientists in the early twentieth century were no doubt aware of the moral questions surrounding experiments on animals. Antivivisection campaigns first developed in the mid-nineteenth century, and though psychology would not become a major target of the movement until some time later, it is quite likely that the scientists whose research involved animals had some sensitivity to the ethical issues involved. By the 1930s, public debate about animal experimentation—and its relation to experiments on people—had been raging for decades.[15] Bazett and Penfield deal directly with the question of pain (and by extension, cruelty), explaining how their procedure prevented the animal from experiencing pain: "The question of any conscious sensation could be completely excluded, and in addition the preparation would remain quiet during experimentation."[16] It is therefore all the stranger that both teams of scientists—Forbes and Sherrington, Bazett and Penfield—spent some of their time barking like dogs, crumpling paper to simulate mice, yowling like cats, and either bringing birds into the lab or chirping like birds themselves (I can't help but picture them doing all this while dressed in white lab coats). Therein lay grand metaphysical questions about the seat of consciousness and subjecthood in the animal's brain and body, and specifically, the researchers' answers to those questions. Why did they write about the surgical procedure as if it left the animal with no consciousness and then approach the animal as if it had some vestigial consciousness in their subsequent experimentation? In the studies of decerebration for the purposes of hearing research, authors did not know where auditory perception took place, and so they were doing their best to ascertain how much of audition was left when the mind was gone. But a deeper answer still might lie in the ambiguous social status of the animal as it lay on the surgical table, or afterward as it lay in a bath of warm water (since the part of the brain that regulates body temperature had been removed).

Researchers at that time used cats because they were cheap, convenient, similar enough to people, and pliable for the experiment. They were more or less interchangeable with other animals.[17] Georg von Békésy's approach to animals is an instructive counterexample, since his research on cochleas led him to experiment on as many different animals as possible to view the movements of the cochlea's hair cells. Békésy believed the key to understanding hearing lay in the basilar membrane—a membrane that runs the length of the cochlea. He first constructed a mechanical model and later—using tools of his own invention—studied the movements of the basilar membranes of dead animals and in cadavers through a microscope. He would drain the cochlea of its fluid and replace the fluid with a salt solution that contained powdered aluminum and coal, and then flash light off the powdered solution to view the interior working of the cochlea. As sound entered, a bulge would run along the basilar membrane, stopping near the base for higher tones and traveling all the way up for lower tones. He would later win a Nobel Prize for this work.

Perhaps emblematic was Békésy's study of an elephant. When he heard that an elephant had died at the local zoo, he sent one of his assistants to retrieve the head. After his assistant made a trip to the zoo and two trips to a glue factory (the first time he did not cut far enough into the elephant's skull to get to the inner ear), he retrieved the animal's giant cochleas and Békésy was able to witness the movement of the hair cells firsthand. Though the incident is recapped in full detail in later histories of auditory research, all he would say in his own published writing was that "by good fortune, the head of an adult elephant became available for study."[18] Békésy's ecumenicism toward his animal subjects suggests an instrumental sensibility. In the universal traffic among the ears of the animal kingdom, a cochlea is a cochlea is a cochlea. Though much could be made of the elephant's captivity in life before its death in the zoo (and subsequent journey to the glue factory), Békésy's research continued the nineteenth-century physiological tradition of gaining knowledge of the hidden processes of life through the dissection of the dead. Thus, auditory research followed dual paths, and the realm of the dead was perhaps the zone in which the meanings of the animals and the people were most fully reduced to the workings of their body parts.

The realm of the living, meanwhile, was more fully suffused with ambiguity and contradiction. Beyond the stated intentions of the decerebrators and their protestations of disinterest as to the significance of the animals they used, we should ask why cats figure so prominently in this tale. As Robert Darnton writes, "There is an indefinable *je ne sais quoi* about cats." In many cultures, cats straddle animal categories. Western European cultures (and presumably North American ones as well) divide animals into at least four categories: the very close (pets), tame but not very close (farm animals), field animals (game), and remote wild animals (not subject to human control and inedible). Cats shuttle around among these categories. In the decades before they became subjects of decerebration, cats were used for their mousing skills and their companionship, but they were also routinely tortured and killed (sometimes eaten), often because they were thought to have magical powers in both life and death—for fertility, luck, power, and protection. According to Jody Berland, cats occupy both sides of binary divisions like "human enterprise versus nature and wilderness, companionate animals or 'familiars' versus wild and edible animals, and domestic (feminine and familial) space versus public space."[19]

Wever and Bray were not the first people to use cats to illustrate an important concept in acoustics. Athanasius Kircher built a "cat piano" in 1650 (see figure 10). It was designed to amuse a bored Italian prince. Pressing one of its keys would launch a spike into the tail of trapped animals selected for the tone of their cries. This particularly cruel instrument reappeared in 1725, when Louis-Bertrand Castel announced his ocular harpsichord. Castel used the cat piano as a joke to show that sounds were not beautiful in themselves; they could only be made beautiful through sequence and harmony imposed by human intent. In a subsequent series of letters, Castel argued that music was something one learned to appreciate and that music-making was a purely human faculty. Thus, the cat piano showed that there is no such thing as cat music, only human music made through tormented feline cries.[20] Two centuries later, the tables would turn as inquiry moved from thought to perception. For Wever and Bray, cat consciousness was to

10. Kircher's cat piano. Source: La Nature, Part 2 (1883): 519–20.

some extent interchangeable with human perception, or for that matter electromechanical transmission.

If there is any historical relationship between the kinds of cat torture and murder one might find on Saint-Jean-Baptiste Day in eighteenth-century Paris (or in a cat piano) and the cold scientific "humane" vivisection of the physiologists and psychologists, it is through a relation of cultural descent. Max Weber describes the pursuit of wealth losing its dimension as a religious calling and becoming an empty activity in itself over the nineteenth century; one might say the same thing about people killing cats. Decerebrated cats were the unfortunate occupants of modernity's iron cage. Stripped of their magic and mystique, they were still killed for human access to their powers. If there is a path from cat massacres to scientific study, it is a path of Weberian disenchantment. "As working and farm animals were forcibly ejected from the domestic sphere," writes Jody Berland, "machines replaced both people and animals. The cat switched from being a hybrid worker-spirit-demon to being a pet as a consequence of changes in the social landscape that redefined the roles of animals as violently as those of people."[21] Thus, one of the ways to understand decerebration is precisely as a kind of disenchantment of both cats and the process of hearing itself. Though they share a trajectory, Wever and Bray differ from the decerebrators who came before them because throughout their experiments, they treated their cats as *anything but cats*. Previous decerebrators communicated with their cats as others. Wever and Bray communicated with their cats as instruments. It is the difference between "communicating with" another person or a pet—where the "with" suggests an object of the communication—and "communicating with" a telephone—where the "with" suggests the means of communication.

Living Circuits and the *Camera Silencia*

Wever's and Bray's cats were microphones, transducers, and transformers for the simple telephone system rigged up between the lab and the soundproof room. They remained category violators, but in a thoroughly modern way. Consider the circuit diagram for their cat telephone: The cat head at the upper left is simply part of the larger telephone system. In contrast to the human ear hanging off the end of the system in the soundproof room, the cat head is literally built into the system as a vital component. There is more than a little bit of Alexander Graham Bell and Thomas Watson mythology in the story of their experiment. While I have been unable to find a record of the words spoken into the cat's ear, it is easy to imagine a reenactment of the telephone's primal scene: "Bray, come here, I want to see you!"[22] A hallway separated the two researchers. One was situated in a room where no outside sound could reach, unless it came through the phone. The sound that came out of the speaker rendered the cat head just another node in the circuit (see figure 11). As Henning Schmidgen writes of soundproof rooms, they are not simply or primarily acoustic technologies; they are social technologies designed for particular configurations of researchers and research subjects. Soundproof rooms accommodate "a many-headed, human *and* animal subject that is in the process of dissolution and that, by means of accelerations and decelerations, by movements in place, continually recreates anew the borders between the inside and the outside."[23] In this case, the soundproof room and phone circuit crystallized the relationship between modern ears and the phone system.

The combination of a phone circuit and a soundproof room in the image renders concrete the institutional arrangements behind the collapse of ears and media. Soundproof rooms were a "laboratory fractal" through which modern psychological research emerged, first through the desire to isolate subjects in time-reaction studies and only later for the purposes of psychoacousticians. Soundproof rooms were a logical extension of twentieth-century academic psychologists' scientific aspirations. If their goal was to

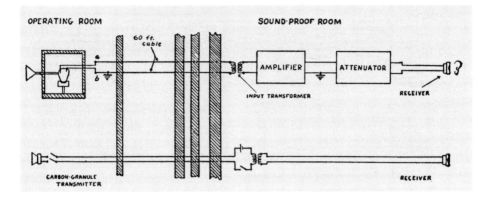

11. Wever's and Bray's cat telephone. Source: Wever and Bray, "The Nature of Acoustic Response," 378.

study a particular subject's reaction to a particular stimulus, it only made sense to isolate that subject from as many other stimuli as possible. By 1929, when Wever and Bray made use of the room at Princeton, the *camera silencia* was a hallmark of the modern psychological laboratory because it was understood as a means to effect the level of isolation necessary for scientific observation of subjects.[24] The installation of a phone system was also not accidental. Wever and Bray got their phone system, audiometer, loudspeaker, and other equipment on loan directly from R. L. Wegel and C. E. Lane, two researchers at Bell Labs. This was the first such installation in North America and as Mara Mills explains, this relationship between Bell Labs and Princeton's psychology lab "facilitated the emergence of modern psychoacoustics."[25] Indeed, Wever's and Bray's own diagram partakes of the iconography of telephone circuits developed at Bell Labs. Compare their cat telephone pictured in figure 11 to the image of a phone circuit in Fletcher's and Wegel's article on normal hearing, published in 1922 (figure 12). Reading this image anachronistically and inappropriately, one could imagine that Wever's and Bray's circuit simply replaces the vacuum tube oscillator with the psychologist's voice and the space between the "exciting receiver" and "capacity transmitter" with a cat's head. If that were not enough, Fletcher and Wegel also explained the role of soundproof rooms in this process: "It is extremely important that all noise interference be eliminated in making measurements near the threshold of audibility. In a room having the ordinary noises from the street the threshold point may be shifted to ten times its value in a quiet place, and in very noisy places

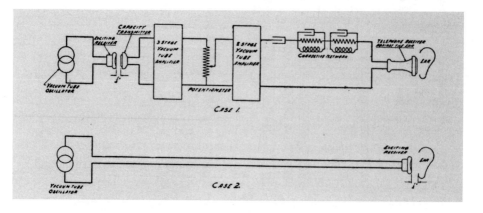

12. Telephone circuits for testing hearing, ca. 1920. Source: Fletcher and Wegel, "The Frequency-Sensitivity of Normal Ears," 558.

this may be increased to 1,000 times."[26] Thus, the assembly of a cat, two psychoacousticians, a phone system (with loudspeaker), and a soundproof room materialized the institutional and epistemological interconnection of early twentieth-century psychology, the telephone industry, and architectural practice. This heady mix nourished the soil beneath our contemporary blends of ears and technologies in music, sound media, communication industries, and hearing science.

In the fractal space of the camera silencia, Wever's and Bray's cats were also model brains, as Wever and Bray attempted to make sense of the connections between the different parts of the ear, the nerves, and the brainstem. Their cats stood in for people, and for the faculty of hearing as such. Wever and Bray were clear on the matter: "The cat was selected as a suitable animal for this investigation. It stands fairly high in the animal scale, and we have casual and experimental evidence indicating its hearing to be comparable to that of man."[27] They were not the first to reach this conclusion. In his discussion of the telephone theory of hearing, published in 1926, Edwin G. Boring suggested the same experimental approach, right down to Wever's study of cat hearing that immediately preceded his work with Bray. Boring wrote that the answer to questions about the relationship between nerves and sound frequencies "will necessarily lie in the experimental determination of the refractory period of the acoustic fibers with a sound stimulus, an experiment which, I am told, is perfectly practicable since the action current can be measured by placing electrodes upon the medulla. The experiment should be performed upon a mammal (e.g., cat) and ultimately the upper limit of hearing should be determined for the same kind of animal."[28] Boring was primarily interested in *human* hearing. The cat was merely an "e.g.," an example, a detour from the main event—the human brain and the human mind.

What would it mean to take the psychologists at their word, that the cats were nothing more than stand-ins for people in these experiments? The substitutionist logic in Boring's and Wever's and Bray's writing, and the work of the decerebrators before them, suggests a warrant for doing so. Another warrant comes from the paired histories of vivisection in medical research and antivivisection campaigns. In both cases, the progress from animal to human, and the consequent construction of the animal as stand-in for people, is quite clear. Opposition to human vivisection was an outgrowth of the anticruelty movements of the nineteenth century because antivivisectionists believe that it was a short step from experiments on animals to experiments on people. The record shows that their fears were justified given the number of documented cases of experiments on the mentally ill, criminals, children (especially orphans), and poor populations during the late nineteenth century and the early twentieth. Some of the experiments were quite close in approach to the cat decerebrators' experiments.[29] As John Durham Peters explains, "The condition of animals has long served as a political allegory of the treatment of humans."[30]

So let us consider the condition of Wever's and Bray's cat head as a political allegory. After opening up the skull and severing the connection between the upper and lower parts of the brain, the "active electrode, usually in the form of a small copper hook, was placed around the nerve, while an inactive electrode was placed elsewhere on the body, in most cases on the severed cerebral tissue or on the muscle of the neck. In most experiments the left pinna was removed and a rubber tube sewed into the external meatus. The tube led to a funnel into which the stimulating sounds were delivered."[31] A close-up of the cat head in their telephone diagram renders the scenario quite clearly (see figure 13).

A telephone wired directly into the brain, a mouthpiece with a tube sewed directly into the head. What a perfectly coercive propaganda model! Here is a head, physically connected to a communication system, from which it cannot disengage itself and which it cannot turn off. Certainly, Wever and Bray were working in the point-to-point world of telephony and had completely discounted the cat's consciousness and so perhaps this sounds extreme or far-fetched. But Wever's and Bray's contemporaries in social psychology

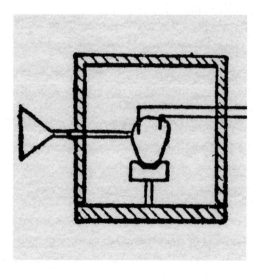

13. Detail of cat head in cat telephone diagram.

and other related fields were already drawing out the implications of this arrangement for mass media like radio or film. Regardless of whether the motivation was social reform or more effective advertising and public relations, media were often conceived in this age in terms of their effects on users and audiences, and those effects were considered to be substantial and direct. Later writers would criticize the findings and methodology behind so-called hypodermic needle effects models, but the fantasy remains a powerful cultural script that has been replayed throughout the twentieth century around comic books, television, video games, and the world wide web. More important for our purposes, this hypodermic model also conceives of communication as primarily a function of transmission, an assumption it would share with the then-emergent metascience of cybernetics.[32]

It is perhaps obvious to cast the cat head as something of a cyborg, a cybernetic organism, "a creature of social reality as well as a creature of fiction"[33]—or not *exactly* fiction, as it turns out. The fusion of cats and communication systems appears as a recurrent theme in cybernetic thought. A robot cat was the case in point in a classic paper by Arturo Rosenblueth, Norbert Wiener, and Julian Bigelow, titled "Behavior, Purpose and Teleology" and published in 1943: "In future years, as the knowledge of colloids and proteins increases, future engineers may attempt the design of robots not only with a behavior, but also with a structure similar to that of a mammal. The ultimate model of a cat is of course another cat, whether it be born of still another cat or synthesized in a laboratory."³⁴ The physical fusion of cats and communication systems would be tried again two-and-a-half decades later in the Central Intelligence Agency's ill-fated "Acoustic Kitty" operation in 1967, in which they surgically inserted microphones and radio transmitters into cats. The CIA researchers came to the conclusion that while cats could be trained to move short distances, "the environmental and security factors in using this technique in a real foreign situation force us to conclude that for our (intelligence) purposes, it would not be practical." Perhaps the agents were convinced when the CIA's first (and probably last) trained feline spy was run over by a taxi early in its first mission.³⁵

There is some reason to consider Wever's and Bray's cat head as an early iteration of the cybernetic thinking that became fashionable later in the century and was so central to packet switching and other technologies crucial for the development of digital media. Cybernetics has had several different iterations, but its core meaning has been consistent. The name *cybernetics* comes from the Greek word for "steersman," and Norbert Wiener defined it as the science of control over the animal and machine through communication. There were several crucial innovations in cybernetics. Theorists in the field promoted the premise that control is a form of communication (or more accurately transmission, since there are other kinds of communication); the premise that insofar as we consider the operation of a system to be the desired outcome, the difference between people and machines is not particularly important; and the premise that the goal of systemic organization is maximum efficiency and maximum control. Alone, none of these premises were particularly revolutionary at the time. Taken together, they constituted a new, general, and interdisciplinary metalanguage for the sciences. Even now, cybernetic thinking is striking in its cool rhetoric of the collapse of human and machine, though such elisions have been quite common since the industrial revolution subjected the rhythms of human labor to the rhythms of machines.³⁶

Psychoacousticians now often discuss the human brain in terms of electrical circuits and information processing. Cybernetics and information theory have to a great degree suffused the language of psychoacoustics. For instance, in the first chapter of Brian C. J. Moore's standard textbook on the psychology of hearing, he frequently refers to the perception of sound as "information" and uses the language of circuitry in casting the middle ear as an "impedance-matching transformer."³⁷ There is more than a little echo of Wever's and Bray's way of thinking in Moore's characterization. Thus far, we have focused on the literalization of the ear-as-telephone metaphor, but it also goes the other way. As we saw with the development of perceptual technics, the telephone circuit's theoretical extrapolation into communication in general also occurred during this period. Just as AT&T developed psychoacoustic research to facilitate maximum use of the bandwidth in its phone lines, the company also developed the math that would become central to the theories of information and communication in cybernetics. Psychoacoustics and cybernetics drank from the same well. They were nourished with the same water.

The standard histories of cybernetics consider the field, and information theory in particular, to be the outgrowth of two Second World War technological complexes: calculating machines and war machines. Manuel De Landa makes the connection explicit. He calls Claude Shannon "the creator of the elementary 'cells' in the body of modern computers," which, with the aid of military motivations and resources, moved "from switching relays to vacuum tubes, then to transistors, finally to ever-more dense integrated circuits." In examining cybernetic theorists' own justifications for the field as a new "universal strategy" for the sciences, Geoff Bowker cites the physicist Niels Bohr, who wrote that the "development of automatic control of industrial plants and calculation devices" was the material context within which it was now possible to think of organisms and machines in fundamentally similar terms. Peter Galison's history argues that the key moment was when Norbert Wiener, having spent a portion of the Second World War trying to predict proper trajectories for antiaircraft guns, realized that he could expand his model to "the vast array of human proprioceptive and electro-physiological feedback systems." The equivalence between human and machine came, at least in part, from an effort to calculate the best way to shoot down enemy gunners.[38]

To these contexts and motivations, we must add another: the phone company. As Paul Edwards notes in his history of the Harvard psychoacoustics laboratory, there was ongoing conversation between Bell Labs (and its infrastructural interests) and the military-funded projects at Harvard during the Second World War. Among other things, the psychoacoustics lab was an important early audience for Claude Shannon's communication theory.[39] If cybernetics was to collapse the distinction between signal-processing equipment and the processes of life itself under the rubric of a general science, one of its crucial antecedents was AT&T's incorporation of human capacity into circuit design in its mix of perceptual technics and a broader social program.[40] Donna Haraway has already noted a connection where "nature is structured as a series of interlocking cybernetic systems, which are theorized as communications problems. Nature has been systematically constituted in terms of the capitalist machine and market."[41] Although the technological descendants of cybernetics and information theory have suffused military thought and practice, they are also constituent features of the everyday media economy. Cybernetics and information theory do not just promote technical logics; they also promote fundamentally economic and economistic logics. This is why the phone company's economic interest in information is a crucial companion to a military history of communication.

As with the psychoacoustics that grew out of phone research, cybernetics sought to study the traffic between animal and machine, and in so doing the field tended to elide the difference between brains and media. Norbert Wiener writes, "When I give an order to a machine, the situation is not essentially different from that which arises when I give an order to a person. . . . The theory of control in engineering, whether human or animal or mechanical, is a chapter in the theory of messages."[42] For him, all of social and material life was abuzz with the movement of messages through systems of transmission. Those messages were sometimes perceived by consciousness and sometimes were unconscious, moving through living or physical bodies by their own motivations. Since the distinction between animal and machine was irrelevant, so was the distinction among conscious, unconsciousness, and preconsciousness. Life itself was a kind of electrical machinery. Perhaps not coincidentally, one of his central examples of life-as-messages involves a kitten:

> I call to the kitten and it looks up. I have sent it a message which it has received by its sensory organs, and which it registers in action. The kitten is hungry and lets out a pitiful wail. This time it is the sender of a message. The kitten bats at a swinging spool. The spool swings to its left, and the kitten catches it with its left paw. This time messages of a very complicated nature are both sent and received within the kitten's own nervous system through certain nerve end-bodies in its joints, muscles, and tendons; and by means of nervous messages sent by those organs, the animal is aware of the actual position and tensions of its tissues. It is only through these organs that something like a manual skill is possible.[43]

The kitten's consciousness is mysterious and inaccessible to Wiener, but so are the signals in its nervous system, calibrated as they are to the animal's digestive system and the movement of a swinging spool of yarn. Despite the electronic language of circuits and messages, Wiener is not saying the kitten is a machine (although I imagine it as a machine every time I read this passage). He is saying that the distinction between a kitten and a machine does not matter for the purposes of control and communication. And therein lies a line of descent from Wever and Bray to cybernetics. A cat is a phone is a cat, and a message is a message is a message. Or rather, a message is information.

The Concrete in the Abstract: Parochializing Information Theory

Another defining innovation in cybernetic theory was the definition of information as a pure quantity without content.[44] This is especially important to note in a book on the MP3 format, since the binary math that cyberneticians developed after the Second World War would become the basis of modern computing, packet switching, and digital code itself. While it might at first seem like a big leap to argue that the phone company is behind a set of innovations more often associated with research for war and weaponry, I will illustrate this possible alternative history of cybernetics through a close reading of one of its own founding documents, *The Mathematical Theory of Communication*. Psychoacoustics and cybernetics, perhaps the two discourses most central to the workings of the MP3 format, thus both have their roots in the phone company's efforts to send less signal down the line, to incorporate users' bodies into the workings of the phone system, and to rationalize the process of communication for the purposes of maximizing profits.

Claude Shannon's master's thesis, completed in 1937 and published in the following year, was titled *A Symbolic Analysis of Relay and Switching Circuits*. It used George Boole's algebraic system for manipulating the numbers zero and one to explain electrical switching. At the time, Boolean algebra was relatively unknown. Today it is the basis of digital computers and telecommunications systems. Shannon argued that binary values in symbolic logic and electric circuits were essentially identical, and that it would therefore be possible to build a "logic machine" that used switching circuits according to the principles of Boolean algebra. After his doctorate, Shannon went to work for AT&T's Bell Labs, where he did some of his most famous work. He was hired to help Bell increase the efficiency of transmitting signals down telephone lines. By applying the principles of Boolean algebra to telephone switches, Shannon provided a major breakthrough in resolving this problem. He developed *The Mathematical Theory of Communication* in relation to this work, as well as work on cryptography and fire detection during the Second World War.[45]

As Warren Weaver writes in the introduction to the book version of *The Mathematical Theory of Communication*, "Information is a measure of one's freedom of choice when one selects a message" — specifically, it is the logarithm (to the base 2) of the number of choices one has. If one were to have two choices, "the transmitter might code these two messages so that 'zero' is the signal for the first, and 'one' the signal for the second; or so that a closed circuit (current flowing) is the signal for the first, and an open circuit (no current flowing) the signal for the second. Thus the two positions, closed and open, of a simple relay, might correspond to the two messages." In other words, information is best represented through binary math, and the quantity of information in any given scenario is measured in bits, which is a contraction of "binary digit." Thus, if one had three binary relays instead of only one, there could be eight possible coded messages: 000, 001, 011, 111 and so on.[46] It thus draws a direct connection from Boole to circuits.

Already in the nineteenth century, scholars interested in logic and math were designing machines to compute complex problems. Most famous was Charles Babbage's failed attempt to construct a "difference engine," but many smaller machines were in fact built. Among them was a small logic machine fashioned by Allan Marquand, a student of Charles Sanders Peirce. Marquand's machine could handle the relationships among four separate true-false propositions through a system of rods and levers (connected by catgut strings, it should be noted). In a letter to Marquand, Peirce suggested modifying the machine: "I think electricity would be the best thing to rely on. Let A, B, C be three keys or other points where the circuit may be open or closed. As in Figure 1 there is a circuit only if *all* are closed; in Figure 2 there is a circuit if *any one* is closed. This is like multiplication and addition in logic." Like Charles Babbage's work on the difference engine (which was largely forgotten until well after the first computers were built), Peirce's letter was not known at the time that Shannon and Weaver published *The Mathematical Theory of Communication*.[47] But the basic premise is the same.

Shannon's contribution was to actually work out the binary math in a form substantial enough to have general applicability to communication engineering.

The connection between information theory and AT&T's corporate strategy is clear in the first two sentences of Shannon's *The Mathematical Theory of Communication*. In his first sentence, Shannon cites the exchange of signal quality for bandwidth as his primary motivation: "The recent development of various methods of modulations such as PCM and PPM which exchange bandwidth for signal-to-noise ratio has intensified the interest in a general theory of communication."[48] Perceptual technics offered a method of negotiating this trade. Shannon was to propose another, drawn from telegraphy. Before AT&T figured out how to quadruple its phone-line capacity in the 1920s, Western Union had searched for a "multiplex telegraph" that would allow for multiple signals to be sent down a single line. The research Western Union funded in the 1870s and 1880s helped lead to the development of the telephone. The idea of multiplexing had been applied to telephony as early as 1891, but no workable model had been devised. One of the first breakthroughs was to find a way to divide a single line so that it could be used for telegraphy and telephony. By 1910 Bell System engineers had discovered that telegraph signals occupied the lower portion of the frequency spectrum—up to about 60 hz, whereas the important frequencies for the telephonic transmission of speech were between 200 and 3200 hz. A means for dividing up the signal was needed, and between 1910 and 1915, Bell Labs engineers worked on various kinds of bandpass filters; that is, filters that would allow for the transmission of only part of the frequency spectrum (by heavily attenuating signals outside the chosen band). These bandpass filters allowed for transmission of multiple signals on a single phone line, with the provision that two signals could not occupy the same frequency bands. Additionally, they restricted the pulses of the telegraph signals to their fundamentals only (eliminating harmonics) and introduced a tiny time lag into the signal. In other words, the motivation for a "general theory of communication" descended directly from AT&T's attempt to extract more profit through more efficient use of its infrastructure in a limited-competition environment. The mosquito of corporate capitalism was embalmed in the amber of information theory.[49]

Shannon's second sentence locates the genealogy of his "general theory" in telegraph and telephone research: "A basis for such a theory is contained in the important papers of Nyquist and Hartley on this subject."[50] In a line of inquiry parallel to the work of Harvey Fletcher's group, Harry Nyquist and R. V. L. Hartley had been studying ways to make the telegraph more efficient for the Bell System. This was crucial, because after AT&T acquired Western Union in 1909, their engineers set out to find a way to better automate telegraph traffic. Until 1910 Bell's telegraphs were run manually and used Morse code, meaning that a telegrapher sat at each end of the line. One would transmit the message; the other would hear the dots and dashes coming through the sounder, translate them back into English, and write them down. This allowed for a maximum speed of about eighteen "dots" a second and also meant that malfunctioning lines could be tested and calibrated through a simple system.[51]

Now that it had an essential monopoly over telegraph lines, Bell System management decided to move from Morse code and operators to teletypewriters. Teletypewriters could theoretically communicate at a speed much greater than eighteen dots per second, and they could be more fully automated, thus saving the Bell System a great deal of money—both because more messages could move through its system and because it would need to employ fewer telegraphers. But much more sophisticated tests needed to be designed for telegraph lines. Telegraphers traditionally described distortion on telephone lines as having bias—signals could be lengthened (positive bias), shortened (negative bias), or the telegraph "hung" in stuck position either on or off. But with teletype, many other kinds of bias could affect the signal. The engineers' goal, then, was to ascertain the tolerable levels of distortion in the channel and make sure distortion was low enough for the signals to get through. While telegraphers could operate at levels of up to 50 percent distortion (sometimes more), management set the requirement for teletypewriters at a maximum of 35 percent distortion, which comes out to about one erroneous character per hour assuming a transmission rate of sixty words per minute. This low level of distortion was necessary because AT&T's main market for teletypewriters was financial institutions, other large corporations, and government agencies that required a high level of accuracy in written messages. Its early teletype customers would include the United Press Association, the Bureau of Aeronautics of the Department of Commerce, the New York Police Department, and countless banks and brokerages.

Unlike a telegraph operator, a teletype was not self-correcting and could not be counted on to recheck the transmission to make sure the error was fixed.[52] This line of research required the development of sophisticated mathematical theories of digital signal pulses in relationship to bandwidth, a preoccupation of Bell researchers throughout the 1910s and 1920s. By 1910 AT&T's telegraph researchers conceived of telegraph signals as "groups of discrete digital pulses arranged in suitable time sequence and duration to enable trains of such pulses to define suitably each letter, figure, or other symbol within the message being transmitted."[53] Hartley, Nyquist, and others built on that basic premise to calculate the maximum number of symbols that could be transmitted in a given time over a given bandwidth, and the results of their work were gospel in the Bell system until the late 1930s, when Shannon published an early version of his *Mathematical Theory*.

To readers only familiar with information theory in the 1940s and after, Hartley and Nyquist's papers might seem surprising because they contain many of the premises for which Shannon is later credited in standard histories. In their papers published in 1928, both Nyquist and Hartley posit telegraphy as a binary system. For instance, Nyquist divides the Morse code letter A into six separate time units: a dot, a space, a dash (equal to three dots), and a space.[54] By 1911 all Morse code letters could be represented in this fashion. As the *Cyclopedia of Applied Electricity* (1911) explained, "The entire scheme of the Morse Code, with its dots, spaces and dashes, their combinations and their relative time values computed according to the unit of time—the dot—and the letters, figures and characters they represent, is shown graphically in the accompanying chart [figures 14a and 14b]."

There would be a series of conceptual steps from the *dot* as a fundamental unit of time in telegraphy to the *bit* as a fundamental unit of communication in Shannon's writing, but the path was already taking shape in the 1910s.[55] Already in the *Cyclopedia*, communication was understood in terms of a binary on-off operation. What was true for telegraphy in this diagram would, forty years later, be taken as true for that quintessential twentieth-century entity, information. In the figure below, although the dashes are two units (rather than three as in Nyquist and Hartley), the basic premise is the same, that Morse code is a binary form so long as we divide time into discrete, equal durations. Nyquist followed the same approach. He divided the time into equal units (noting that they need only be *approximately* equal) and within each time unit posited a finite number of possible conditions. From this basic premise, he was able to mathematically model a wide range of possible telegraphic scenarios, select "a criterion

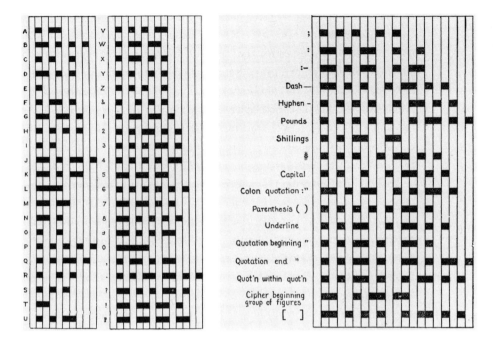

14a and 14b. Morse code as a binary system, as it appeared in the *Cyclopedia of Applied Electricity* (1911). Source: American School of Correspondence, *Cyclopedia of Applied Electricity*, vol. 6, 151, 177.

of perfect transmission," suggest conditions that would meet this requirement, propose several alternative scenarios, and offer an elaborate discussion of the differences between different kinds of telegraphs and the effects of interference on them. In short, the move from Morse code to math allowed Nyquist to extrapolate from specific telegraph systems to a general theory of telegraphic signals.[56]

R. V. L. Hartley took this one step further by expanding Nyquist's general theory of telegraphy to a general theory of all communication systems. He aimed to "set up a quantitative measure whereby the capacities of various systems to transmit information may be compared" and applied it to "telegraphy, telephony, picture transmission and television over both wire and radio paths."[57] Hartley's "quantitative measure" would later become the cybernetic definition of information found in Shannon and Weaver. The similarity is striking. Hartley believed that it was "desirable therefore to eliminate the psychological factors involved [in communication] and to establish a measure of information in terms of purely physical quantities." Like Shannon after him, Hartley realized that if information was to be conceived as a quantity, the "amount" of information is conditioned by—but not simply equal to—the range of possible selections at any given moment. Specifically, his measure of information was the logarithm of the number of possible sequences in a message. Hartley's model presumed that the "capacity of a system to transmit a particular sequence of symbols depends upon the possibility of distinguishing at the receiving end between the results of various selections made at the sending end."[58] This model of system capacity corresponds exactly to the questions that drove the contemporary work of Harvey Fletcher and Georg von Békésy in hearing research. If the ear was less sensitive than the telephone receiver, if it could not discern certain elements of the sound "encoded in a message" on the transmitting end, then there was no point in transmitting those sounds. Twenty years later, Shannon would take Hartley's definition at face value. "The fundamental problem of communication," wrote Shannon, "is that of reproducing at one point either exactly or approximately a message selected at another point."[59] Like the psychoacousticians before them, the information theorists and their mathematician progenitors at Bell Labs wanted to isolate the minimum threshold of communicability—or rather, the minimum threshold of intelligibility to people and to machines—in any system of transmission. The universal ambitions of this theory were clear from the start.

Hartley's model was not just a model of telegraphy or telephony. In sequence, he applied his informational approach to the study of telegraphy, telephony, radio, sound recording, picture transmission, and television—all in the course of a twenty-eight-page article.[60] Shannon's examples were largely the same:

> We may roughly classify communication systems into three main categories: discrete, continuous and mixed. By a discrete system we will mean one in which both the message and the signal are a sequence of discrete symbols. A typical case is telegraphy where the message is a sequence of letters and the signal is a sequence of dots, dashes and spaces. A continuous system is one in which the message and signal are both treated as continuous functions, e.g., radio or television. A mixed system is one in which both discrete and continuous variables appear, e.g., PCM transmission of speech.

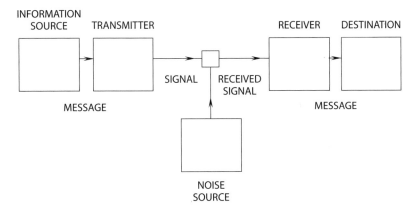

15. Claude Shannon's "Schematic Diagram of a General Communication System."
Source: Shannon and Weaver, *The Mathematical Theory of Communication*.

> We first consider the discrete case. This case has applications not only in communication theory, but also in the theory of computing machines, the design of telephone exchanges and other fields. In addition the discrete case forms a foundation for the continuous and mixed cases which will be treated in the second half of this paper.[61]

Shannon proceeds through exactly the same sequence that Hartley had followed in 1928, a sequence ordered not only by mathematical convenience, but also by the needs and interests of Bell Labs as it sought to make AT&T's enterprise more efficient and more profitable. Figures 15–18 show how this process looks when we map social relations, media industries, and communication infrastructures back onto Shannon's signal-flow abstractions.

Hartley is therefore a historical source of two of the central premises for which Shannon is most often credited today—the idea that information is a pure quantity (the logarithm of the number of choices in a message) and the idea that even though this model was derived from telegraphy, it is applicable to all communication systems. The central tenets of the cybernetic theory of information thus emerged from research on phone systems in the 1910s and 1920s. Shannon acknowledged as much when he wrote that Nyquist and Hartley formed "a basis" for a general theory of information. However, they were its core, not its precondition.

Shannon's true innovations can be better understood now that the debts named in the first two sentences of *The Mathematical Theory* have

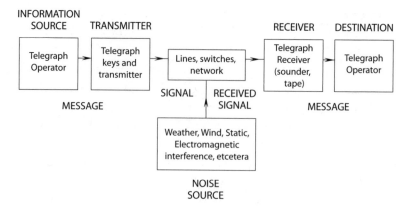
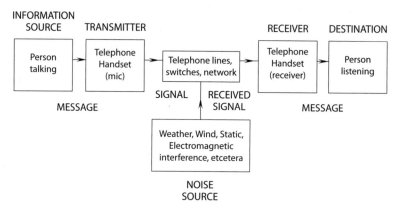
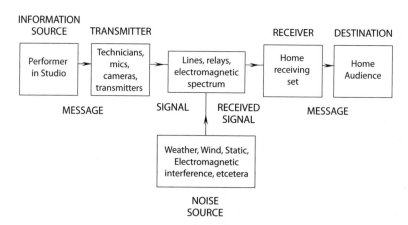

16. Shannon's model, mapped to telegraphy. Image modified by Liz Springate.
17. Shannon's model, mapped to telephony. Image modified by Liz Springate.
18. Shannon's model, mapped to broadcast. Image modified by Liz Springate.

been explored. As already noted, he showed that the problems preoccupying Bell's engineers could be represented not only logarithmically (as had been demonstrated by Hartley), but also through binary math or bits. While the telegraphic *dot* could simply approximate equality of duration, the mathematical bit was precisely a value of zero or one. His mathematics of communication was more properly probabilistic. He abstracted the problem socially to correspond with his mathematical abstraction of the problem. Weaver especially, in his introduction to Shannon's paper, argues that there is no functional difference between the systems that Shannon, Nyquist, and Hartley studied and any imaginable form of communication, including face-to-face speech. Derived as it was from the specific problems of the telegraph and telephone system, the information theory model of communication could be applied to *any* area of human conduct. As Weaver proudly noted, it was not *a* mathematical theory of communication; it was *the* mathematical theory of communication. In other words, the communication system was the ideal type of communication, and a given communication event was some kind of approximation of this system.

There is an uncanny resemblance to another contemporary definition of communication, one tied to propaganda research and mass psychology: who says what in which channel to whom with what effect? Although originally attributed to Harold Lasswell, the idea emerged from John Marshall, the associate director of the humanities division at the Rockefeller Foundation. As William Buxton points out, Marshall's purpose was not so much to offer a general theory of communication as to explain how the emergent field would play a role in the war effort, and how its problems might be divided among different researchers.[62] Although the problem was different, the basic question of transmission was connected to the relationship between encoded and received messages.

Shannon also added another crucial factor to his calculations: noise, or entropy within the system. In 1928 Hartley had written that "external interference, which can never be entirely eliminated in practice, always reduces the effectiveness of the system. We may, however, arbitrarily assume it to be absent, and consider the limitations which still remain due to the transmission system itself."[63] Shannon turned Hartley on his head and considered noise as *part* of the communication system in order to find ways to deal with it. As Weaver explains, noise was part of the system itself: "In the process of being transmitted, it is unfortunately characteristic that certain things are added to the signal which were not intended by the information source. These unwanted additions may be distortions of sound (in telephony, for example) or static (in radio), or distortions in shape or shading of picture (television), or errors in transmission (telegraphy or facsimile), etc. All these changes in the transmitted signal are called *noise*." Weaver went on to point out that the greater the number of possible messages in a system, the greater a chance for error—with increased information comes increased noise.[64] Shannon's contribution, which was also a conceptual cornerstone of cybernetics, is that the goal of information theory was to find ways to minimize noise in an environment where there were many possible messages.

And thus emerged a group of engineers and scientists who understood a fundamental homology among the processes of hearing, electrical communication systems, and computers. They were concerned with the process, rather than the content, of messages. They looked for ways to trade in signal definition for better bandwidth so that they might send more messages down a single line. In so doing, they sought maximum efficiency in their transmissions and looked for satisfactory, or in Shannon's words, "approximate," symmetry between the moment of encoding and the moment of reception.

Reverse Engineering the Napster Cat

Efficiency, symmetry, the privilege of process over content, the homology between mind and media: these were embodied ideals of mid- and late-twentieth-century engineering culture.[65] With the rise of big science alongside R&D, it is safe to say that the ideals of bureaucracy and efficiency so central to the business corporation of the early 1900s became values held dear for the more noble goals of scientific collaboration in the service of some abstract notion of progress.[66] Over time, the proximate motivations for extracting that extra last bit of efficiency from communication systems became more abstract and more diffuse. Few engineers would cite anything like perceptual technics or maximum profitability in a context of stagnant prices as a motivation for improving communication systems. The engineers came by these values honestly because they drive the professional ethos of engineering from top to bottom. Psychoacousticians talk about the human auditory system as an information-processing device because it has been the most effective way to make hearing "work" in a world full of media. But an individual's motivation is not the same thing as a historical or structural explanation. Just as my desire to educate students and promote the cause of knowledge does not explain the historical role or cultural significance of universities, the same can be said for the engineers and psychoacousticians, or members of any profession for that matter. People's investments in their own fields—what Pierre Bourdieu calls the *illusio* or "investment in the game"—come from a wide range of motivations; the effects of those investments often have nothing to do with participants' motivations. "It is precisely the social aspects of scientific practice that are systematically excluded from practitioners' discussions about methodology."[67]

If we follow the history, the core ideas about hearing and communication written into the code of the MP3 emerged from a concrete relationship among interested parties early in the twentieth century—the phone company, its customers, its engineers, its competitors, its regulators. The administrative mentality of the modern corporation shaped research that gave us some of the twentieth century's central concepts of hearing and communication. That same mentality and those same ideas, along with decades of sedimented history beneath them, now dwell among the abstract words one so often encounters in writing about communication technologies: progress, efficiency, power, control, choice, economy, and freedom. Perhaps those words are thrown around so easily and with such frequency because the origins of the values and interests behind their association with communication technology are now so obscure. Two other words explain so much more: administration and exchange.

Perhaps we might encapsulate the story of this chapter in a simple picture (figure 19). A green-eyed cat with a sly smile listening to headphones: the Napster logo stares back. Called the "Kittyhead" by Napster executives, this is the second version of a logo designed by a friend of Shawn Fanning, the founder of Napster. Fanning chose it simply because "it looked really cool." The manager of a branding firm puts it in more conventional marketspeak: "It's challenging and aggressive. It says 'we're the underdog. We're the new kid on the block and we are the antiestablishment.'"[68] Mixed metaphors aside, who really *can* tell a cat what to do, anyway? The Napster logo nicely encapsulates the processes of sedimentation, forgetting, and ambiguity so central to the history chronicled in the last two chapters. Here is a picture of a cat connected to a sound system that is supposed to be a mark of agency (for both company and consumer) and rebellion. And yet it bears

19. Napster Kittyhead. Source: Electronic Frontier Foundation, http://w2.eff.org/endangered/list.php.

an uncanny resemblance to Wever's and Bray's cat head, unconscious, decerebrated, wired, and sewn into a system that it cannot comprehend or choose to leave. Of course, the "Kittyhead" is both everything Napster says it is and a reminder of the violence in the Wever and Bray diagram. In either scenario, we see beneath the visual metaphor of a cat with headphones the homology of interior and exterior systems—rebellious cats, rebellious listeners; communication networks, neural pathways. The homology between micro and macro is already presupposed, and the fundamental question of what it means to listen is already settled. However you read the Kittyhead, listening is above all else about one's position in the world of media, an attempt to negotiate it. It is about the balance among phenomena of administration and exchange, and the place of listening in that configuration. There is no "outside the system," at least not for these cats, because media are already inside them.[69]

The telephone and the panoply of communication problems that sprung from it helped to organize a field of ideas about hearing, information, and communication that are still widely used today. They appear without much thought or history in everyday speech, psychology textbooks, schematic diagrams, and new media technologies. But they are motivated. Inasmuch as we continue to imagine perception and cognition as forms of information processing, we are in part extrapolating from very specific concerns of Bell Telephone in the first half of the twentieth century. What we claim to know scientifically of hearing-in-itself—and how we know it—passes through the modern technological formation of sound. Psychoacoustics and information theory presuppose a vast interconnected system of media and speak in their terms, even as they reach back toward conceptions of nature—human, feline, or otherwise. "Nature builds no machines, no locomotives, railways, electric telegraphs, self-acting mules," wrote Marx as he contemplated the changing configurations of people and machines under nineteenth-century capitalism. "These are products of human industry; natural material transformed into organs of the human will over nature, or of human participation in nature. *They are organs of the human brain, created by the human hand*; the power of knowledge, objectified."[70]

3. PERCEPTUAL CODING AND THE DOMESTICATION OF NOISE

Today, we think of perceptual coding as a digital process, but it did not necessarily have to be. As a concept and technology, perceptual coding was developed in fits and starts, in places all over the world and by people who were not in touch with one another. There is no Alexander Graham Bell or Thomas Edison of perceptual coding. There was no single "a-ha!" moment of discovery or invention, though by some measure it should have been invented in the 1950s, rather than in the twenty-year period between 1970 and 1990. By 1950 AT&T researchers had constructed an artificial basilar membrane, essentially, an electrical model of the inner ear.[1] Shannon's mathematical theory of communication was readily available in the 1950s, and a variety of psychoacoustic models could have been put to work and combined with the analog equipment of the time to make a crude perceptual coder. If the goal was to design a communication system that excluded frequencies in the audible range that users wouldn't hear, it should have been a short step from AT&T's 1920s-era strategy of only passing a limited band of audio for speech to a strategy whereby AT&T went into that band to pull out further unnecessary frequencies. But of course, it did not happen this way. Retrospection is useful to highlight connections that were not always apparent at the time. Even my choice to apply the term *perceptual coding* to technologies and ideas from before the late 1980s is anachronistic, because the phrase had not yet been coined.[2]

"The pages of the history of science record thousands of instances of similar discoveries having been made by scientists working independently of one another," wrote Robert K. Merton.[3] One fable of invention illustrates Merton's point well. JJ Johnston tells a story of a poster session in 1988 at the International Conference on Acoustics, Speech and Signal Processing (ICASSP—a division of the Institute of Electrical and Electronics Engineers, the IEEE). It was a lucky coincidence that he was there at all. "The only reason I got to go to the '88 conference was because it was in New York and they didn't have to pay for travel." Johnston was presenting a poster on his perceptual-coding research at Bell Labs. For several years earlier in the decade, he'd worked on perceptual coding as a method to reduce the bandwidth of digital audio, and as a test of Bell Labs' new computers. His bosses could imagine no practical application, however, and moved him over to video research while they filed for a patent. Once Bell had the patent, the company allowed him to publish his findings. Next to Johnston's poster was one by the German Fraunhofer Institute that discussed work on perceptual coding, presented by Heinz Gerhäuser. "His poster is like the other half of my poster. I'm looking at it and Gerhäuser's looking at it and he's like 'I could give your poster.'" Like Johnston's, Gerhäuser's poster represented several years of work in the area.[4]

Simultaneous discovery occurs when "prerequisite kinds of knowledge and tools accumulate" and "when the attention of an appreciable number of investigators becomes focused on a problem, by emerging social needs, by developments internal to the science, or by both."[5] While interviewing people involved in the development of perceptual-coding technology, I heard several explanations for why the technology developed when it did, usually in some combination. There was a cluster of conceptual breakthroughs that provided routes around existing roadblocks in other methods of audio compression. Increases in computing power allowed for more sophisticated models of the human ear. New techniques in signal processing allowed for more efficient coding. Researchers sometimes made discoveries simply through a lot of trial and error experimentation. And by the end of the 1980s, emergent commercial applications in radio and telephony helped to propel and sustain interest in perceptual coding. These were the accumulated requisite knowledge and tools, the relevant social needs and scientific problems. In this chapter and the next, I will recount some of the stories behind these developments, but I wish to set them inside a larger cultural and historical constellation.

To Merton's motivating social needs and internal developments, we should add a third axis: culture. Culture animates and suffuses research in the broader sense of ambiance, the "whole way of life" in which the researchers live outside their labs. It also subtends research in the more specific sense of the particular historical contexts that shape researchers' practical understandings of sound, hearing, communication, noise, and technology.[6] Perceptual coding emerged at the intersection of ideas about sound, voice, hearing, signal, and noise. The core developments in psychoacoustics and engineering that led to perceptual coding were part of larger historical constellations of sound culture. Early discussions of techniques that would eventually become integral parts of perceptual coding have striking resonances with discussions of sound, noise, and listening that were happening around the same time in psychoacoustics, aesthetics, architecture, music, computer engineering, and dentistry (again, in different countries), even though those discussions happened in different languages and were understood as relevant to different problems and ends. To be realized as a digital process, perceptual coding required a substantial amount of computing power, coding skill, and a workable psychoacoustic model. But three related conjunctures made perceptual coding thinkable in the first place:

1. The ongoing project of seeking to overcome the subjectivity of listening, especially by separating the formal content of sound from any meaning or connotation it might have. The desire to overcome subjectivity was as old as psychophysics itself, and information theory and informational models of hearing further developed it. In the 1940s and 1950s, psychoacoustic research into masking—a phenomenon where one set of frequencies hides another from the ear—began to take off. A predictive theory of masking called *the theory of critical bands* extended the idea that the ear did not simply reflect vibrations out in the world, but instead actively shaped them. The theory of critical bands offered a new way for psychoacousticians to deal with the subject-object split in sound. Humanists like Roland Barthes also sought new ways to think about sound that could get around the thorny problem of subjectivity.

2. The emergence of computers as potential components of sound media, or sound media in themselves. This is the period in which computer engineers and avant-garde musicians first began to use computers to intervene in processes of perception, rather than just as musical instruments or as models of perception (an unsurprising coincidence given the ongoing traffic between these two fields).

3. The domestication of noise. Historically, noise had been understood as something to be eliminated, the object of abatement. But a group of approaches developed over the twentieth century that sought to domesticate noise in one way or another, either to render it useful—in environments like manufacturing, avant-garde music and sound art, dentistry, or office

work—or irrelevant, as was the case in communication engineering. The domestication of noise is perhaps the biggest shift recounted in this chapter, because it recontextualizes the other two major developments I chart—the theory of critical bands and computers' emergence as potential sound media. Instead of considering noise as a problem that masked other wanted sounds, engineers began to imagine that they could move noise underneath other more desirable kinds of sounds. Noise could be masked and put in its place; it did not have to be eliminated. Similarly, if computers could be imagined as sound-reproduction technologies in their own right, perceptual coding could be imagined as a natively digital process.

These are broader developments in mid- to late-twentieth-century sound culture, developments in which engineers and psychoacousticians were caught up and to which they contributed. Thus, the chapter follows two paths to explain why perceptual coding emerged when it did. First it proceeds from an internalist explanation of the history, which focuses on the development of the theories of masking and critical bands, and their application to engineering, especially the coding of digital audio. It then opens out into a broader discussion of two themes: the changing attitudes toward computers as sound technologies and the cultural status of noise, both of which were essential conditions for the form that perceptual coding took at the turn of the 1980s.

One reason that ideas flowed across fields is that people moved across fields. There were movements of people between signal processing, avant-garde composition, psychoacoustics, and communication engineering; between avant-garde music and critical theory; and between psychoacoustics and architecture. The career of Marina Bosi is an excellent example. Bosi was heavily involved in MPEG, eventually chairing it. Before that, she spent two years at the French avant-garde music institute IRCAM (Institut de Recherche et Coordination Acoustique/Musique) doing her doctoral work. She has also worked for Digidesign, manufacturer of Pro Tools, the industry standard in digital multitrack recording. She is now in residence at Stanford's Center for Computer Research in Music and Acoustics (CCRMA), which, like IRCAM, combines interests in experimental music technology and experimental composition. Similarly, we can find architectural acousticians citing the work of psychoacousticians like S. S. Stevens. We can find all sorts of links between Bell Labs and avant-garde composition, especially in computer music, since Bell Labs' Manfred Schroeder (an important figure in the story I tell) worked with Pierre Boulez in founding IRCAM (and other key members, like Jean-Claude Risset and Max Mathews, also worked at Bell Labs). Also, as Georgina Born has argued, the institute was at the center of debates around modernism, postmodernism, and the arts.[7] This traffic among people could explain how ideas in radically different fields might have cross-fertilized one another. It would also be possible to offer an explanation for the emergence of perceptual coding in different places grounded in the globalization and standardization of engineering curricula, since if a generation of engineers is socialized into the set of paradigms, it may well come to the same problems and solutions simply by exploration. But the history of perceptual coding is not only about the history of perceptual coding. It is also about the history of sound. A contextualist history tracks individuals, instruments, and institutions, but it also steps back to consider the broader cultural currents in which they swim.

Taken together, the three conjunctures I have identified resist easy naming. The fields in which they happened were not strictly speaking isomorphic. Thus, I turn to a tradition in cultural theory that gives names to clusters of contingent connections that exert some force and cogency as a group but lack a clearly defined center. Theodor Adorno used the term "constellation" to mean "a juxtaposed rather than integrated cluster of changing elements that resist reduction to a common denominator, essential core, or generative first principle." I find the evocations of the term appealing, and so adopt it here. History writing demands a certain level of interpretation, and writers must make choices in how they represent patterns of events, just as naming patterns of stars in the sky makes them recognizable, endows them with meaning, and renders them useful for navigation. "We need no epistemological critique to make us pursue constellations," wrote Adorno, "the search for them is forced upon us by the real course of history."[8]

Gaps in the Mind's Ear: Masking and the Theory of Critical Bands

The most important psychoacoustic concepts for perceptual coding were the theories of masking and critical bands. Masking proposed that a louder sound could "hide" a quieter sound of similar frequency content from the ear. The theory of critical bands proposed that masking effects could be conceived in terms of frequency regions somewhat like highways in the ear. With knowledge of critical bands, masking response could be predicted. Although masking does not represent a single physiological process (but rather several), it is an operational term, which is to say that it offers a way of simplifying the complex operations of the auditory system, rendering them knowable and predictable. Theories of masking attempt to explain how the limits of human hearing change in the presence of different kinds of sound. They aim to get at the most basic potentials and absences in auditory perception. For this reason, masking research has been one of the central themes in modern psychoacoustics.[9]

As a concept, masking has been understood in terms of the relationship between ears and technologies. This goes back to the earliest published references to masking, such as an article published in 1876 by Alfred M. Mayer. Mayer quoted from his friend Alexander J. Ellis, who made use of clocks in his tests. Though many clocks are silent today, they were most certainly common auditory technologies in 1876:

> Several feet from the ear I placed one of those loud-ticking spring-balance American clocks which make four beats in a second. Then I brought quite close to my ear a watch (made by Lange, of Dresden) ticking five times in a second. In this position I heard all the ticks of the watch, even those which coincided with every fourth tick of the clock. Let us call the fifth tick of the watch which coincided with one of the ticks of the clock, its fifth tick. I now gradually removed the watch from the ear and perceived that the fifth tick became fainter and fainter, till at a certain distance it entirely vanished and was, so to speak, "stamped out" of the watch.[10]

In some ways, this passage seems to sit several miles distant from modern psychoacoustics. Ellis used his own senses as the basis for a scientific experiment, a practice that was already in decline in some scientific fields at the time he was writing.[11] The approach and language of presentation is quaint: comparing the noise of a loud clock (one of those noisy *American* clocks) with a no-doubt quieter German pocket watch by means of his arms. And yet the passage is rich with themes central to subsequent discussions of masking phenomena. Mayer chose a familiar, technologically produced sound as the means by which to study the masking phenomenon. From the very beginning, masking was studied as it occurred in relationship to mechanically produced sound. Subsequent masking studies would use tone generators to generate measurable and repeatable frequencies.

Key among these was a study by R. L. Wegel and C. E. Lane in 1924 at Bell Labs (the same Wegel and Lane who would later lend phone equipment to Wever and Bray for the cat experiments). Using an air-damped telephone receiver and two vacuum tube oscillators with filters designed to eliminate effects due to harmonics (an early form of analog synthesizer), Wegel and Lane produced two tones at variable pitches and volumes: the first tone varied from 150 to 5,000 cycles per second, and the second "masking" tone was set at 200, 300, 400, 600, 1,200, 1,800, 2,400 and 3,500 cycles per second.[12] They had subjects listen to the phone receiver while they varied the tones, asking when one tone would become audible or inaudible. While Ellis's and Mayer's study identified the existence of masking, Wegel and Lane conducted the first large-scale quantitative study of masking. Their article offers simple frequency and amplitude curves that predict when one tone would mask another, rendering it inaudible. They concluded that tones were most likely to mask those most similar to them in pitch (though two tones close together could also produce a "beating" sound) and that the effect was more pronounced at higher volumes.

In the 1920s, masking research asked after the mechanics of the inner ear. It found that the inner ear's workings were situationally variable. Wegel and Lane spent some time in their article on a "dynamical theory of the cochlea," where they hypothesized that masking may be a result of an inability of the basilar membrane to register vibrations that are too close together. "The brain can detect very small changes in position or altitude of the vibration curve of the basilar membrane, but cannot distinguish between these changes unless they exceed a certain definite amount." In other words, small changes in intensity can be mistaken for small changes in pitch. Wegel and Lane supposed that if this was the case with changes in a single tone, it may also obtain in the brain's ability (or inability) to distinguish between two very similar tones. Shortly thereafter, Harvey Fletcher would summarize it thus: "When the ear is stimulated by a sound, particular nerve fibers terminating in the basilar membrane are caused to discharge their unit loads. Such nerve fibers then can no longer be used to carry any other message to the brain by being stimulated by another source of sound."[13]

In masking research, sound was an effect of the changing relationships among vibrations, ears, and minds. Sound was not simply something out there in the world that human and animal ears happened to encounter and faithfully reproduce; nor were human ears solipsistically creating sound through the simple fact of hearing. At its core, the faculty of audition was limited and imperfect. Even if in some ideal conception sounds were understood as a pure quality of vibration out in the world, the theory of masking held that humans did not have the ears to hear it. Hearing was itself a medium and could therefore be understood in terms analogous to the media that were being built to address it. If it were possible to design sound technologies, such as telephones, that could anticipate the behavior of the mind's ear, they could be much more effective (in any number of ways) than if designers simply attended to the physical properties of sound in the air or the sound of transducers, since the physical properties of sound did not neatly correspond to what could be heard. Wegel and Lane presupposed the priority of sound reproduction and the centrality of human speech in their research.[14] Although it was presented as a general theory of hearing, the concept of masking had become medium-specific.

In his subsequent discussion of masking, Harvey Fletcher makes a more sustained attempt to consider its applications. He notes that a theory of masking helps to explain the effects of room noise on perception of telephonic speech and further offers a way of understanding the relationship between ambient noise in the spaces of modern life and the intelligibility of speech in general. Fletcher goes so far as to quantify noise levels and speech frequencies in order to offer a hierarchy of thresholds of intelligibility. At the top is the soundproof room, and his scale then descends through "country residence, quiet office, average office, noisy office or department store, railway train or automobile, New York subway, boiler factory."[15] Though Fletcher doesn't offer any direct application of the theory of masking, already in the 1920s sound researchers considered the audibility and character of reproduced sound in relation to the noisy landscapes of modern everyday life. Fletcher's list of typical sonic environments and their various noise levels suggests that masking might be one key to tuning sound-reproduction technologies to the environments in which they are heard.

Masking theory thus prioritizes a particular concept of hearing in the conceptualization of sound technology. It renders hearing as a process, one that is situated and temporal, since one sound masks another in a moment of coincidence. Crucially, masking also spatializes hearing. The exterior spaces of home, work, travel, and consumption in Fletcher's list suggest that hearing changes depending on where it happens and in relation to sounds out in the world. In a few years' time, this exterior spatialization would be supplemented with a series of interior spaces called *critical bands* (the German psychoacoustician Eberhard Zwicker would later call them *frequenzgruppen*—groups of frequencies). The first use of the term *critical band* appears as "critical bandwidth" in an essay by Fletcher titled

"Auditory Patterns" (1940). In it, Fletcher was interested in the relationship between three tasks: "how to describe and measure the sound reaching the ears; then we need to know how to describe and measure the sensations of hearing produced by such a sound upon a listener. To do this quantitatively, we must also know the degree and kind of hearing ability possessed by the listener."[16] Fletcher's thesis, based on his knowledge of masking, was that there was a good deal of difference between the vibrations in the air and the sounds perceived by listeners. His goal was to find a way to predict or at least describe the ways in which the ear acts as a filter. Different parts of the basilar membrane respond to different frequencies, so Fletcher reasoned that masking tones would effectively occupy parts of the membrane, rendering them unable to hear additional frequencies in the same region until such sounds reached a certain volume. By mapping out the positions of the nerve endings and their responses to sounds, it would be possible to determine the zones of sensitivity and insensitivity to sounds in the presence of a masking tone.

Figure 20 shows a diagram of four notes of a bugle playing "Taps." Fletcher mimicked the snail-shell shape of the basilar membrane in the cochlea for his diagrams. Since each note is made up of a complex of frequencies and since different parts of the basilar membrane are receptive to different frequencies, it is possible to map the resonance of a note on the basilar membrane. Low frequencies are closest to the center and high frequencies are closest to the edge. Of this diagram, writes Fletcher, "for each note the fundamental carries only a small fraction of the total loudness. This is due both to the mechanism of the coronet and also to the mechanism of hearing. So it is the interpretation of these continually changing auditory patterns in the ear which brings to us all of our information about the acoustic world about us."[17] By graphing out the ear's responses to complex tones, Fletcher showed that it was possible to identify a set of "critical bands" for masking, and that each of these bands would correspond to a "filtering" of sound by the ear.[18] By the 1960s, critical bands had become an important concept in the psychoacoustic literature. There was debate as to exactly how they worked, how best to divide up the audible range, and how they related to one another. Fletcher's original "critical bands" were redefined as "critical ratios," and the critical band was then considered to be the frequencies *modified by* the filtering action of the inner ear. The empirical referent of critical bands is debated to this day. Researchers do not agree as to whether critical bands actually exist or are simply useful analyti-

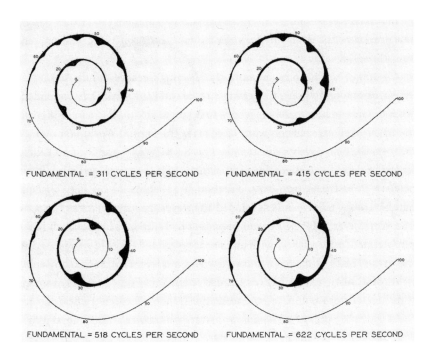

20. Auditory patterns for four notes of a bugle playing "Taps." Source: Fletcher, "Auditory Patterns," 63.

cal constructs.[19] But because of the way it spatialized sound and hearing, critical-band research would be crucial to the development of perceptual coding.

Veit Erlmann argues that the spatial conceptions of audition were one of the most important innovations in the history of ideas of hearing in the twentieth century. They dealt a blow to notions of a resonant hearing self that had been in development for centuries and aided "a quest for a more privileged role for hearing in twentieth-century thought."[20] Erlmann's main case rests on Georg von Békésy's account, published in 1928, of the workings of the basilar membrane. While Helmholtz and his followers believed that different hair cells in the inner ear responded to different frequencies, Békésy showed that zonal divisions in the basilar membrane (in the inner ear) were actually the basis of sensitivity to different frequencies. The basilar membrane contained within it different zones of receptivity. Fletcher's snail-shell diagrams relied on the model of hearing developed by Békésy. Both partook of emergent constructions of spatialized hearing. Békésy's theory imagined how zones of embodied receptivity on the basilar membrane interacted with sound in the world. The theory of critical bands imagined how the ear created divided highways for sonic information and then directed traffic.

Masking research blossomed in the ensuing decades. While American researchers tend to point to Harvey Fletcher and the work that descended from his research, researchers whose first languages are German or French are more likely to focus on work from after the Second World War and to cite two German psychoacousticians: Eberhard Zwicker and Richard Feldtkeller.[21] Feldtkeller was employed at Siemens (which maintained an electroacoustic laboratory) from 1933 on, though he claims in his unpublished autobiography to have only developed an interest in acoustics after the Second World War.[22] Zwicker began his career in Feldtkeller's lab, which happily blended "pure science" with industrial research. He would go on to write important texts of his own, and to develop models relevant to many subfields of acoustic research. Zwicker also spent some time at Bell Labs and thus helped cross-pollinate American and German research traditions.[23]

In 1967 they published a book whose argument was encapsulated in its title, *The Ear as a Communication Receiver*. The book was not so much a revolution in psychoacoustics as it was a synthesis of work that had been going on for some time in their lab and elsewhere. It was also, crucially, a work of translation and popularization. The authors enumerated four goals for the book: to act as a "school of hearing" to translate psychoacoustics for engineers, to inform psychoacousticians about mathematics and physics relevant to their work, to prompt otologists to repeat experiments with hearing-impaired listeners to develop diagnostic tools, and to get musicians interested in psychoacoustics.[24]

Masking was one of their central concerns, and they did much to popularize ideas about masking among people outside the psychoacoustics community.[25] At the core of their explanation of masking was the distinction between stimulus and sensation. As we have already seen, psychoacousticians had long known that external measurements of sound pressure level in the air (a stimulus) did not neatly correlate with the experience of perceived loudness (a sensation). The ear does not respond like an electronic measurement instrument. Central to this research was a concept called the *just noticeable difference*, which was originally developed by the nineteenth-century psychologists Ernst Heinrich Weber and Gustav Theodor Fechner (who popularized Weber's approach and applied it to stellar magnitudes in astronomy). Rather than measuring a physical stimulus and assuming its intensity was the same as its perceived intensity, the Weber-Fechner law described the relationship between the physical intensity of a stimulus and the perceived intensity of a sound, since the two were not the same thing. Psychoacousticians developed new tools and approaches to measure just noticeable differences in hearing. Concepts like the decibel, a basic measure of loudness, resulted from this method. Decibels are logarithmic, which means that a doubling of sound pressure does not result in a doubling of loudness.[26] S. S. Stevens's differentiation of *sone* and *phon* in 1936 was another attempt to construct an absolute scale of perceived intensity (rather than a relative scale where one sound is said to be louder than another).[27]

Zwicker and Feldtkeller extend this project by distinguishing between "stimulus quantities" and "sensation quantities." A stimulus quantity is something like a tone or other sound out in the world that confronts the ear. It can be measured in terms of sound level, frequency, duration, and angle of approach to the listener. A sensation quantity, meanwhile, attempts to overcome the problem of ambiguity in description:

> the listener can only describe the sensation only vaguely. There are only a few words available to describe them: loud and soft, shrill, roaring, murmuring, etc. Nonspecific words from other areas of perception are used to further detail the description: high and low, bright and dark, hard and soft, etc.
>
> The accuracy of this collection is not satisfying. To overcome this limitation, methods have been found by which the relations between stimulus and sensation can be described by curves and mathematical expressions, as is common practice in science. This is possible because the listener can concentrate on a single component of the sensation and abstract it from all the others. . . . It is possible to judge which of two consecutive sounds is louder, even if their pitches differ. Listeners can focus their attention on the loudness and ignore the pitch. The experiment can also be reversed by asking the listener to judge which of the two tones is the highest-pitched one without regard to any difference in their loudnesses.[28]

While there is clearly a relationship between the stimulus and the sensation, one cannot immediately deduce the quality of a sensation from the attributes of a stimulus. Psychoacoustics tries to find a way around that condition when it is applied in perceptual coding. Sensation quantities, then, are Zwicker's and Feldtkeller's attempt to get around what Roland Barthes called the problem of the "adjective" in his essay "The Grain of the Voice" (1972).

Barthes's complaint was that "if one looks to the normal practice of music criticism, it can be readily seen that a work (or its performance) is only ever translated into the poorest of linguistic categories: the adjective.... No doubt the moment we turn an art into a subject, there is nothing left but to give it predicates.... The predicate is always the bulwark with which the subject's imaginary protects itself from the loss which threatens it."[29] From a different direction, Barthes's psychoanalytic language points to the same problem as Zwicker's and Feldtkeller's psychoacoustic language. In both cases, the closed-off world of subjective experience is a fundamental problem of knowledge and power. The goal was to get beyond conscious experience as it is perceived to consider instead a preconcious level of sound (quantities for Zwicker and Feldtkeller, qualities for Barthes) that transcends—or subtends—individual subjective experience. Zwicker and Feldtkeller aimed for repeatable, verifiable, scientific knowledge that transcends any particular individual in the form of statistical aggregates and probabilities. But the only possible route of access to this knowledge was to reach inside and beneath the subject, to get the listener to split his or her sonic experience into the subjective and *something else*. Hence the concept of the "sensation quantity," an abstracted aspect of a sonic experience on which the subject can report. Getting to the sensation quantity required listeners to focus on loudness, pitch, or duration while studiously ignoring the other aspects of the sound. In so doing, Zwicker and Feldtkeller sought to coax their test subjects to reveal aspects of the listening process that were not otherwise accessible to the listeners themselves but that would, after many tests established an aggregate, be rendered accessible. We cannot normally hear our basilar membranes at work. We cannot hear hearing. But we can listen hard in order to tell if we can hear something.

Sensation quantities are numbers made up by psychoacousticians that are based on the reports of listeners who are properly trained for the right kind of hard listening, listeners who already know the relevant qualities of sound and are able to abstract them from one another. The fact that the results are an artifact of the experiment itself, a combination of technologies, forms of knowledge, and highly cultivated techniques of listening, does not make them less real or necessarily arbitrary. On the contrary, the sensation becomes a thing that one can study. The sensation gains density, repeatability, and a measure of materiality because it retains a certain level of opacity and "surplus" for researchers, because something can be known about it but there remains a good deal which is not yet known. This ambiguous status is precisely what allows it to hold the attention of researchers and even, perhaps, become the basis of a technological system.[30] Five years apart, then, writers in two completely disparate fields identify the listening subject as something *that must be gotten through* while traveling the path to the truth of hearing. They proposed routes both above and below the listening subject to get at the truth of hearing, and proposed techniques of listening for vehicles.

Zwicker's and Feldtkeller's other key ideas—such as *absolute and masked thresholds*—follow from the basic separation of stimulus and sensation. An absolute threshold is the point at which a sound becomes audible or inaudible: "Even listeners who are not experienced in hearing experiments can reliably determine their absolute thresholds. Anyone can judge whether a tone is audible of not." But absolute thresholds don't tell us much about hearing in the real world, because sounds always occur in the presence of other sounds. Hence the need for the concept of the masked threshold. Because louder sounds mask quieter sounds, "the masked threshold lies above the absolute threshold, at least in the frequency range encompassed by the spectrum of the masker."[31] This point is not fundamentally different from Wegel's and Lane's claim in 1922 that masking changes the frequency response of hearing, but the analytic is more advanced. In figure 21, the curved line "threshold in quiet" marks the frequency response of ears in silence. The curve "masking threshold" shows how that sensitivity changes in the presence of sound. The "masked sound" would be audible by itself, but in the presence of the "masker," a louder sound, it is inaudible. Zwicker's and Feldtkeller's move to consider masking in terms of thresholds rendered the process quantifiable and therefore easier to translate as an engineering problem. The masked threshold asks after the precise conditions under which a sound is audible or inaudible in the presence of other sounds.

Though it wasn't translated into English until 1999, *The Ear as a Communication Receiver* had an immediate impact in Germany and was shortly thereafter translated into French. It was hugely successful as a work of translation across disciplines, and it had all the necessary conceptual building blocks for a theory and practice of perceptual coding. But *The Ear as a Communication Receiver* does not offer a theory of perceptual coding. Although it is possible to read it retrospectively as hinting at such an approach, we should be wary of giving the authors too much credit in this one

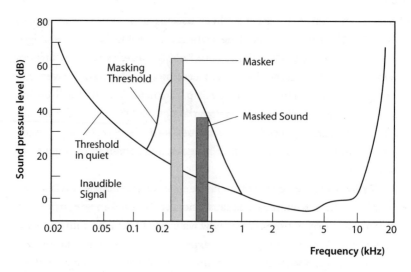

21. The measured frequency response of hearing changes in the presence of sounds. Source: Pohlmann, *Principles of Digital Audio*.

regard. For all the raw materials at hand, there was no extant discussion among psychoacousticians or communication engineers of using the theories of masking and critical bands as the design basis for something like perceptual coding. Their research was useful in countless other areas, and Zwicker went on to develop a method of evaluating the subjective loudness of loudspeakers that became the industry standard.[32]

From Mechanical Mouth to Digital Ear

Bell Labs' work in digital audio descended, at least in important parts, from work in speech synthesis. Since at least the 1930s, Bell Labs researchers were interested in speech synthesis for a number of reasons: it was a key to speech recognition technology (for instance, when you speak a name into your mobile phone and it automatically dials the number for you); it would be useful for reading to the blind; it offered a means to encode signals for security purposes; and it was conceived as a way to save bandwidth by taking speech apart at the transmission end and recompiling it at the receiving end, thereby allowing for the transmitted signal to be of much narrower bandwidth than if it were to contain the full spectrum of speech.

The device that synthesized speech was called a *vocoder*. Vocoder speech sounds intelligible but not altogether human. Vocoders reproduced speech by modeling the mechanism of speech. As Homer Dudley (widely credited as the vocoder's inventor) wrote in 1939, the goal was to build "an electrical synthesizing circuit that would be a reasonably good analogue of the vocal system" and then to provide "the necessary controls for operating this synthesizer by currents obtained from the analysis of a talker's speech." Although vocoders grew smaller in size over the next thirty-seven years, the basic idea behind the vocoder remained the same. Bishnu Atal, who began working on coding of speech in the 1960s, recalls that "the philosophy was that since we produce speech by moving our articulators and modulating the air flow from the lungs through the vocal tract, the information contained is at a much lower rate because our vocal organs cannot move very fast. So it was driven primarily by how we produce voice and since we produce voice in a way that doesn't take much bandwidth, we thought we could compress the voice by somehow mimicking the way we produce it. That was the philosophy for speech compression from the 1920s to the 1970s." Applications of vocoders in the 1970s included Texas Instruments' "Speak and Spell," talking cameras, and the use of vocoders in music.[33]

Although vocoders sounded intelligible, they didn't really sound human. Manfred Schroeder, the director of acoustics and speech research at Bell Labs, became interested in ways to code speech that would allow for some error and variation in the character of speech. In conjunction with Atal, Schroeder developed a method called *adaptive predictive coding*, that would, "as speech is being encoded, continuously predict, on the basis of preceding samples, the subsequent speech samples; compare the prediction with the actual speech; and transmit the prediction error, which is called prediction residual. In decoding, use of the same prediction algorithm and knowledge of the prediction residual allow accurate reconstruction of the actual speech." In other words, adaptive predictive coding only transmitted the difference between a predicted signal and the actual signal, leaving most of the rest of the signal behind. Since the computer at the receiving end would make the same prediction as the computer at the sending end, one would only need to transmit the parts of the sound that hadn't been predicted.[34] The theory was elegant, but in practice it left something to be desired. Though it was a definite improvement over its predecessors, adaptive predictive coding did not produce a very high-quality signal. Its uses were restricted to military applications. In fact, the fate of voice compression at the time looked somewhat dim. Frustrated by decades of slow progress in vocoders, AT&T no longer considered voice compression important for telephone services. The research from the mid-1960s until about 1980 was "driven by government needs for secure voice communication."[35]

One of the problems with adaptive predictive coding was that it generated a lot of noise. As Mara Mills has pointed out, acousticians had tended to define noise in terms of its frequency characteristics: nonperiodic, irregular, or otherwise not behaving like pitched or recognized sound. This was in contrast to a broader subjective and social definition of noise as "unwanted sound" that at its extreme could be a threat to the social order, often because it was tied to unwanted populations, or to the discomfort of relative elites. Harvey Fletcher redefined noise in the 1920s as "unwanted disturbance" in electronics quite broadly, thereby blending the two definitions. "Fletcher reframed noise within the context of intention and control. For him the term signified any 'extraneous sounds which serve only to interfere with proper reception.'"[36] As noise-abatement campaigns had often been about the management and even removal of populations or activities from particular areas of cities, it is perhaps unsurprising that the definitions circulating through communication engineering, information theory, and cybernetics led to an approach that emphasized the elimination of noise or entropy in a channel. This was also a fundamental tenet of Shannon's information theory and Wiener's cybernetics. For Shannon, noise was that which interfered with communication in a channel (extrapolating from the telegraph and telephone to a general theory of communication). Wiener, meanwhile, characterized noise as "extraneous disturbances." To use Michel Serres's turn of phrase, the two ends of a communication channel "battle together against noise."[37]

In concordance with this construct of noise as an interfering element to be eliminated, engineers who confronted noise in adaptive predictive coding originally believed that they could solve the noise problem by finding ways to get rid of the noise, or at least minimize the noise produced by the process. But this solution proved unsatisfactory because listeners' *perceptions* of noise do not neatly correlate to the intensity of a signal—this was the problem of knowledge identified by Zwicker and Feldtkeller (and Barthes). Bell's computers could measure signals, but they could not tell which parts of the signals listeners would hear as signal and which parts as noise. "We were getting saturated," Atal recalls. "Progress was not being made, speech quality was not high, it was not good enough for commercial use." In conjunction with Joseph Hall, a psychoacoustician, Schroeder and Atal started looking for ways to reduce the perceived loudness—rather than simply the signal strength—of the noise in vocoder speech. As Schroeder explains, "We started work on the masking of noise by tones. The results of that were then applied to adaptive predictive coding, with the aim of making the quantizing noise as inaudible as possible, minimizing the loudness of quantizing noise. But basic knowledge about hearing that goes into this work was not available in 1973 when we started this." What they needed was a mathematical model of how the ear worked.[38]

Schroeder had been interested in models of auditory perception since at least the late 1960s. When Hall arrived as a new employee at Bell Labs in 1966, he was given free rein to research "distortions" in the cochlea—in other words, differences between the measurements of sounds that showed up on instruments and the behavior of the basilar membrane in conveying sound. Hall quickly got interested in model building, a process that was greatly enhanced by the availability of new, more powerful computers at Bell Labs. For him, the computer provided "a way of solving a bunch of equations that would take you forever to do it if you didn't have the computer. The thing is running as a sample data system. Rather than an analog model it's a digital model, so you break time up into little intervals and you put in signals representing the acoustic signals." Over time, Hall was able to develop a computerized model of the cochlea, probably the first in the world.[39]

Hall had also read the work of the psychoacoustician Rhona Hellman. Where prior research had studied how noise masked tones, in an article published in 1972 Hellman turned the scenario on its head, examining how tones masked noise. Although the masking effect she discovered was much smaller, she was able to document it. Hellman had studied with Eberhard Zwicker, and there was some traffic between her lab in Boston and Bell Labs in New Jersey. The exact path from Hellman's ideas to the proposition of actually using tones to mask noise is not well documented. Hall probably translated her ideas for Schroeder and Atal, and the three of them then further developed the concept together.[40]

The result of this collaboration by Schroeder, Atal, and Hall is a widely cited article, published in 1979, titled "Optimizing Digital Speech Coders by Exploiting Masking Properties of the Human Ear." In it, they argued that the best way to get rid of noise in speech systems "should not be to reduce the noise power as much as possible, but to make the noise *inaudible* or to minimize its subjective loudness. 'Hiding' the noise under the signal spectrum is feasible because of human auditory masking." Hall's model used masking thresholds identified by researchers like Wegel and Lane, Fletcher, Zwicker and Feldtkeller, and Hellman. His computer model of the cochlea proved that these mathematical models could be computed and applied to digital audio signals. The paper was largely a review, digest, and synthesis of existing psychoacoustic research for engineers. However, Hall did some additional masking tests that used essentially the same methods as those developed by Wegel, Lane, and Fletcher in the 1920s. He played a tone and added noise in the same critical band. He would then increase and decrease the levels of noise relative to the tone and ask subjects at what point they could (or could not) hear the noise. The insights and experiments were simple, but in retrospect, the paradigm shift was huge. Where previously speech researchers and communication engineers treated masking as a problem, as when noise would mask speech on a phone line, Schroeder, Hall, and Atal turned the paradigm on its head. Extrapolating Hellman's research method to a technical principle, they argued that rather than fighting noise, one could simply distribute noise in such a way that it would be inaudible in the presence of a signal.[41]

Bell Labs was not the only organization working on the coding of speech and the use of masking to reduce perceptible noise. In addition to a group of Japanese companies (which would later dispute some of AT&T's patents in speech coding), there was research going on at MIT, in Germany, and in Argentina. The same year that the article by Schroeder, Hall, and Atal appeared, a doctoral student at MIT's Lincoln Labs named Michael A. Krasner submitted a dissertation titled "Digital Encoding of Speech and Audio Signals Based on the Perceptual Requirements of the Auditory System." (It does not appear that the Bell researchers were aware of Krasner's work; Krasner was, of course, well aware of the literature on masking and cites an early paper by Atal and Schroeder on predictive coding.) Krasner began from the premise that audio systems which performed well according to traditional measures like signal-to-noise ratio could still be "judged as annoying to listen to" and listeners may prefer some kinds of systems that exhibit "lower levels of intelligibility" according to traditional measures. Like Schroeder, Hall, and Atal, Krasner tried to build a coding scheme that would use the limitations of the human auditory system to make the system "efficient, using only the bit rate necessary to maintain its [perceived] quality" because "the degradations introduced through its processing are not audible when presented along with the audio signal."[42] And like the Bell

Labs researchers, Krasner turned to computer modeling of auditory masking as the key to the system.

Chapter 3 of his thesis, "Perceptual Requirements of a Digital Encoder," is a lengthy survey of the theory of critical bands, citing the standard psychoacoustic literature on masking of the time. Yet Krasner noted an important gap in the psychoacoustic literature. Previous psychoacoustic researchers had used relatively narrow bands of tones or noise to run their masking tests, but Krasner was hoping to design a processor that could work for speech or music, which are "wideband" phenomena since they are composed of many complex frequencies from the low to the high range of human hearing. His solution was to divide up the wideband signal into twenty-four narrower bands that roughly corresponded to the critical bands identified by Fletcher, Scharf, and others, and so long as within each band the noise was below the audible range within each critical band, no noise would be heard. Krasner implemented his model, built a prototype, and tested it out on listeners with some success. While his system wasn't perfect, Krasner concluded that "by exploiting the limited detection ability of the auditory system as determined by masking experiments, the system achieves performance that is comparable to or better than other encoders at the same bit rate." Although further research was necessary on masking and on the implementation of a masking-based model for audio coding, he concluded that "within a few years, it should be possible to implement multi-channel encoding systems as real-time systems."[43]

Krasner's thesis reads impressively today. He identified the need for a filterbank that corresponds to the theory of critical bands; he privileged auditory perception over engineering specifications; and he even identified issues that still plague perceptual-coding researchers to this day.[44] Yet it didn't find immediate application, and Krasner is much less widely cited than Schroeder, Hall, and Atal (though Krasner does now have a place in the Wikipedia entry on MP3). Perhaps because Krasner's research was funded by the United States Department of Defense's Advanced Research Projects Agency it seemed less relevant to their priorities than to those of the telecommunications sector, where perceptual coding would yield immediate commercial advantages—a point not lost on Krasner, who lists broadcasting as one of the major possible applications of his device. Perhaps it was simply an issue of publication. Krasner's dissertation was never published, and research papers growing out of it remained in relatively limited circulation. Nevertheless, through his approach to music and his testing method, Krasner's work would be an important antecedent for the research at Fraunhofer in the late 1980s.

Fraunhofer's research also had its genesis around the same time that interest in perceptual coding developed at Bell Labs and MIT. Dieter Seitzer, a professor at the University of Erlangen, approached the German government in 1977 to fund a project on the digital transmission of music over phone lines (some aspects of the phone system have been digital since the 1960s). Seitzer was refused funding, but he formed a research group that studied the possibility. Their work soon turned to the problem of data compression, and shortly thereafter they turned to psychoacoustics as the key to the process. One of Seitzer's students was a man named Karlheinz Brandenburg. By the late 1980s, Brandenburg was part of a research group at the Fraunhofer Institute in Germany that was intensely interested in perceptual coding. After JJ Johnston and Brandenburg's colleague Gerhaüser met in New York in 1988, Brandenburg traveled to the United States to work with Johnston at Bell Labs. Between Bell Labs and Fraunhofer, most of the key ideas for the MP3 codec were developed, and the key North American patents still refer back to this collaboration in the late 1980s and early 1990s.

Yet for all this genealogy, we are still left with the question of why perceptual coding emerged when it did. Researchers in at least three separate locations in the 1970s who were not in touch with one another realized that there was something to be gained by employing the ear, and specifically its perceptual limitations as outlined in the theory of masking and critical bands, as the basis for digital sound reproduction in limited-bandwidth environments. At the time, researchers did not necessarily understand perceptual coding to be a paradigm shift or a major new approach. For instance, in a conference paper presented in 1978, Atal and Schroeder noted that the ear registered noise differently in different parts of the frequency spectrum, but their approach to noise was still considered part of the larger project of predictive coding. Hall wasn't fully aware of the connection between his work on masking and MP3 until many years later, when there started to be lawsuits about the patents. Krasner conceived of the problem within the broader rubric of perceptual technics. For him, it was an issue of efficiency in digital channels.[45]

In AT&T's case, it initially struck me as uncannily similar to the shift that occurred in the nineteenth century immediately before the invention of telephony and sound recording. Leon Scott's phonautograph, which was invented in 1857 and led to Bell's telephone and Edison's phonograph, was based on a model of the eardrum, the tympanic membrane, which is more or less still mimicked today in every microphone and speaker that transduces signal to sound or sound to signal. Before Bell and Edison, inventors were more likely to model their attempts at sound reproduction on the means of sound production, either the mechanism of the voice or the means by which music was made. The collision of acoustics, physiology, ear medicine, engineering, and commerce created the first generation of analog sound-reproduction technologies in the 1870s.[46] A similar shift from mouth to ear as a model for sound reproduction occurred at the moment that perceptual coding was first conceptualized. But it is not a replay of the same exact story. Part of the conceptual centrality of the voice for researchers at AT&T no doubt had to do with the economic centrality of speech in the telephone system. Speech was the main sound transmitted through telephony, and a working phone system only required that speech be intelligible, not that it have any particular high definition. Compression was important to save money and also, as we have seen, for security purposes in military or corporate applications. The idea of vocoders and later digital speech synthesis made sense in the mid-century Bell System because they had already been taken a certain distance by ongoing research into hearing (which was more or less continuous from the 1910s on), since the phone already only produced those frequencies most important for speech. Speech was not always the first shore reached as new sound technologies came aground in the twentieth century: electric sound generators, tape machines, and analog synthesizers made tones and music and still only approximate the voice; computers made music long before they spoke.[47]

Speech may have still enjoyed an ideological privilege in some quarters as the most important or meaningful kind of sound, but music figures prominently from the very first conceptualizations of perceptual coding in both Krasner's and Seitzer's work. Thus, the history of audio coding is best characterized as an ongoing process of shuttling between source and destination models. The rise of perceptual coding reflects one movement in this longer historical arc. After reducing the frequency bands for speech as far as they could in the 1920s based on studies of hearing, Bell researchers turned to models of speech production as a way to further reduce bandwidth through technologies like the vocoder. When predictive coding, based on speech models, turned out to be too noisy, they turned back to the ear to solve the problem. The shift from model to model is an ongoing element of innovation in sound-reproduction technology. Take one model as far as it will go, then shift to another to solve a problem raised by the first. The approach continues today in the design of digital signal processing.[48]

Computers as Audio Technologies

It would be easy to read the development of perceptual-coding research as a natural outgrowth of research on masking and critical bands, combined with a long-term interest in efficient signal transmission by the telecommunications industry. But by that measure, perceptual-coding research, or something like it, should have existed more or less at the moment that Fletcher first advanced the theory of critical bands in the 1920s—or at least shortly after the article Fletcher published in 1940 on critical bands. Several factors came together to make possible the ferment of the 1970s and 1980s in this area. Most obviously, there was the matter of computer power. If perceptual coding is to be understood as a digital process, computers would not have been powerful enough to do that kind of processing until sometime in the 1970s, and even then, only the most well-funded research sites, like Bell Labs and Lincoln Labs, would have had computers that could achieve something like perceptual coding. In the 1960s, the IBM 7094 computers at Bell Labs could process at a rate of about 500 to 1, which meant that it would take 1.3 hours to process ten seconds of bandwidth-limited speech, which was fast by the standards of the time. But that did not factor into the clumsy process of analog-to-digital conversion, which would take even more time. Even by 1984 and 1985, when JJ Johnston was tasked to try out the Alliant FX-Series UNIX supercomputer at Bell Labs, it still took hours to process audio. "You understand, there were no CDs at this point. The way I got data was I took an LP, played it on my turntable and then into my cassette deck, carried my cassette deck into work and got a 12 bit A-to-D to spool it onto disk. It took about two hours to get about 10 seconds worth of music. So, I didn't have a big repertoire to work with." Johnston's research into perceptual coding at Bell Labs was, among other things, an attempt to simply give the new Alliant Computer a workout. He'd read the paper by Hall, Atal, and Schroeder and realized that it would be possible to actually build the system that they proposed (Atal apparently mapped out the system but it was never built). "I was the UNIX geek in my department— these things are in UNIX—so it's sort of like, 'Here! You're the administrator. Oh, and by the way, test it.' So I talked to Joe [Hall] and I literally drew up on a board their model—their masking model. And I went back and I programmed it in Fortran and I broke the compiler a few times. Brand new computer, brand new optimiser, you know, that happens, there wasn't something wrong with it."[49]

During the same time period, Fraunhofer used a Hewlett-Packard minicomputer (what we would today call a desktop computer) with a 128 mb disk drive, which was considered high end and expensive for the time. But it was considerably less powerful than Bell's. Karlheinz Brandenburg recalls, "I had to re-write a driver to get audio in and out of the computer in real-time. There was no audio interface; we had to build one ourselves and modify the systems for ourselves." Harald Popp, Brandenburg's colleague at Fraunhofer, recalled that when he started working with perceptual coding as a hardware problem in 1987, it took eight hours to process twenty seconds of music. Oscar Bonello, who invented an early perceptual-coding system in Argentina in the late 1980s, used consumer-grade IBM-PC computers and also designed his own sound card. Curiously, Michael Krasner's thesis does *not* indicate that analog-to-digital conversion was a significant issue—conversion equipment appears to have been ready-to-hand. For speech, "four speakers, two male and two female, were recorded with an Electrovoice 667 microphone in a soundproof room, directly in digital format through a 16 bit A/D converter. For testing, the speech was played back directly from digital storage through a 16 bit D/A converter and Star SR-X electrostatic headphones in the soundproof room." Despite Krasner's interesting exception regarding getting sound into the computer, even highly funded labs were working at the limits of their processing capacity when perceptual coders were being developed in the 1980s.[50]

But additional computing power was not enough. Perceptual coding also required a shift in attitude toward the relationship between computers and sound, and specifically music. Georgina Born's insights about the status of computers in psychoacoustic research at IRCAM in 1984 are an interesting foil in this respect, because they highlight the ways in which computers were moving from analysis to synthesis not only in engineering but also in experimental music. The shifting attitude toward computers also helped make perceptual coding conceivable and possible. A French institute dedicated to avant-garde musical composition with computers, IRCAM had long had traffic with communication engineering. Manfred Schroeder worked with Pierre Boulez in founding the institute, and from the beginning IRCAM researchers were very interested in psychoacoustics and cognitive research (a thread that remains alive and well today in university music schools).

> There has been a development from music analysis as a purely analytic field to one that, employing the computer and in conjunction with the rise of cognitive music studies and AI, aims to provide both computer analysis of musical structure and also computerized models of 'musical knowledge' or 'rules' as aids to composition. The computer has therefore come to be seen as a tool for analyzing the deep structures or 'cognitive rules' characteristic of certain musics, but equally for simulating these rules—and indeed for generating entirely new abstract structures as frameworks for composition.[51]

Born writes that analysis and composition become ever more entwined in this period, and that computer approaches to composition had become more and more autonomous from existing musics. She calls these developments part of an "extreme contemporary expression of musical modernism, the tendency for theory to become prior to, prescriptive of, and constitutive of compositional practice." To this end, IRCAM composers became more and more interested in the "scientific study of cognitive universals" as a basis for their new approach to music.[52] Of course, the stated goals of the IRCAM composers were entirely different from those of engineers at AT&T, MIT, Fraunhofer, and elsewhere. The composers at IRCAM wanted to create music that had never been heard before, while the communication engineers wanted to reproduce familiar speech and music. But the two fields experienced similar shifts in their attitudes toward computers. In both cases, computers moved from analytic tools to modeling devices, and specifically devices that modeled the process of perception. Hall's cochlear model was initially just that, until Johnston tried to build a revised version into a perceptual coder. Perceptual models became programs for composition at IRCAM. The process of sound analysis—as a surrogate for the ear, or at least the mind's ear—thus becomes a core component of the computer's status as an emergent sound technology.

It is possible to conceive of the increase in computer power and the change in attitudes toward computers as sound technologies as twinned processes. The quantitative increase in computing power was such that it effected a qualitative change in the ways that engineers, psychoacousticians, and composers related to computers in this period. But the change in computer power does not explain the particular qualitative changes that happened. Computers had been conceived as universal machines, which can model other processes—including human processes—at least since Alan Turing.[53] So the idea was available but not in dominance. Further, although perceptual coding is generally understood as a digital process, it is possible to also conceive of it as an analog process. Since Fletcher's paper on critical bands, and certainly since Zwicker's and Feldtkeller's detailed enumeration of masking thresholds and just-noticeable differences, one could imagine an analog perceptual-coding device containing a filterbank of twenty-four filters (which correspond to twenty-four critical bands of hearing), an envelope detection system not all that different from envelope filters then in use in synthesizers and vocoders, and a filtering system to remove some signal or subject it to audio compression (which would reduce its dynamic range). Full-bandwidth signal could enter it and it could transmit only the audible parts of that signal, based on some preset parameters derived from available studies of masking at the time.[54]

Given the history of knowledge about masking and the development of analog signal processing (filters and transistors, especially), such a machine could have been built sometime in the mid-1960s, should anyone have seen fit to do so. It is unlikely that our fictional analog perceptual coder would have worked very well. Its ability to analyze incoming signals would have been primitive compared to the kinds of Fourier analysis computers could do even in the 1970s. It might not have reduced bandwidth enough to be useful, and even if it did reduce bandwidth enough, it would likely have been too crude to produce useful speech or pleasing music. But then AT&T's many analog vocoders didn't work very well either, even when they were digitally controlled. Nevertheless, the company spent piles of money and countless hours trying to make them work. Computing was an affordance that made perceptual coding more of a possibility and easier to accomplish, but faster computing does not simply cause the kinds of conceptualizations we find in the late 1970s as people turned to masking to solve communication problems. Other points in the constellation afford a richer view of the changes happening at this moment.

The Domestication of Noise

Masking also has a history as a solution to problems outside of what we normally consider communication engineering. The idea of masking machine noise with recorded music appears to go back at least to 1915, and the Second World War spurred an increase in studies of music to aid in the manufacturing process, as did programs like BBC's "Music While You Work." It was not possible to compete with loud machinery in practice until the development of more powerful amplifiers and more reliable audio compressors in the 1940s. Karin Bijsterveld reads the masking of noise by music as part of "a long-standing symbolism of sound that associated noise with chaos and rhythm with order." Certainly rhythm connoted order, but the practice of covering industrial noise with music also operationalized the ear's masking behavior, even if at the time, the practice wasn't described in terms of masking (or the theory of critical bands, which was still in its infancy in the 1940s). Music in factories was a writ-large example of one of the two goals of perceptual coding: it hid unwanted sounds (noises) underneath more desirable sounds (music).[55]

Another chapter in the domestication of noise might begin like a bad joke: "a man walks into a dentist's office with white noise generator." In 1958 J. C. R. Licklider took a cavity, an idea, and a white noise generator to his dentist, Wallace Gardner, in Cambridge, Massachusetts. Licklider had spent time at Harvard's psychoacoustics lab. By 1958 he was working on human-computer interaction at MIT.[56] Like many psychoacousticians at the time, Licklider was interested in masking. Licklider used his white noise generator to mask the sound of the drill. With just that and no other anesthetic, he claimed to feel no pain as a cavity was drilled out and filled. Exhilarated by their initial success, Licklider and Gardner immediately sought out another patient:

> The office secretary was immediately pressed into service as a patient and observer. This young woman had a history of extreme fearfulness in the dental chair. She had never before had cavities prepared without a local anesthetic or an analgesic. Nevertheless, she agreed to forego the use of conventional agents. She put on the earphones, adjusted the volume of the masking sound and sat quietly through the preparation of three large cavities without appearing to experience any pain. At the end of the session, she reported that she had felt none.[57]

As one episode in a long history of male managers subjecting their secretaries to various schemes around new technologies, the story has all the usual ingredients: men testing out their new toys and then calling in a female experimental subject. Her timidity proves the power of the technology ("extreme fearfulness . . . agreed to forego"), and the success of its

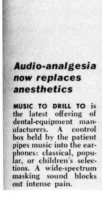
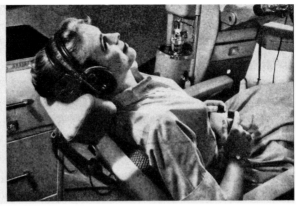

Audio-analgesia now replaces anesthetics

MUSIC TO DRILL TO is the latest offering of dental-equipment manufacturers. A control box held by the patient pipes music into the earphones: classical, popular, or children's selections. A wide-spectrum masking sound blocks out intense pain.

22. Audio analgesia. Source: Maisel, "Who's Afraid of the Dentist?" 61.

use on her paves the way for generalization of the research and the results. Licklider and Gardner thus went on to conduct a series of experiments. They determined that patients needed to be able to control the volume of the sound, and that they needed a choice of music as well as white noise. They named their machine the "audio analgesiac" and offered a selection of tuned noise or music recordings. They called the tuned noise the "waterfall" sound.[58]

In an article on new technology in dentistry titled "Who's Afraid of the Dentist?" the August 1960 issue of *Popular Science* pictured relaxed, smiling women (with lipstick well applied, of course) ready for all manner of dental treatments (figure 22). The answer to the titular question was clear, "with the help of music and masking noise, she's not."[59]

Over the next decade, controversy persisted in the dental community about the effectiveness of the machine and others like it. Some claimed it was a sham; others found it worked in a surprisingly large proportion of their patients. There was no agreement as to why it worked when it did. A review in 1969 of a decade's research concluded that "audio analgesia is not a panacea. While its limitation in use must be recognized, at the same time, its advantages must not be ignored. All pain-relieving drugs are systematic irritants which must be administered to very close tolerances. All have side effects which are often unpleasant and sometimes dangerous. Audio analgesia seemingly has no dangerous side effects if proper limits are followed as to sound intensity and duration."[60] This may be a tempting origin story for the diffusion of Muzak in dentists' offices. But more than that, it is a story about the domestication of noise and the primacy of perception. Licklider, Gardner, and their audio analgesiac enacted a new approach to noise as something potentially useful. For them noise was localizable, knowable, and could be tamed.

The music and dentistry examples appear to have been highly specific. Despite the commercial success of the Muzak corporation, masking music wasn't extrapolated as a general principle of industrial design or architectural acoustics until the 1960s and 1970s. Quests for a dampened, echoless, clean, or "one best" modern sound guided architectural acoustics in the earlier part of the twentieth century. But by the 1960s and 1970s, the field had abandoned its vice-like grip on echo, and a plurality of sounds and sonic signatures were now possible for buildings.[61] In this context, masking noise became useful and productive, rather than something to be eliminated. Leslie Doelle's *Environmental Acoustics* (1972) provides an unusually explicit treatment of the subject. The principle of applying masking noise to mask *other*, unwanted noise appears throughout the book, and the approach is intriguingly analogous to the approach of Schroeder, Hall, and Atal to "hiding" noise via masking:

> In many situations noise-control problems can be solved by drowning out (or masking) unwanted noises by electronically created background noise. This artificial noise is often referred to as *acoustical perfume*, although the term *acoustical deodorant* would be more appropriate. The process suppresses minor intrusions which might interrupt recipients' privacy.
>
> Noise from ventilating systems, from a uniform flow of traffic, or from general office activities contributes to artificial masking noise.
>
> In designing landscaped offices the provision of a relatively high but acceptable degree of background noise (from the ventilating or air-conditioning system) is essential in order to mask undesirable office noises created by typewriters, telephones, office machines, or loud conversation and to provide a reasonable amount of privacy.[62]

Elsewhere, Doelle lists "appropriately selected and well-distributed background music" and "the sound of a water fountain" as other examples of "acceptable" background noise. To be sure, such background noise had to meet some basic standards: masking noise will "beneficially mask other disturbing noises," provided that it is continuous, not too loud, unobtrusive, and "carries no information, such as intelligible speech or identifiable music. An excessive masking noise will impair audibility or intelligibility by drowning out faint speech, soft music, or other very low-intensity sounds one might want to hear."[63]

Doelle's *Environmental Acoustics* proposed a measurement called a *noise criterion curve* (NC) that specified the maximum amount of "preferable" background masking noise, a quantity that would "provide a satisfactory environment for listening to speech and music or for any other activity." The goal of the NC was to help designers facilitate the incursion of useful background noise into built spaces, a project that was sufficient but not excessive since an insufficient level of background noise "often provides inadequate masking and will not secure sufficient privacy against interfering noises from adjacent rooms" and too much would drown out "wanted sounds" like conversation. One might object that Doelle was merely talking about sound here, rather than noise, but it is interesting that he repeatedly emphasized that masking noise must be "meaningless." In his discussion of masking noise in offices, he wrote that "acoustical privacy" was the goal, which includes the eradication or obfuscation of "intelligible speech originating from an adjacent or nearby area."[64]

Doelle's NC descended from a line of research on noise in offices by Leo L. Beranek and his collaborators. Using research questions and methods borrowed from the psychoacoustician S. S. Stevens, Beranek and his collaborators studied the base lines for acceptable noise and quiet in office environments. In studies like "Criteria for Office Quieting Based on Questionnaire Rating Studies" or "Apparatus and Procedures for Predicting Ventilation System Noise," Beranek quantified and classified the sources of noise in office environments and considered those sources as they were perceived by people in those environments. He tried to establish maximal noise levels for different kinds of offices and to distinguish between the kinds and frequency content of sounds that were acceptable to workers in the offices (both executive and clerical staff). But Beranek stopped at establishing acceptable noise levels for office work. For him, masking remained a problem to be avoided, and noise was primarily understood as a problem for abatement.[65] In contrast, Doelle saw masking noise as a potentially positive force that could be used in shaping the acoustic environments of buildings. Thus, Beranek and Doelle's NC measurement is clearly of a piece with Zwicker's and Feldtkeller's masking thresholds. Doelle's combination of quantification and a view of masking as potentially productive set his approach apart from attempts earlier in the century to simply soften the noise of machinery with music. Doelle relied on the quantification of sound and the isolation and objectification of the masking function itself, so that the interaction of sounds within the mind's ear is the guiding principle of sonic design. In Doelle's *Environmental Acoustics*, some types of noise become resources in the attempt to obliterate other types of noise through the systemic application of masking principles. This approach would prove especially useful in open-plan offices (see figure 23). Herman Miller's "Action Office" arrangement, first released in 1968, was a predecessor to the familiar modern cubicle-farm format. Although open-plan offices had existed for decades, this layout was now moved into fields where workers on large floors would need to talk with one another. Doelle's architectural acoustics used masking to create an acoustic space that was smaller than the physical space inhabited by corporate workers and machines (see figure 24). One form of distraction enables or facilitates a second form of productive distraction—for instance, when the hum of a HVAC system smoothes out the noise of conversations in the next cubicle.[66] Like an animal previously thought to be wild, noise was domesticated and put to use in Doelle's architectural acoustics. In the process, noise took up a position as part of the communicative order, rather than as a form of entropy that threatened it.

Decompositionism: Sound Reproduction after Noise

In 1977, the year Dieter Seitzer made his unsuccessful application to the German government to transmit music over phone lines, Jacques Attali published *Bruits: Essai sur l'economie politique de la musique* in France (translated into English as *Noise: The Political Economy of Music* in 1985). *Noise* is an intoxicating and romantic book, and its heavy use of cybernetic language in the guise of social theory makes for fertile reading when set against the history I have laid out in this chapter. Attali offers a sweeping twentieth-century history of the political economy of music. He argues for a new way of thinking through music as a political-economic force, conceived in relation to noise. Attali traces the political economy of music "as a succession of *orders* (in other words, differences) done violence by *noises* (in other words, the calling into question of differences) that are *prophetic* because they create new orders, unstable and changing." For Attali, noise was a threat to order, but also a force for change: because noise challenges social order, he reasoned, power seeks to control noise, silence it, or make it so ever-present as to be banal. Yet he further understood noise and music as also prophetic,

23. An open office (pre-cubicle-farm). Workers, separated by furniture but not by walls. Drawing by Liz Springate.

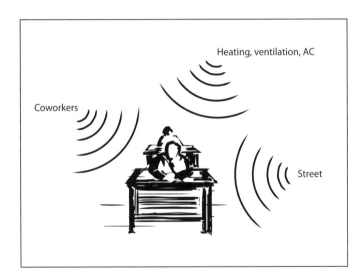

24. Doelle's scheme. Ambient noise reduces the phenomenological sound space surrounding each person. Although the office is open, nobody can hear very far, and the sound space closes in on the listener, establishing some conversational privacy. Drawing by Liz Springate.

able to anticipate emergent forms of social organization in their codes.[67] Attali's conception of noise as chaos, entropy, and generative force no doubt grew from many influences. In the single proposition that noise threatens order and social orders seek to control noise, one hears echoes of Romantic poets, Nietzsche, Marx, twentieth-century antinoise crusaders, and communication engineers and information theorists.

Attali's treatise posited the control of noise as a metaphor for all social control precisely at the moment when communication engineers began to articulate a paradigm where noise no longer needed to be eliminated or reduced if it could simply be rendered imperceptible to the ear. It is possible that Attali's analysis of noise as a total social phenomenon and the hypotheses put forward by Seitzer, Schroeder, Hall, Atal, and Krasner all fit a larger pattern of intellectual history where a discursive or social formation only comes into relief as it begins to go out of existence.[68] For Attali's part, he adopted cybernetic language as a social theory precisely at the moment that communication engineering exhibited a new attitude toward noise. Once you can use signal to hide noise, the game is up. Noise ceases to matter as a perceptual category. Sure, it will register on your instruments, but in a perceptually coded world, you can only see the noise on a screen or a printout. It is no longer for the hearing.

Sound reproduction after noise privileges hearing, but as a striated component of an imperfect communication system. It hides its flawed workings in the gaps and shadows of its imagined listening subject. In perceptual coding, noise beyond the hearable is irrelevant, and so is signal beyond the audible. Perhaps Seitzer, Schroeder, Hall, Atal, Krasner, and Attali all saw the same thing from different directions. The mutual antagonism between noise and control that had been outlined in telephony, cybernetics, and information theory, and which Attali and historians projected back to at least the 1600s, was coming to an end.[69] The engineers confronted the hard limits of communication systems and a paradigm of noise reduction and elimination that, as Bishnu Atal suggested, had run its course both economically and analytically. Dentists explored noise as an antidote to pain rather than as its source. Architectural acousticians turned to masking noise as a productive force in shaping auditory environments. No longer did acousticians seek total eradication of noise; now they merely hoped to arrange it. The same could be said for avant-garde composition, sound art, and audio recording. Noise became a source to be utilized in the service of creative expression, and acoustic space became ever more plastic with multitracking and new forms of reverberation that proliferated from the 1960s onward.[70] Attali may have had just enough critical distance to pen a book that for all its absurd hubris—or more likely, because of it—articulated one of the central sonic problems of Western culture up to that moment.

Attali says as much himself: "We see emerging, piecemeal and with the greatest ambiguity, the seeds of a new noise, one exterior to the institutions and customary sites of political conflict." Attali calls this new age one of "composition," and his musical examples in this section are drawn from improvisational jazz and new directions in art music. "In composition, stability, in other words, differences, are perpetually called into question. Composition is inscribed not in a repetitive world, but in the permanent fragility of meaning after the disappearance of usage and exchange." His language is strident and utopian, in the socialist tradition. For Attali, composition is a utopia because it figures (or prefigures, since for him sonic codes herald coming social orders) the end of the separation between use and economic exchange.[71]

Contemporary writers tend to associate this age of composition with the rise of digital audio, the boom in sound art, the growth of sampling and recombinant music, and the wave of music piracy online and on the streets of many major cities.[72] Yet despite their attractiveness to many people (this author included), none of those practices has quite taken on the utopian tack that would be necessary to fulfill the dreams that come at the end of *Noise*, even if each of them issues its own challenge. Digital audio reduces the cost of recording, but it remains firmly rooted in the ideologies and practices of several communication industries. Digital recording technologies may do just as much to standardize the sound of music—through the proliferation of standards and presets and the tastemaking done by mastering engineers[73]—as to challenge those standards. The boom in sound art challenges the revered status of the museum, the musician, and the composer but it also upholds those notions, if only in the negative. Far too often artists still fetishize noise as transgression or a challenge. Sampling, turntablism, mashups, and remixing all challenge the contemporary order of intellectual property, but they have not undermined it. MP3s may be pirated, but they are products of a money economy and they still operate inside it. A quick look at the available recordings on popular file-sharing sites reveals that the most available recordings are also frequently among the bestsellers.

It may be just as much the case that Attali saw the beginning of the end of his whole conceptual edifice, a set of changes that unsettle the orders of noise he put forward before him. At the very least, that is how I have come to see his moment. To honor Attali's questions, if not his answers, let us use the neologism *decompositionism* to describe the new malleability of sound and noise across cultural domains that emerged in the 1960s and 1970s. That epithet is vague enough that we don't have to see it as the spirit of a whole age or a total social condition (or even a "school" of engineering or artistic practice). We can conceive it merely as a way of thinking and doing things with sound.[74] Its origins clearly lie in perceptual technics, as in Homer Dudley's claim that in his vocoder's "synthesizing process, only the specifications for reconstructing the sound wave are directly transmitted," rather than the sound itself.[75] From psychoacoustics to oil drilling to architecture to sound installations to DJ battles to granular synthesis to cellular telephony to noise-canceling headphones, sound-reproduction technology renders sound meaningful and effective through processes that analyze it, decompose it, and reassemble it.

But decomposition goes beyond the desire to decompile and recompile sound, to analyze it and synthesize it. Decompositionism also has a managerial streak, where all sound and noise is potentially useful and possible to organize. Decompositionism demystifies noise, but it also demystifies sound and hearing in the service of perceptual technics. Through vast technical infrastructures and a new common sense, sound and noise became audio, and audio could be disassembled, isolated into its components, and then administered. Instead of allowing noise to endure as a threat to order, decompositionism gave noise its place within the world of sound and signal. Sometimes noise is hidden away; sometimes it is endowed with meaning and portent; and sometimes it is simply let be.

Decompositionism was just emerging across many fields in the 1960s and 1970s, and remained far from a total social condition or a dominant theory of sound of any kind. Even today, noise is still a threat to many people in many situations, and its elimination is still a principle in many arenas of practice, from a never-ending procession of noise-suppression-strategies engineering to the politics of urban zoning laws. Many people still believe that taking sound apart does some violence to it. Decompositionism merely opened up the possibility of a plurality of relationships to noise for engineers, for listeners, and for many others through the total disassembly of sound. Decompositionism heralded neither a utopia where people would be freed from exchange nor a "grey world" of endless, meaningless exchange, to use Attali's term of dystopian derision for mass culture and mass production.[76] Thus, instead of Attali's utopian premonitions for the end of the battle between noise and order, we should instead consider one of his more sober propositions as useful for understanding the next phase in the history of perceptual coding: "Every code of music is rooted in the ideologies and technologies of its age, and at the same time produces them."[77] As perceptual coding moved from a theory of signal processing to a codified institutional practice and then a mass phenomenon, its history intertwined with a cluster of changes in sound cultures, media systems, technology industries, and policy regimes. Decompositionism may appear avant-garde or otherwise radical in my telling, but it would come to stand among the most ordinary features of contemporary cosmopolitan sound culture.

4. MAKING A STANDARD

Perceptual coding was part of a larger shift in thinking about sound media, from considering audio in terms of transduction and noise suppression to terms of technological decomposition and recomposition. These conceptual shifts would have remained purely theoretical if not for burgeoning industrial interests that helped propel the development of audio data compression in the 1980s. Between 1979 and 1989, perceptual coding moved from an experimental project to a practical solution to problems in a range of industries. By 1989 at least fifteen companies and institutes went public with perceptual-coding schemes, while others quietly worked on their own schemes. Ten years earlier, only one publicly available record of a perceptual coder existed—Michael Krasner's at MIT—and there were only two published English-language articles on the subject. In the late 1980s and early 1990s, a set of policy developments would help synthesize emergent perceptual-coding standards into a set of coherent industrial practices.

After finding some success marketing to both media industries and consumers, the makers of digital audio technologies sought ways to get around existing limits. The video compact disc (VCD), a precursor to the DVD that found wide use in Asia and other markets around the world, was based on the same technology as a compact disc, but it contained both audio and video, meaning that it could not accommodate full-bandwidth CD audio. Full-bandwidth CD recordings also couldn't fit on phone lines or be broadcast over digital satellite radio, but compressed audio could. Hard-disk and solid-state storage was still limited by today's standards and expensive, so the more audio that could be crammed into a smaller space, the better. Lower-bandwidth digital audio afforded opportunities to create new services that didn't exist before, and to make more efficient use of materials and services already in existence. But each new application posed a concrete problem for its developers: what standards would they use for perceptual coding, and how would their technology work with other related audio technologies?

This chapter considers the emergence of the Moving Picture Experts Group in the technical and industrial context of the 1980s and early 1990s. While the next chapter offers a sonic history of MPEG audio, this chapter considers it as an industrial and regulatory development, exploring the problems it hoped to solve, the agencies involved in its development, the decision-making behind the standard, and the compromises that were built into it. One of my main struggles in writing this chapter is how to tell the prehistory of the MPEG format. We know its outcomes today, but at the time they were totally unpredictable. If we project backward from the success of MPEG audio in a variety of industries, we miss the messy part of standards-making and sell short its political significance.

We normally think of the MP3 format's primary policy implications in terms of intellectual property politics, but the MP3's role as a standard is also central to its history. Standards simultaneously engage technical and political problems. Companies need to figure out how to get different audio technologies to "talk to" one another, while at the same time, most aspire to control the terms of the technical conversation. Because standards usually apply to international industries, governments have not been heavily involved in their regulation. This has had several major effects. Communication-policy scholars have often lamented the lack of public debate around media standards that are set by governments. However, when companies or industries set standards, there is not even a possibility of public debate. Industry-set standards also tend toward conservatism, since they aim to keep the big players big and to render emerging markets stable and predictable. Because companies' participation in standards organizations is voluntary, any standards imposed are nonbinding in any legal sense, though once a standard reaches a certain level of suffusion in a given market, participation may be coercive in other ways. (For instance, imagine the grim market forecasts for a new portable media device today that could not play MP3 files.) The emergence of MPEG was a response to a particular set of problems in a particular commercial environment. As a standard object, the MP3 format is thus a creature of policy as well as of technology, economy, and culture.

Over-the-air digital audio broadcasting provided a major motivation for research in perceptual coding. The earliest digital audio broadcasting technologies were developed for satellite delivery. These systems worked, and the technology required no compression, but it did not allow for mobile reception or local broadcasting, thus cutting out two cornerstones of radio industry practice at the time. It also did not work in the FM band, meaning that listeners' sets would have to be replaced (today, the consumer electronics industry would consider the need for replacement to be a virtue). The Institut für Rundfunktechnik (IRT) in Germany had been working on data-compression techniques for radio in collaboration with German broadcasters, and this approach generated enough interest that the German government commissioned a study on the problem of digital radio in 1984. The study concluded that it would be possible to develop terrestrial digital radio with the aid of data-compression technologies. However, this initiative would require a good deal of research and international collaboration.

By the end of 1986, a consortium of nineteen organizations from Germany, France, the Netherlands, and Britain signed an agreement and applied for funding to the European Commission as a "Eureka" project. Eureka, founded in 1985, was a pan-European initiative for market-oriented research and development, and the group's proposal to the commission pitched this as a necessary step forward for the radio industry. The consortium's stated goal was to develop digital audio broadcasting as a successor to AM and FM broadcast bands. It received funding in 1987 as project number EU-147 (hence the name Eureka 147), and by 1988 money was pouring into research in at least four countries. The goal was to develop formal standards for digital audio broadcasting, then to implement them in both technology and worldwide policy.[1]

At the time, it was not widely believed that digital technology might be part of a major reorganization of the broadcast industry; the future commercial applications of digital audio weren't immediately apparent. Early applications of perceptual coding sold by companies like Fraunhofer in Germany and Solidyne in Argentina focused on the existing needs of broadcasters. Pitches to venture capitalists in this period went over like lead balloons because new technical applications and new industry niches were not readily apparent to most potential investors.[2] One finds the same attitude in media theory. Writing about radio at approximately the same historical moment that the MPEG project was unfolding, Friedrich Kittler understood radio as a so-called mature medium, one whose important historical development had already happened: "Media that have reached their levels of saturation are hard to write about. They disappear at the juncture of high technology and triviality. They reach your ear from neighbouring yards as only the nightingale once did. . . . The normal scenario however (to quote the engineers) is that transmission is constant although, or just because, it is wireless. Due especially to being the first electronic medium to program day and night, radio has become a Platonic substantiality that causes it to vanish irresistibly as a technological medium."[3] The idea of a media form as "Platonic substantiality" nicely captures the work that communication-industry organizations undertake in their efforts to manage technological change and to control markets. A new technology, protocol, or use can be employed in projects to either extend existing industrial relations and cultural forms or to transform them. If we cannot imagine alternatives, then extension seems only natural.

The Politics of Standards

In December 1988, a group of digital audio experts met in Hanover, Germany, as part of a conference of the brand-new Moving Picture Experts Group. Their mission was to establish some industry-wide standards for the compression of digital audio data. In addition to radio and the plans for a video compact disc, there were calls for a standard in other industries. As more and more communication equipment became digital, manufacturers worried about a bottomless soup of competing standards and protocols creating a market that was too unpredictable. Companies from around the world were represented at the meeting, and each one had already been developing some kind of data-compression scheme for audio, sometimes oblivious to the work done elsewhere. Other groups had attempted to deal with the problem, but MPEG was part of a coordinated effort. Formed by Hiroshi Yasuda and Leonardo Chiariglione in 1988, it was an offshoot of JPEG—the Joint Photographic Experts Group—that set standards for still images. Chiariglione was especially invested in the idea of industry-wide standards. His account of the project mentions the VHS-Betamax conflict in analog video recorders and points out the wastefulness and redundancy of European countries that allowed mobile phone providers to build multiple versions of the same network. Chiariglione's story of the motivation for the meetings is idealistic in tone and contrasts his sense of the norms in his industry—computers—with those of telecommunications and broadcasting. From his perspective, telecommunications industries had a long history of incompatible standards (one need only consider the bizarre range of analog television formats that were in place for most of the twentieth century), while computer engineers were more interested in the value of interoperability, where something that worked in one place, on one platform, would work on another. Chiariglioni casts the history of computer standards in a sunnier light than is probably warranted (consider incompatibilities in operating systems or networking protocols), but his idealism is certainly part of the reason MPEG worked at all.

The history of standards for audiovisual formats is not usually told as a story of diplomacy, peaceful negotiation, and rational parsing. Far more conventional is a tale like that of Edwin Howard Armstrong, the inventor of FM radio who jumped to his death from the thirteenth-floor window of a Manhattan apartment on the afternoon of 1 February 1954. In the 1920s, Armstrong developed a mode of radio transmission that was superior to the systems then in use. It sounded better and was less susceptible to interference. It also allowed for more space in the spectrum to be used for radio, thereby giving listeners greater choice. Armstrong was granted patents in the United States in 1933 and within a few years, FM receivers were available for sale and FM radio stations were broadcasting. Yet Armstrong's version of FM operated in a part of the frequency spectrum that RCA had been planning to use for television. RCA launched a massive anti-FM public relations campaign and began lobbying the Federal Communications Commission. In January 1945, the FCC moved FM to another part of the electromagnetic spectrum, eliminating an inconvenient conflict for RCA and rendering all the sets built to Armstrong's specifications completely useless. Armstrong sued RCA for royalties and patent infringement. RCA stalled in the courtroom, buying time to find workarounds that would allow them to claim they were not using Armstrong's patents and pushing the decision toward 1950, when some of Armstrong's patents would expire. Outmaneuvered by RCA's legal team and exhausted from a fight that had consumed his life and destroyed his marriage and career, Armstrong chose suicide over continuing to pursue his case. Apart from the sad and sensational outcome for Armstrong, the combinations of secrecy, coercion, public relations, litigation, legislative lobbying, and attempts to control the market are common elements of the process through which standards come to dominance.[4] One can find similar stories to that of FM in telephone, radio, sound recording, and consumer audio standards.[5]

For bottom-line oriented companies and for engineers who preferred research to acting as delegates in some contrived legislative body,[6] standards had in the past often appeared as "a necessary evil," or worse, "a complete waste of valuable resources, as something hampering progress." But by the time MPEG convened in the late 1980s, this attitude was joined by another one, that standards might be useful industrial levers for companies, another way to pursue their own objectives of market conquest and rational expansion. Initiatives like MPEG emerged in this climate.[7] Yet even a tale of companies instrumentally seeking to bend standards to their wills is too rosy a picture: as one of the industry's great clichés puts it, the great thing about standards is that there are so many to choose from. The "possibility of influencing the allocation of long-term returns makes *de facto* standards competitions such fertile ground for corporate strategy."[8] With the wide range of standards organizations in existence, companies can essentially go shopping for a standard that they like. It is therefore only with hindsight that we can understand MPEG as destined to matter at all. Private standards can come to dominate a market just as easily as those sanctioned by standards organizations. Proprietary standards like .doc or .pdf files, or Blu-ray video are bids to turn their owners into market regulators of a sort.

Even though perceptual coding did not develop in that way, it very well could have. If any company had had a market-ready perceptual coder with the backing and leverage to dominate the market in December 1988, it is unlikely the MPEG meetings would have happened, or mattered if they did. For instance, if a private company had developed a commercially successful VCD that relied on a proprietary format rather than MPEG's specifications, MPEG would be irrelevant to that history. If another company unaffiliated with MPEG had developed a software program to perceptually code audio from compact discs, it is possible that MP3 would not even rate as a bay leaf in the alphanumeric stew of format history. At the time of MPEG, both Sony and Philips were pursuing independent lines of research into perceptually coded media that would eventually yield the MiniDisc and the Digital Compact Cassette, respectively. In addition to its use of different storage media, each technology made use of a proprietary perceptual codec, one that was brought to market with full knowledge of the MPEG exercise then in effect. If one of those codecs had come to market dominance, we might have yet another story of perceptual coding, and this book might be called *MiniDisc: The Meaning of a Format*. But despite some commercial success for both (eventually MiniDisc "won," at least in the portable recorded market—though it has largely been replaced by solid-state storage), neither format was of industrial or cultural significance beyond specialized market niches.[9] All this is to say that there was not a single industrial line that leads inexorably from the research at AT&T and Lincoln Labs in the 1970s to the flood of MP3s on the internet at the turn of the twenty-first century.

The MP3 is thus a very unusual kind of standard object in audio. It is a major audio format that is the result of an international standards exercise, complete with rules for participation and debate, elaborate testing systems with well-documented results, and enough of a paper trail (and enough compliant interviewees) that I am able to write this history after the fact. It is a widely available standard, free to end-users but privately owned. With policy scholars' vocal advocacy of all things democratic, this would appear to be a development worth celebrating, an example where the system worked. And yet, things are not so simple. As we will see, even the people who are credited with the development of the MP3 consider it a technological compromise at best. For instance, employees of Fraunhofer, which owns the patents, will freely laud the superiority of their AAC (Advanced Audio Coding) codec, which was developed after the MPEG-1 standard was set.

The MPEG standard emerged at the intersection of several standards organizations and several different standards' interests. It was at once a standard for computers, consumer electronics, and broadcast, all of which traditionally had very different standards cultures. Because of this, it partakes of a bit of each of these cultures. It has the centralization of a telecommunications standard, the stability of a consumer electronics standard, and the modularity of a computing standard. It has the near-universal reach of a telecommunications standard, and yet different parts of it are owned and administered by different companies. If you build a piece of software or video game system with one kind of MPEG decoder, you pay royalties to one set of companies. Another kind of decoder obligates you to an entirely different set of patent holders.

Telecommunications standards were often structured around the value of *interconnection*, which involves a dominant network to which "foreign" devices may be connected. In the case of landline telephones, a central telephone network operates according to a single protocol, and telephones and answering machines (for instance) can be connected to it on the user's end. Broadcast radio or television work the same way: a nation decides how the electromagnetic spectrum will be divided up, and then radios and television sets are built to receive over-the-air signals in those frequencies.[10] Conversely, computer standards were usually structured around the ideal of *interoperability*, where different machines can run the same software, exchange files, and work together even if they do not work according to the same protocols or operating systems. Thus, Windows, Mac, and Linux systems can all connect to the internet and exchange files with one another, and each one can properly encode, decode, and make use of files that conform to widely used standards, like the MPEG. In the United States and many Western European countries, another important difference between standards cultures is that governments intervened quite actively in telecommunications standards but did not make as much explicit policy regarding standards in computing and consumer electronics. Though some would characterize a lack of governmental or international regulation as "leaving these standards to the market," it might be better to say that standards were simply managed by industrial interests, since oligopolistic market conditions existed just as often as free market conditions. In fact, the relatively laissez-faire approach of governments to regulating consumer electronics and computers has led to a situation where companies that own major standards have a vested interest in maintaining the largest possible locked-in user base. Privately owned standards may therefore seem competitive, but the owners of standards often pursue their interests by doing everything they can to limit economic competition as much as possible, or by making it as difficult as possible for a competing standard to capture part of the market.[11]

The politics of standards is therefore a crucial point of entry into debates about the future of media regulation and policy. As Tarleton Gillespie writes, the regulation of media is now "as much technological, commercial, institutional, social and ethical as it is legal."[12] If we are moving into an age in which important technological shifts in communication happen as often in software as in hardware, if new formats can sometimes eclipse new media in importance, then the politics of standards may eclipse the governmental regulation of broadcast or telecommunications as a crucial site where policy happens, and where the everyday affordances and constraints of media are shaped for the end-user. One could certainly argue that the MPEG standard is less crucial to the everyday operation of the internet than packet switching or domain-name conventions and allocations, file compatibility and extensions on computers, and so forth. But the MPEG case illustrates the ways in which different values influence the development of a standard, and how different interests are represented in an otherwise closed conversation. At the same time, any history of MPEG would be incomplete without a critical discussion of standards politics. If we want to understand why an MP3 sounds the way it does, we need to understand how technical decisions regarding MPEG technologies were bound up with the politics of nations, cultures, and industries.

The Moving Pictures Experts Group evolved within a complex ecology of standards organizations. MPEG's largest parent organizations were the International Organization for Standardization (ISO) and the International Telecommunications Union (ITU). The ISO is a nongovernmental body that publishes standards for a wide range of industries, and it descends from international standardization initiatives that have been going on for over 150 years, often motivated by innovation in communications infrastructure. Founded in 1947, the ISO is a meta-organization, overseeing standardization practices that had previously been split between the International Electrochemical Commission (IEC) and the International Federation of National Standardizing Associations (ISA). Countries join the ISO, and one member organization is allowed to represent each country within it. The organization's work is entirely decentralized, with technical committees, subcommittees, and working groups developing standards that are then approved and published by the organization. Despite its approach resembling that of the United Nations, the ISO has to compete for influence, and some authors have suggested that its actual policy influence over corporate practice—as well as the influence of the bodies with which it competes—is now minimal. Founded in 1865 as the International Telegraph Union, the ITU was perhaps the first modern standards organization, and has long worked to mediate the different telecommunication protocols in different countries.[13]

The ISO got involved in the MPEG project for a number of reasons. It had just recently founded Joint Technical Committee 1 (JTC1) in 1986 in collaboration with the IEC. The committee was specifically tasked with the management of standards for emerging computer and so-called information technologies, and its goal was to centralize standardization in the field and avoid the creation of competing standards by bodies with overlapping jurisdictions (as was the case with ISO and IEC; over time, ITU also became more involved with this committee). The committee was supposed to be guided by the purported special needs of IT manufacturers, most notably the reduction of bureaucracy so that standards could be implemented more quickly, for instance, by taking a single company's standard and ruling it as a standard for a whole industry. The ISO was also a logical home for a project that was in part the result of an international consortium of companies and governments who sought to set a single protocol to ensure a wide range of consumer electronics technologies might work together.

Because ISO had already set up the Joint Photographic Experts Group (JPEG), which set standards for still images online, MPEG was something of a natural outgrowth for them.[14] Still, it would be wrong to think of the meetings around the MPEG standard as themselves being a significant media-historical event. While we can look back with certainty and say that the MPEG audio meetings were important for shaping the sound of compressed digital audio and the standards by which it would circulate worldwide less than a decade later, this is only possible because of hindsight. The MP3 standard only became important after the fact, and somewhat in spite of the business plans of its developers. At the time that MPEG came together, there was no world wide web, no internet as we know it today, so the MP3 developers' version of interoperability was quite different than the one we might now imagine. They were thinking about hardware boxes that might go in professional studios or in people's homes, and of the portability of content within and across industries.

Imagined Industrial Landscapes of Digital Audio

MPEG's imagined electronics world was still divided into categories of producers and consumers, as defined by the post–Second World War consumer electronics and media industries. If there was a dream of universality, it was the old universality of the multinational conglomerate, not the fantasy of information that wants to be free, or a vast lateral network of users. One analysis of standards politics during the 1980s calls this tendency "common systemness," which was not only important for users, as we might think today, but also for manufacturers who hoped to be able to compete in a predictable market. It is only through massive processes of imitation and broad dissemination, where "vast markets are conquered by various competitors, that mass production can bring down the costs of production." Global standards help facilitate this kind of market conquest precisely because they settle a conflict in one industry—digital audio coding, in this case—in order to assist other industries, such as radio, consumer electronics, and the recording industry. The development of the MPEG format was therefore an example of "transectorial innovation," a process whereby "new technologies are no longer confined to a single application, to a single sector; they are disseminating and interpenetrating the whole economy."[15]

Even if JTC1's effects were clearly transectorial, the committee was designed and staffed with a deliberate bias toward the growth of the information technology industry, or at least industry representatives' own views of how best to attain that goal. Among other things, this meant that at the level of standards-setting, a very specific professional ethos dominated. "Many members of standards working groups consider 'technical sophistication' a major prerequisite for participation in the standards setting process. People perceived as lacking this attribute are likely not to be taken [seriously]." While this is certainly a generalization, it fits the MPEG case fairly well. The main players in MPEG's audio group were all representatives from major research institutes or corporations. Some would promote particular corporate agendas, while others would promote particular technical values, but in most of the discourse around MPEG's audio group, the standardization of compressed audio was first and foremost a technical problem. If politics entered into the equation, they were seen as something that interfered with the quest to improve the technology. Though communications engineers did not all share the same values, it was their values that would dominate the formation of the MPEG standard.[16]

The starting point for the group was the digitization of audio that had already happened. Parts of telephone systems in various parts of the world had been digital for decades by the time the MPEG audio group first met, and digital networks that had previously found closed, cozy homes in the military, banking, and education sectors were becoming increasingly intertwined and connected with other industries.[17] These networks already had a set bandwidth, and MPEG's original goal of 1.5 mbps for signal transmission was based on the network technology of the time. The most important standard for digital music was the compact disc, itself the result of industrial cooperation between Sony and Philips, which gave up competing to establish a single standard for digital audio.[18]

In contrast to the compact disc, nobody knew exactly what MPEG's new formats would be for. A survey of the range of industrial and technological models proposed for digital audio at this moment is quite instructive, because the schemes do not presuppose the separation of hardware and software to which we are now accustomed. The MPEG audio group meeting in 1988 ended with a call for proposals for developing a digital compression standard that would work for "various applications like CD-ROM for audio and video, DAT recorder, [and] Digital Audio Broadcasting." Although the list seems perfectly reasonable, it gives a clue to the level of confusion at the moment as to what, exactly, the transmission of compressed digital audio was *for*. In a single sentence, we have consumer and professional technologies mingled together and put alongside the revitalization of radio, an entire medium. CD-ROM and DAT were specific technologies, with their own specifications and standards. Computers were only beginning to be understood as multimedia devices, but CD-ROMs were imagined at the time as a new consumer format (like the compact disc) that would have wide dissemination, like videotapes and compact discs. DAT (digital audio tape) was a professional recording format used in recording and broadcast studios. Although some consumer DAT machines would come on the market (with more commercial success in Europe than North America), prerecorded DATs were not available for sale, and DATs were already able to store a fair amount of data in a compact space. Meanwhile, digital audio broadcasting was an application, one might even say a dream for a medium, and represented a complex of technologies rather than a standardized storage medium like a CD-ROM or a DAT cassette.[19]

A report for the IEEE authored by Hans-Georg Musmann in 1990 suggests some other applications for compressed digital audio. The original

MPEG audio mandate specified that compressed audio should "enable digital storage media like the CD to be used for interactive video," what became the video compact disc (VCD). But the authors of the original VCD specification had forgotten about audio entirely. This left an impression on the Fraunhofer engineers who joined the project in progress. Karlheinz Brandenburg said "normally audio takes up all the data on an audio-CD but they chose to begin with videos in the absence of sound—obviously movies without sound were out of fashion so they needed to find a way to compress audio." Bernhard Grill joked, "Video without audio is only good for surveillance. That's why MPEG was named 'Moving Pictures Expert Group' and then added the subtitle 'and associated audio.' For a long time we felt like the poor little 'associated audio' guys."[20] In addition to VCD, Musmann's report proposed other concrete applications for compressed digital audio like "a digital audio recorder without movable components" (which most MP3 player and smartphone users hold in their hands today) and the transmission of digital audio over phone lines or ISDN lines. The commercial implementations of these technologies weren't clear at the time. Probably the portable audio player was imagined like a Walkman and the ISDN like a version of cable radio, but these two applications have in subsequent years become central to a variety of media industries.[21]

Alongside the European Eureka 147 initiative discussed above, digital radio was a recurring topic in the United States at meetings of the National Association of Broadcasters (NAB). The Federal Communications Commission authorized experimental broadcasts as early as 1987, and there was considerable interest in Eureka 147. Audio made up about a third of the $10 billion consumer electronics market in the United States, and digital audio was becoming more important. The industry needed economic models for the new digital audio media. Perhaps not surprisingly, those models were based on broadcasters' points of view. Consider the diagram from an NAB report, published in 1990 (figure 25).

The paths in the NAB's diagram follow the rough technical divisions of broadcasting at the time: terrestrial broadcast, cable, and satellite. The industrial models follow as well: advertiser-supported broadcast and subscription-based services, all provided by professionals to audiences. Reading backward from the present, it is interesting to note that terrestrial broadcasting is still largely an analog operation between stations and audiences. But we can imagine "satellite services" as roughly analogous to XM and Sirius radio, and "cable networks" as the audio-only channels that cable

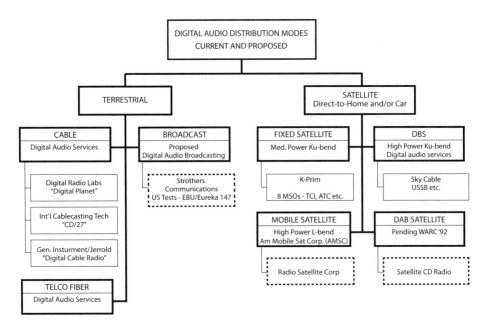

25. "Digital Audio Distribution Modes: Current and Proposed," ca. 1990.
Source: De Sonne, *Digital Audio Broadcasting*, 3.

networks routinely pipe in with their video content. There was no broad consciousness of the world wide web in 1990, but it is interesting to note the "TELCO FIBER: Digital Audio Services" in the lower-left-hand corner of the diagram, which is the closest analog available.[22] It is the furthest off and the least interesting for the report, and receives no discussion at all. There is no slot in the graph for lateral exchange of recordings, nor even an inkling that it might be an issue, no discussion of mass customization, and little discussion of portable media beyond car radios. Broadcasters were settled enough in their industrial structures that they imagined new technological forms as extensions of existing forms, despite the fact that both television and FM had been parts of major shakeups in radio broadcasting decades before. The assumption underlying the entire report is that existing industrial players are the relevant ones for thinking about the development of digital broadcasting. One can even see this in their citation of "consumer demand" as the reason for the shift to digital audio, ignoring the troubles that the consumer electronics and recording industries had in getting people to buy CD players and CDs. Consumer demand for digital content had to be created—it did not simply exist out there waiting to be tapped. In the case of compact discs, sales lagged in the United States until record labels made it very financially difficult for stores to continue stocking LP records. In other words, one sector of the media industry believed another sector's public relations.[23]

It is quite likely that if we were to unearth memos in the telecommunications industry, we would find a similar marginalization of radio in its dreams for digital audio, and the consumer electronics industry would no doubt provide another blueprint. To a great extent, MPEG, which is the technical basis of so many important developments in the sound culture of new media, had its roots in a number of sectors of the communication industry's attempts to extend its status quo into the digital domain. Even though the capacity for decentralization and lateral exchange may have existed, it appears that all of the major institutional players expected that MPEG audio and video content would continue to follow the producer → distributor → consumer model on which the vast majority of the media economy was conceptualized at that point.

Regardless of competing visions of the future, players in each of these industries were conducting at least rudimentary research into the compression of digital audio by the time the MPEG audio group was formed. This is clear because the fourteen proposals for audio coding received by the MPEG audio group in June 1989 cut across communications industries and continents. Musmann's report listed "companies from the fields of telecommunications, broadcasting, computers, consumer electronics and VLSI [chip] manufacturers." Federally funded research institutes like France's Centre Commun d'Études de Télédiffusion et Télécommunications (CCETT) and Germany's IRT submitted in competition with telecommunications companies like British Telecom, AT&T, and Nippon Telegraph and Telephone; and those groups were also competing against consumer electronics manufacturers like Sony and Matsushita. The fourteen proposals were clustered into four competing groups based on the technology that each proposal used, and three of the groups contained companies from at least two different sectors. The two codecs that would eventually form the basis of the MPEG standard were MUSICAM (Masking Pattern University Sub-band Integrated Coding And Multiplexing) and ASPEC (Adaptive Spectral Perceptual Entropy Coding). MUSICAM would later be enshrined as the technology behind layer 2 of MPEG-1 and ASPEC would be the basis of the specification for layer 3 (MP3). MUSICAM was a collaboration between IRT, Philips, the

CCETT, and Matsushita (a subsidiary of Panasonic); ASPEC was a collaboration between AT&T, Thomson, Fraunhofer, and France Telecom. Interestingly, the other two codecs, which would be less successful in the tests, were less multisectorial collaborations: Adaptive Transform Aliasing Cancellation (ATAC) was a collaboration between consumer electronics manufacturers and chip manufacturers; Sub-Band Adaptive Differential Pulse Code Modulation (SB/ADPCM) was a collaboration between two telephone companies.[24]

The fourteen competitors who gathered together in the MPEG audio group were building a standard that had no guarantees regarding industrial use and no clear or obvious industrial path to realization. Coming from so many different fields, their interests and expectations differed. Of course, companies lobbied for their own schemes, but broader assumptions that reflected industrial priorities were built into those schemes. So part of building the standard involved specifying its relationship to other technologies, like ISDN and VCD, and specifying what mattered, in particular, about an audio codec.

The ISO set a wide variety of requirements that balanced sonic factors against technical factors such as compatibility and processing demands. Each proposal was scored according to a complex scale that weighed sonic factors against technical factors. Compromise was built into the standards-making process as well as the standard itself. The list of "weighting factors" offers a glimpse of the competing values that went into the MPEG standard (see table 1). Sonic values brushed up against standard digital virtues like random access, and robustness brushed up against errors, metadata (though the "ancillary information" was probably not understood in the same fashion that we understand metadata today), simplicity, and speed. The values were at once aesthetic and technical. A *good-sounding* codec was also a good *sounding-codec*—in the evaluation, there was no difference. MPEG audio was a delicate dance between form and content. The technology presupposed the separation between hearing and content that one finds in psychoacoustics and information theory. At the same time, the process by which the technology was made created an arithmetic holism where the total score for a codec was the sum of its (weighted) performances in each category: "$S = S_1 \cdot W_1 + S_2 \cdot W_2 + \ldots S_{11} \cdot W_{11}$." This logic is analogous to the decomposition and recomposition of audio, except here it is applied to the whole standard. MPEG breaks sound, sonic experience, and its technical

TABLE 1. Weighting factors for evaluating competing algorithms for perceptual coding

PERFORMANCE	WEIGHTING FACTOR W
1 Sound quality of forward audio playback	121
2 Sound quality fast forward audio playback	067
3 Random access	118
4 Ability to encode in real time	055
5 Data capacity for ancillary information	093
6 High quality stereo	086
7 Intermediate quality audio	096
8 Robustness to bit errors	089
9 Encoder complexity	059
10 Decoder complexity	117
11 Short decoding delay	072

Source: Musmann, "The ISO Audio Coding Standard," 2.

facilitation down into manageable, measurable, and relatable components in the service of creating new arrangements of technology, sense experience, and industry.

The scheme might appear arbitrary, but it demonstrates how a particular standard is situated among technical, industrial, and aesthetic concerns all at once—and attempts to negotiate them. That's what an MP3 does as a standard object. If one digs down into the references of the various qualities "Sx," there is an apparent incoherence, since presumably an evaluation of audio quality such as stereo coding at higher bitrates is a completely different matter than the evaluation of a processing quality like random access. The weighting factors and common numbers impose a false arithmetic equivalence, but live practice touches on all these qualities in a way that makes separating them difficult. An MPEG audio file is at once about sound quality, compatibility, manipulation, storage, data intensity, and distribution. The arithmetic abstraction here is necessary to, as Alfred North Whitehead put it, "confine your thoughts to clear-cut definite things, with clear-cut definite relations." The abstraction in a standard is a "productive device," in Matthew Fuller's words. It helps effectuate a "naturalized landscape of nouns, things, homogeneities" in which it operates.[25] MPEG's weighted list of criteria has its biases in kind and in weight. Radio and postproduction need fast-forward playback, while DJs might have wondered about something like "slow-forward" playback. To read the weighting scheme, decoder complexity is only slightly less important than overall sound quality—and what counts as decoder complexity is clearly based in the available computing technology around 1990. The scheme also has its aporias, since the sound of coded audio once it is edited is never considered.

After clustering the original fourteen proposals into four groups, MPEG was ready to set up its tests. Each approach was represented by a hardware device that could encode and decode audio—a codec ("coder-decoder"). The codecs were then subjected to demanding performance and listening tests in 1990 and 1991, and the results were rated according to the ISO's weighted measures. Two codecs emerged as preeminent in the tests: MUSICAM (backed by Philips, Panasonic, and others) and ASPEC (backed by Fraunhofer, AT&T, and others). Competing industrial interests led to competing interpretations of the test results and outright conflict in meetings. MPEG's layer scheme was thus an attempt to mediate industrial competition and conflict by creating a standard with options built in. These options corresponded to different interests. Layer 1 is not in wide use. Layer 3 is the specification that we now call MP3 (descended from ASPEC). Layer 2 (descended from MUSICAM) is less well known, but it is an important character in the history of perceptually coded audio. Layer 2 was less complex, and therefore less taxing on computers (a concern in 1992). It was also less susceptible to transmission errors in digital audio broadcasting, but it did not compress data as efficiently as layer 3, which meant that a layer-2 file of the same sound quality as a layer-3 file was bigger. The abstractions and weights set out by the ISO now became the subject of hot debate: should lower complexity be more important than sound quality at lower bitrates? Given that particular industrial interests backed each layer, these were political, economic, and industrial matters, not just engineering decisions.

The layer scheme was therefore a compromise that allowed MPEG to sidestep the problem of actually settling matters of industrial competition.[26] Despite the lack of an "MP2" epithet, layer-2 audio has found wide use in satellite television broadcast, digital audio broadcast, video compact discs, DVDs, cassette-based high-definition video cameras, and elsewhere. Because they were originally designed to handle signal processing in different ways, layer 3 is compatible with layer 2 but layer 2 is not compatible with layer 3. As part of the compromise, layer 3 was literally built on top of layer 2, which means that the filterbank in layer 3—the part that analyzes the audio before it is perceptually coded—is unnecessarily complex. It introduces a technical flaw that did not exist in the original ASPEC design.[27] While the sonic differences were small ("there's a little bit more smearing at times in the MP3 filterbank. But it's not a great difference, really," said Bernhard Grill), the engineering was inelegant. Still, the technical compromise allowed for political compromise. Layer 3's advocates got their technology included, at the cost of building it on top of layer 2 to preserve that layer's relative simplicity. There was also a technical expediency to political hedging on MPEG's part. The layer scheme preserved some dimensions of the ASPEC algorithm, which performed better at lower bitrates, while also preserving the simpler MUSICAM algorithm, which required less computing power. The standard's flexibility also hedged against the impossibility of knowing what uses would come to prevail, since MPEG's management of rivalries to get a standard did not extend to actual industrial applications.[28]

The MPEG standard devised in 1993 did not put to rest matters of industrial competition, nor did it ultimately settle questions of sound quality in perceptually coded audio. But it marked an emergent, crystallized set of understandings, practices, protocols, and industrial relationships. The standard allowed for the proliferation of standard objects that could move between countries, media, operating systems, and protocols. It provided grounds for the subsequent adoption of standards for a whole range of broadcast, recorded, and transmitted audiovisual media still in use today. These adoptions and uses did not happen overnight. The standards process offered no guarantees that it would happen at all. MPEG had no power of enforcement and no power to govern once the standard had been set. Yet cross-industrial self-governance is still often the closest thing to any kind of format policy.

The MPEG process was far from perfect. The standard explicitly rendered industrial compromises as technological compromises. MPEG's criteria for decision-making were skewed to the questions brought to the table at the moment of its founding in 1988, as well as to the kinds of technical values and knowledge that are so privileged in organizations like the ISO. MPEG's process is not in any simple way a model for future standards—whether in video, audio, or elsewhere. Yet at the same time, the MPEG story points to the need for better governance of audiovisual standards and formats, one guided by a notion of collective good. The MPEG standard blended the ideals of interoperability and interconnection that guided different standards cultures. While these ideals are debatable, they will only be subject to debate for as long as companies are responsible to something other than the bottom line. Without international standards policies, we are simply left with patent holders trying to lock in as large a portion of their market as possible, or at best collective cross-industrial self-governance like the ISO. The story of MPEG poses standards as an as-yet-unresolved issue of political representation in the development of new communication technologies. That undecided issue of representation also resounded as a sonic problem in MPEG's listening tests.

5. OF MPEG, MEASUREMENT, AND MEN

The classic questions of representation in liberal political theory—who speaks for whom, to what end, and under what circumstances?—find their sonic counterparts in listening tests. Almost every dispute over sound technologies eventually revolves around analogous questions: who listens for whom, to what end, and under what circumstances? In 1990 and 1991 the Moving Picture Experts Group had to confront these questions as it tried to mediate among competing industrial claims and paradigms, and as it conducted three massive, intensive listening tests to reveal the point at which different perceptual-coding schemes for audio would break down. Through that revelation, MPEG could adjudicate among competing technologies to come up with a single international and cross-platform standard. Although engineering discourse generally casts sound reproduction in terms of problems of definition, transmission, and operation, the listening tests are fascinating for how they mingled technical and industrial concerns with directly aesthetic questions of what kind of system might be pleasurable to listen to. The tests dramatized a form of political representation, in which one group of testers and listeners listened in place of a future public who would eventually hear the technology, and in so doing helped reveal something about the sensibilities inscribed in close to a century of audio tests.

Ears did not settle political or industrial disputes over compression formats and they still don't. But they provided a political threshold through which all contending coders had to pass, and thereby set the terms through which technics and aesthetics might be negotiated through one another.

This chapter examines the history and epistemology of listening tests, both in MPEG and in the broader swath of perceptual-coding technologies that led up to the standard now known as MP3. It also ruminates on the place of aesthetic judgment in the design of sound technologies. The two magnetic poles of universalism and particularism orient debates around the listening tests, the interpretation of psychoacoustic data, and how we think about the composite listening subject written into the MP3—or the ideal audience of any sonic technology. Engineering and standards-making parlance presents listening tests as largely practical and technical procedures, but the delegation and construction of listening in the tests have a much richer political and aesthetic texture.

Stories of tests are epistemic stories. They tell us how knowledge is produced, managed, and negotiated. The authority of test data, derived from their performance of objectivity and publicity, explains why MPEG turned to testing as the last stage in making a choice among competing coding schemes. This chapter reads MPEG's listening tests against a longer history of testing perceptually coded audio to explore how particular ways of listening and assumptions about subjects and aesthetics get written into a format. I consider listening tests and the talk around them as cultural documents, situating them against other, alternative approaches to audio testing in a broader discursive field. While I will consider the tests critically, my critique is not intended as a simple debunking, for despite its flaws, the format works remarkably well for the majority of its users, and as we will see, alternative discourses of auditory judgment carry with them their own sets of problems. Instead, this chapter establishes some conditions under which truth-effects can be generated through listening tests, and offers an explanation of why the MP3 format works so well on so many different kinds of recorded audio, despite its well-known flaws and despite the failure of MPEG's listening tests to achieve the universal status they sought.

Initially, to people outside the field, the listening test may appear as an absurd scenario. It uses exceptional listeners in exceptional environments and a limited set of recordings to produce numbers that represent the sonic performance of an audio system. It appears to subject an aesthetic question—which system sounds better?—to an apparently anaesthetic situation. Test subjects aren't even supposed to listen to music or speech. They are told to listen for tiny timbral details in the sound of a system's playback. But MPEG's listening tests were not just about musical taste or preference or even objectifying subjective experience. Testing was the ground on which MPEG thought it would resolve its internal industrial conflicts between competing sets of companies with competing products. Its removal of subjective experience was synecdochic for its attempts to remove private economic interests from its proceedings. In order for that to happen, the process had to appear as objective and disinterested as possible to its participants.

The test results provided the grounds for the inclusion and elimination of codecs in the standard. Between the first and second tests, two of the codecs were eliminated, leaving two codecs to compete for the final standard: MUSICAM—a collaboration between IRT, Philips, the CCETT, and Matsushita (a subsidiary of Panasonic)—and ASPEC—a collaboration between AT&T, Thomson, Fraunhofer, and France Telecom.[1] The tests, and the documents that arose from them, were public performances of the technology (or at least performances for a limited group of delegates who represented a public), where large groups of people involved in the organization could be enlisted as earwitnesses to the subtle differences between coding strategies and protocols. As with all witnessing, these accounts had to be interpreted. They were not simple or straightforward reflections of what happened on a particular day in a particular room. Testing can be said to be a form of technological performance, and the testing environment an elaborate kind of staging and theater. The publicity of this performance is an essential component. Because everyone making the decision has the same access to the same test results, the reasoning goes, decisions about which technologies to adopt can be made in a disinterested fashion. Of course, it is never that simple. As Trevor Pinch writes, "It is during testing that different interests in the development of the technology are manifest."[2]

Testing is the basis for knowledge of any new audio technology that comes from a laboratory research setting. So it is no surprise that in the documents leading up to the establishment of the MPEG standard, there is at least as much written about the conditions of the listening tests as there is regarding the results of the tests. Test data have a certain privilege in discourses about the functioning of technology. The goal of a test is to assess whether or not the technology works as intended, but considerably more is at issue in what initially appears to be a simple matter of assessment. Though "test data are usually thought of as providing access to the pure technological realm, a means by which the immanent logic of a technology can be revealed," tests are by their very nature made up of contrived conditions.[3]

A highly orchestrated and scripted scenario was meant to guarantee that privacy—whether in the sense of idiosyncratic taste, an idiosyncratic room, or private economic interest—had no place in the tests. In spite of the vast range of recorded music and listeners, MPEG wanted a process that would work on any recording; they wanted a system with potentially universal reach. Credibility for the whole process essentially rides on this moment of testing, despite the fact that the tests and their results are not entirely public and in some sense are irrelevant to popular taste or to a format's economic success. The listening test functions in engineering practice as the anticipation of all possible future uses of the codec, which is why it must be a worst-case scenario. Louis Thibault, who helped design a widely used ITU protocol for listening tests published in 1994, describes how the goal of the listening test is to

> quantify the merit of a codec knowing its perceived quality will vary according to that material and that system and the particular listener and the particular playback environment. The philosophy behind all of this is to make everything worst case so that the result you find by applying this method is the worst quality a codec can deliver. If this 'worst case' quality is acceptable, we can infer from that most of the time it will be perceived as CD quality.[4]

This is standard engineering practice, just as bridges are designed to withstand a load much heavier than they are likely ever to bear.

A worst-case scenario for a perceptual-coding system includes high-definition playback equipment that plays back the widest possible range of frequencies; a sonically treated room that allows the broadest frequency spectrum to be heard; recordings that are particularly difficult to encode for one or another reason; and "expert" listeners trained to hear even the subtlest of changes in recorded audio while comparing one version of encoding with another.[5] Early published articles on perceptual coding say relatively little about each of these four aspects, but over the past two decades, engineering's common sense of worst-case scenario testing has been codified in ITU recommendations and elsewhere.[6] Current industry guidelines for testing perceptually coded audio were only published after the major tests for MPEG in 1990 and 1991. But the final MPEG tests, which aimed to determine whether experts' conceptions of coding corresponded with how the format was likely to perform in real-world situations, conformed to the basic idea that perceptually coded audio should be tested in a worst-case situation.

Reading documentation on listening tests means confronting a series of universal and transcendent aspirations. For its users, the listening test is a sonic context that aspires to transcend all sonic contexts. It aspires to universalism. The technology that wins the test should work for as many recordings and as many real-world situations as possible. The rules for running a listening test aim to eliminate listeners' individual interests and predilections—factoring them out of the equation. The very organization of the test is designed to both facilitate and represent disinterest in the choice of any particular contending technology. If the test had three commandments, they would be "Leave aside your interests! Make judgments! Make them as universal as possible!" All this effort toward disinterested judgment invokes the legacy of Kant's aesthetic theory—or at least hints at its flavor. Listening tests are therefore also susceptible to some of the critiques of Kant, most notably Pierre Bourdieu's thesis that aesthetic disposition is intimately and inextricably tied to social position.

In *The Critique of Judgment*, Kant uses the example of music when he distinguishes between that which is merely pleasant and that which is truly beautiful. Music at a dinner party is "meant to dispose to gaiety the minds of the guests, regarded solely as pleasant noise, without anyone paying the least attention to its composition; and it favors the free conversation of each with his neighbor." For Kant, the problem with this scenario is its instrumentality and context. The pleasure of music at a dinner party derives from sensation and location. The music exists as part of an environment and a social experience. But proper aesthetic judgment requires a context that transcends context, a situation where judgment is divorced from the situation of its rendering. "The universal communicability of a pleasure carries with it in its very concept that the pleasure is not one of enjoyment, from mere sensation, but must be derived from reflection; and thus aesthetical art, as the art of beauty, has for standard the reflective judgment and not sensation." The contemplative audience member, seated, silent, and attentive in the concert hall, better approaches Kant's ideal. If we move from music to sound reproduction, the listening test occupies a similar place. It turns listening pleasure (or displeasure) into an object of contemplation for the listener. In MPEG's listening tests, listeners were not meant to listen to the music, or even the speakers or headphones. They were meant to listen for sonic artifacts of the technical process to which the recording was subjected. In other words, they tried to hear the codec at work. Everything in the listening-test scenario was designed to facilitate this kind of disinterested, reflective judgment, in the service of producing a "universally communicable" sonic format through that judgment.[7]

Because the testing context is carefully circumscribed, it is not always clear whether it tests the technology or the user. Listening tests followed this pattern. Perceptual-coding research developed through the 1980s and researchers in different places all ran into the same problem. By the late 1970s a range of people began to figure out that it ought to be possible to use what is known about masking to create better-sounding audio at lower bandwidths. We now turn to the politics of listening, to consider what it means that a small group of engineers with a coherent aesthetic helped to shape the sound of the world's most common audio format.

How Tests Work

The possibility and the practice of perceptual coding were two different things. The people who built early perceptual coders started with tables of masking responses or critical bands taken directly from articles or textbooks in psychoacoustics. But these proved insufficient for building technology that would process music or speech so that files were smaller but still listenable. Michael Krasner had already documented these issues in 1979 in his unpublished dissertation, which includes an experiment on tones masking noise because he could find no scholarship in the area (though in fact Rhona Hellman and others had published on the subject, and Hellman's work influenced some of the thinking at Bell Labs). During the mid-1970s, psychoacousticians also began to reassess masking research, taking an interest in the quality of sonic experience rather than simply in its gaps and absences.[8]

This pattern repeated itself in the mid-1980s as engineers worked to develop perceptual coders. In Germany, first at the University of Erlangen and later at the Fraunhofer Institute, Karlheinz Brandenburg was working on a scheme called OCF (Optimum Coding in the Frequency Domain—a predecessor to MP3): "I literally took Zwicker's [book] *Psychoacoustics* and read through it, and for some of the rules built into the psychoacoustic model of OCF. I looked into the figures, derived parameters from the figures and put them into my code. So it was literally relying on that. Of course, we soon found that some of the figures were a little misleading." At Bell Labs in the 1980s, JJ Johnston ran into the same problem. He began by using data from Harvey Fletcher's research into masking, but "the thing that happened is it didn't do that much good for voice coding." In Argentina, Oscar Bonello's group used masking curves derived from the work of Richard Ehmer and also ran into problems that led them to do extensive modifications followed by listening tests.[9]

Earlier psychoacoustic research didn't immediately translate because psychoacousticians didn't often do masking research with music or speech. They used sinusoid (sine wave) tones or broad or narrowband noise to do their tests. Neither music nor speech—the two holy grails of perceptual coding—behaves anything like sinusoids or noise bursts. As Johnston put it, "Voiced speech is about halfway between tone and noise in terms of the way the ear deals with it."[10] The development of perceptual coding required essentially redoing some basic research into human hearing, at least insofar as it was relevant to getting a codec to render speech functional and music pleasing. The goal of this research was not a universal theory of human hearing or musical pleasure, but simply the creation of an algorithm that would convert CD-quality audio files into much smaller files with little or no perceptible loss of definition for most listeners in most situations.

Early tests in the career of a codec were simple A-B tests in which engineers would listen to their own work to ascertain whether there were obvious problems. The engineer would play a version of the reference recording and compare it with a version subjected to the coding algorithm being tested. Problems were obvious, so there was no need for more sophisticated testing. Although coding problems would be plainly audible, their solutions often weren't, as the engineers were still just figuring out which of dozens of parameters were causing the problems they heard. As Brandenburg explained, "You could change its behaviour completely with a different parameter set and sometimes it would work better with this type of music and then better with that type of music, and so on. . . . You had so many parameters to play with and so little knowledge of how things would work in the end that it was quite a tricky process."[11]

The final tests for what we now call MP3 were performed on panels selected by the Moving Picture Experts Group in July 1990 and May and November 1991. These later tests were *double-blind triple-stimulus tests*—

tests in which subjects would hear three recordings but know the origin of only one. Listeners were given a *known reference*, which was the original recording. They would then be given two more *hidden* recordings, one of which was the known reference, and the other was the same recording run through a codec. The test was *double-blind* because neither the listener nor the test administrator knew which of the last two recordings was the known reference.[12] Everything in these tests was carefully calibrated to render the different protocols for perceptual coding as the only meaningful variables. The documentation for the tests provides lists of recordings, names of test subjects, test procedures and deviations therefrom, exhaustive lists of equipment with signal flow diagrams, and reverberation times for the rooms in which the tests were held. For instance, the following description of Studio 11 highlights all the steps taken to ensure that it follows the principles of good acoustic design: "The studio 11 is a former small drama studio slightly modified in order to improve the acoustics. The listening room has nonparallel walls and floated floor. The absorption is evenly distributed over the walls. The floor is covered with a carpet. Boards with slightly absorbing surface have been suspended from the ceiling in order to achieve a V-shaped cross section. The windows and the wall behind the loudspeakers are covered with curtains."[13]

Even greater detail is available regarding the process of the tests themselves. Consider the signal-flow diagram (figure 26) from the first round of tests in July 1990 for what it reveals about the testing process and its epistemology. Although the double-blind triple-stimulus test is simple in concept, it turned out to be complex in execution, since the technology had to be set up in such a way that neither listeners nor administrators would know ahead of time which hidden recording was which. On the right hand of the diagram, six listeners ("subjects") operate a control terminal that allows them to switch between three different signals. One is the known reference, the other two are hidden—the reference again and the coded audio. A computer with a custom audio interface selects which signal is which for each listener, so neither the listener nor the test administrators know which of the hidden signals is the reference and which the coded audio. The subjects can listen through headphones or through a pair of speakers. As the transducers are the most crucial link in the chain besides the listeners themselves, brand names ("Stax" and "Genelec") are given to signify quality and expense, and an exhaustive list of makes and models for all equipment used is included in the report of the tests.

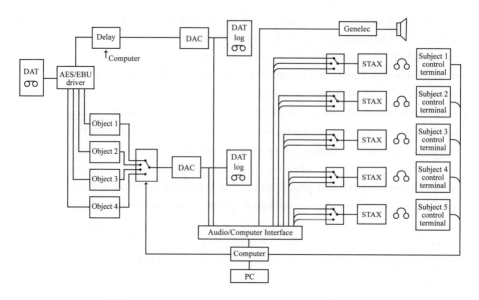

26. General system setup for MPEG listening tests, July 1990. Source: Bergman, Grewin, and Rydén, "MPEG/Audio Subjective Assessments Test Report," appendix 3. See also Bergman, Grewin, and Rydén, "The SR Report on the MPEG/Audio Subjective Listening Test April/May 1991."

On the left of the diagram, a digital audio tape player plays back a CD-quality recording through a digital interface that splits the signal.[14] Taking a right at the fork in the road, the signal is split again as it passes simultaneously through four "objects." Each object is a physical box containing one of the four competing codecs. A computer switches between each of them for each test (but not within a test—codecs were compared to the reference recording on the DAT, and not directly to one another). The coded and decoded audio then would pass through a digital-to-analog converter on its way to speakers or headphones and then to subjects' ears.

Taking a left at that first fork in the road in the diagram, the signal is routed through a computer-controlled time delay, as different codecs took different amounts of time to process the signal, and the test required that subjects hear both signals in perfect sync so they would not know that the delayed signal was the coded audio. After the delay, the signal passed through a digital-to-analog converter on its way to the transducers and subjects. Two other DAT machines recorded the playback from each side of the system in case it would need to be reproduced later.

Everything in the test is undertaken to mediate the relationship between listening subjects and sonic objects. The safeguards and anonymizing features are both technical measures and rhetorical levers designed to install a sense of the test's reliability and credence in its results. Every feature of the process quietly but firmly announces disinterestedness. Just beyond the control terminals on the right side of the diagram lies the mind's ear of each of the five participants in each phase of the test. The competing codecs are just inside the diagram but on the far left. The testing scenario expends most of its energy (and the diagram draws most of our attention) to the middle of the process. The test's real object is the stream of audio the codec releases through the maze of signal paths, converters, and switches. Every step in the network is designed to symbolically or perceptually erase all mediations — of interest, technologies, sound, time, experimental process — and to save the sonic artifacts introduced by the codecs themselves. But the diagram makes all of these mediations readily apparent. If, after many repetitions of a test, listeners could not distinguish between the coded signal and the known reference with a greater frequency than if they were guessing (i.e., more than half the time), the coding was considered transparent. If they could correctly distinguish between the signals, then the level of difference between the signals began to matter. Obvious and annoying differences were more problematic than minor differences, but all differences mattered.[15]

In a way, the whole testing exercise pivots on the word *annoying*. It responds to the need identified by perceptual-coding researchers for a test that makes explicit how music and speech ought to sound and what counts in each case as *good sounding*. This is reflected in the scale used for the test. For much of the twentieth century, one could say that psychoacousticians followed an anaesthetic aesthetic.[16] That is, psychoacoustics made aesthetic decisions about sound but did not present them as such. Perceptual tests used a scale that ranged from the "threshold of audibility" to the "threshold of feeling" or "threshold of pain."[17] But by the 1980s, another scale had come into common use in perceptual listening tests for audio equipment (see tables 2 and 3 and figure 27). In the terminology of the scale, we can see the cool scientific language of psychoacoustics melt in the warm radiance of judgment. It offers listeners only three adjectives but they speak volumes: *imperceptible*, *perceptible*, and *annoying*. The first two correspond to the concept of just-noticeable differences. They translate the subjective experience of hearing into measurable units; you can either hear a difference or you can't. But since the test is also aesthetic and since the discourse

TABLE 2. Imperceptible-annoying grading scale used in the MPEG listening tests

The grading scale

5.0	Imperceptible
4.0	Perceptible, but not annoying
3.0	Slightly annoying
2.0	Annoying
1.0	Very annoying

Source: Bergman, Grewin, and Rydén, "The SR Report on the MPEG/Audio Subjective Listening Test April/May 1991," appendix 1.

TABLE 3. Better-worse grading scale used in the MPEG listening tests

Grading scale used during the fast-forward assessments

3	Much better
2	Better
1	Slightly better
0	The same
−1	Slightly worse
−2	Worse
−3	Much worse

Source: Bergman, Grewin, and Rydén. "MPEG/Audio Subjective Assessments Test Report," appendix 7.

of beauty in engineering is relatively submerged, beauty's opposite appears: repugnance, or at least its quieter cousin annoyance. *Annoying* is an aesthetic category, and one of some refinement, as evidenced by the five-point scale recommended for use within ten decimal places (for a total of fifty increments of annoyance—imagine the range of possible uses for such a scale in daily life!). The scale was already in use for audio by the mid-1980s; it appears in 1986 in a CCIR (Comité consultatif international pour la radio) standard cited in MPEG reports.[18] The scale itself borrows ideas from the field of food engineering, where it had been in use since the 1940s in panel tests of processed and preserved foods. Taste is thus more than a metaphor. It points us back to a mode of transmission and a particular way of engaging

Absolute grade			Difference grade
5.0	Imperceptible	0.0	
4.9		−0.1	
4.8		−0.2	
4.7		−0.3	
4.6	Perceptible but NOT annoying	−0.4	
4.5		−0.5	
4.4		−0.6	
4.3		−0.7	
4.2		−0.8	
4.1		−0.9	
4.0		−1.0	
3.9		−1.1	
3.8		−1.2	
3.7		−1.3	
3.6	Slightly annoying	−1.4	
3.5		−1.5	
3.4		−1.6	
3.3		−1.7	
3.2		−1.8	
3.1		−1.9	
3.0		−2.0	
2.9		−2.1	
2.8		−2.2	
2.7		−2.3	
2.6		−2.4	
2.5	Annoying	−2.5	
2.4		−2.6	
2.3		−2.7	
2.2		−2.8	
2.1		−2.9	
2.0		−3.0	
1.9		−3.1	
1.8		−3.2	
1.7		−3.3	
1.6		−3.4	
1.5	Very annoying	−3.5	
1.4		−3.6	
1.3		−3.7	
1.2		−3.8	
1.1		−3.9	
1.0		−4.0	

27. A subjective quality scale. This was not the actual grading scale used in the tests (a simpler five-point scale was used) but this one has the virtue of showing fifty gradations of annoyance. Source: Pohlmann, *Principles of Digital Audio*, 409. See also "Methods for the Subjective Assessment of Small Impairments in Audio Systems including Multichannel Audio Systems," ITU-R BS.1116.

the senses. John Philip Sousa's famous derision of sound recording as yielding "canned music" finds its twentieth-century analog in the word *processing*.[19] MPEG audio is processed sound for listeners who live in a processed world.

The taste-test researchers used the term *hedonic*, which means "of or pertaining to pleasure." In psychology, the term had been in use from the beginning of the twentieth century to refer specifically to pleasurable or unpleasurable sensations or feelings, considered as quantifiable affects. A book by John Beebe-Center, *The Psychology of Pleasantness and Unpleasantness* (1932), systematized research into the area, building on the work of

Gustav Theodor Fechner.[20] But using hedonic as an adjective here collides with hedonics, the noun, which refers to the doctrine of pleasure in ethical philosophy (which shares a root with the more commonly used English word *hedonism*). The use of a hedonic scale in measuring the performance of audio technologies therefore privileges pleasure as the guiding principle for evaluation, and despite the more staid language (annoying, not annoying), it implicitly puts pleasure as the first principle and purpose of sound reproduction. This is a striking contrast to the way engineering discourse usually represents the goal of sound reproduction, which tends to revolve around a measure of definition (for instance, technical specifications).

The functional replacement of a threshold of pain with a threshold of annoyance in measurement scales is at one level a simple matter of practicality since we are talking about musical artifacts and not loudness. But in charting a historical path from pain to annoyance, we follow the well-worn trail of Norbert Elias, whose notion of the civilizing process relies on senses that make increasingly fine distinctions, retreating behind an ever-advancing "threshold of repugnance." In Elias's account of modernity, as the bourgeois subject becomes more entrenched and actual threats of violence become less and less common in daily life, subtle movements and perceptions take on greater and greater meaning.[21] *Taste* is the name given to the senses so cultivated. Listening tests thus stage the civilizing process for audio. They subject different modes of audio signal processing to various "civilized" sonic sensibilities in listeners. The listening test subjects processes of sonic refinement to subjects' refined sonic judgments; or perhaps it is the other way around.

In the world of MPEG, reference begins its life in the digital realm. The starting point for the MPEG associated audio group was the compact disc (red-book) standard—which was well established by 1988. MPEG's original goal for signal transmission was based on the network and storage technology of the time. So, the listening tests for MPEG referred to compact disc audio as their base line. For instance, the report from May 1991 defines "transparency" as "subjectively indistinguishable from the reference source, a CD player," even though digital audio tape was used in the tests (DAT has the same bit depth and sample rate specifications as CD, though the converters and analog signal paths in the devices may have been quite different depending on manufacturer).[22] The MP3 is a format that takes other formats as its aesthetic and technical base line. It is designed for a world populated with recordings, and it was built with the intent of affecting the conditions in which recordings circulate. It is a mediatic creature. The reference was therefore at once sonic and economic: by establishing CD-quality audio as the reference point, the MPEG tests also established a ground for extracting perceptual capital. If listeners could not distinguish between MPEG audio and CDs, then it would be possible to substitute the latter to enhance the value of transmission networks and recording media that could not accommodate the red-book standard. MPEG audio is thus mediatic in two senses, both affirming sound's and music's existing modes of circulation and seeking to adapt them to the existing digital infrastructure.

Although the testing scale measures from imperceptible to annoying, its object is something else: timbre. The audible differences between perceptually coded audio and source recordings are differences of timbre. As these are quite subtle, listening tests can require extensive training. The Audio Engineering Society, which presents itself as the world's leading body for professional audio engineers, now produces a CD on which listeners are instructed in the various kinds of artifacts one might encounter.[23] But in the early listening tests for a given a codec, it was not always clear what one was listening for. When Oscar Bonello's Argentinian team started testing out their technology, they felt "like Adam the first time he saw Eva.... We spent some time in order to learn that this was a different type of distortion." Bernhard Grill, who was present for the 1990–91 MPEG tests, also questioned his judgment after the fact. While he thought the coded audio streams sounded transparent at the time, he says that he would not think so now. Even though listeners may find them hard to consciously detect, subtle yet audible differences between MP3 and CD-quality audio have become an important part of mediatic sound culture in the intervening years. Despite all the talk of imperceptible differences, the issue was well known during the tests. An account of the tests presented at the AES meeting in London in 1991 referred with some anxiety to listeners having a span of thirty years (or longer) to discover artifacts once a system was standardized: "It can be compared with somebody who moves into a new house. The first time he looks through the window he only sees the beautiful view. After a few days he detects a small flaw in the glass and from that moment he cannot look through the window without seeing that flaw."[24]

By the late 1980s, it was well established that artifacts of perceptual coding could be identified, heard, and reproduced. For instance, one artifact of some perceptual coding is called *pre-echo*. Already in 1979, Michael Krasner had identified a problem with "onset transients" in coded speech, where distortions of percussive sounds like the letter *t* were quite perceptible to listeners. This problem has continued down to the present day. Pre-echo smears a sharp transient like a cymbal hit, so the initial attack has less bite and occurs over a slightly longer period of time than in the reference recording. The recording of castanets used in the tests (and discussed briefly below) will still today yield audible pre-echo when processed through a standard 128k MP3 coder. However, the effect is very subtle and bothers some people more than others. While some listeners will hear it right away and be annoyed, others will take years to discover it, never discover it at all, or simply experience it as part of the sound of the music, like tape distortion or dust on a record. More to the point, without identifying an artifact, it cannot be tested for. JJ Johnston recounts of his tests at Bell Labs: "At that point we didn't know what pre-echo was and we didn't know to have a signal that would cause it to happen, either, so we didn't hear it." Similarly, because of the way MP3 codes stereo data, the stereo image of the sound can change. For some people with some recordings on some systems, this is very noticeable. For others, it is not perceptible at all.[25]

Because of the variability among people, listening tests involve a great deal of repetition in order to produce reliable and predictable results. Listeners are given training on the equipment and on what to listen for, and they are given practice so that they become more reliable over time. Louis Thibault, stressing the importance of reliability and reproducibility, considers listeners as part of the reproduction system and measurement apparatus: "[Listeners] calibrate their internal rating scale and that is important because our listeners are meters, and if you want a meter to provide a reliable measure you must calibrate it before using it—you do it with volt meters and you do it with human subjects. The goal of the training is that their rating of a given audio sequence will remain the same days or weeks after the initial test."[26] As they have for most of the twentieth century, perceptual technics subsume ears and media.

Although the testing scenario ostensibly sorts out codecs, it must first sort out listeners. Even when the listeners are given extensive training, a good listening test strains the auditory palette. Karlheinz Brandenburg recalled, "For larger sets of music, it was very tiring because the test subjects had to listen to each test item twenty times. I remember at AT&T at one point in time we had a revolt—people were just refusing to participate in the listening tests [because they were so awful]." Bernhard Grill called the MPEG listening tests "very hard. If you do listening tests for almost transparent codecs it's really strenuous. Each of the sessions lasts two hours or one-and-a-half hours and you're really exhausted afterwards. It really requires a very high degree of concentration."[27] The signal flow diagram for the listening tests does render the listening subject as part of the reproduction system, and as its most important variable. The image's absenting of listeners' bodies suggests that they may in fact be the hardest part of the test to pin down. This may seem like a circular logic: the test tests the listeners before the listeners test the technology. But as Pinch reminds us, the test scenario is by design a contrivance. It is not entirely circular. Rather, in the testing scenario a series of political modulations occur: subjects become acclimatized both to the peculiar scenario and to the peculiarities of the technology. Their experiences within that circumscribed situation then help shape the further development of technologies that may be in turn inflicted back upon them at a later date. Subjects and machinery thus influence one another in a very particular set of contrived situations.

Who Listens for Future Listeners?

If it matters how perceptual codecs are listened to in their early development, it also matters who listens to them. If listening tests had a political ideology, it would be a modified form of representative government. The subjects of listening tests represent a future public that the codec hopes to one day confront. The test situation is thus not only a worst-case scenario, as the engineers would have it. It is also a moment of listening—and speaking—for others and thereby representing them. Guidelines for listening tests refer to their ideal subjects as "expert listeners."[28] The expert-listener bias mimics the biases toward a certain level of technical sophistication that is more generally present in standards organizations. Facility with the technology—and with the technical language that accompanies it—is a prior condition for representing others in the test scenario: "This attitude on the side of the committee members further hampers successful representation of these groups of stakeholders, as they will in many cases not have adequate resources."[29] The name *expert listener* is therefore not accidental. It represents a structural bias in engineering culture and a political bias that shapes the making of standards and formats.

The first round of listeners for perceptual codecs are usually the engineers themselves, essentially testing the audio on their own ears. While these engineers have a great deal of technical knowledge, they may or may not have "golden ears." In this sense, the engineer building the codec is not an expert listener, merely an expert who listens. Since audio issues are fairly obvious and technical issues fairly opaque early on, the argument goes that no other listener is needed. For more advanced tests, expert listeners are often drawn from the ranks of musicians, recording engineers, broadcast engineers, piano tuners, and audiophiles, all of whom must develop an ear for technology, an auditory virtuosity that facilitates making careful judgments and finely graded distinctions. In essence, the ideal expert listener functions as an extension of the reproduction system, as a meter for it, measuring the audible performance of the codec and reliably reporting back to the engineers. Anyone who can produce trustworthy results can be an expert listener. But in practice, it is easier to begin the search for expert listeners among populations that spend a lot of time thinking about and listening carefully to sound, and especially to sound-reproduction systems. Expert listeners listen for form, not content. They listen to the system, not the speech or music that it transmits. An expert listener needs a good ear for timbre (specifically, the timbres of different codecs) and must be able to make consistent judgments over time. Many testing facilities have coteries of expert listeners to whom they turn (and whom they pay) for the final testing of equipment, and all of the codecs that made it into the MPEG tests had already been through rounds of testing with expert listeners in their own "native" corporate environments.[30] As the list above suggests, these listeners are a highly specialized, selective, relatively elite, and predominantly white European male and bourgeois crowd.[31] The list of participants in the final MPEG tests confirms this. Most were engineers who had worked on one or another of the coding schemes proposed for adoption or had some other interest in the standardization process. It is unclear whether they actually qualified as expert listeners, but they were experts in the technology, and often interested parties.[32]

Here, we directly confront the question of who listens for whom and with what effect. Georgina Born's study of IRCAM in France of the 1980s notes that while the psychoacoustic research being developed at IRCAM made strong claims for its "universal" models of perception, those models were highly specific to the culture that developed them. Thus, much of the research was concerned with pitch and timbre, but there was almost no discussion of rhythm, despite the importance of rhythm and percussion in most of the world's music outside of the Euro-American art music tradition.[33] In the case of perceptual coding, the lack of attention to pre-echo in early testing and coding strategy may evidence a certain bias toward rhythm also evident in music theory, law, and other institutions that have hypertrophied ideas derived from art music aesthetics.

It is not enough, however, to note the particularist universalism of psychoacoustic research practices. For this raises the more fundamental question of what difference *difference* might make in the listening tests or in the broader development of MPEG audio. Although the people who worked on perceptual coders repeatedly challenged existing masking research, they did not often question existing practices of subject selection for psychoacoustic and engineering research. For instance, an article from 1992 by Søren Bech, a researcher for the high-end audio company Bang and Olufsen, sums up the field's common sense regarding the selection of research subjects for listening tests around the time of the MPEG tests. He argues that subjects should be selected on the basis of who can best produce a reproducible measure (on an appropriate rating scale) of a prespecified aspect of the perceived sound being tested. Cultural differences like gender, age, race, class, nationality, and language are not directly named in the article or in the literature Bech cites. Test subjects can produce variability in test results, he argues, but these are primarily on factors like the condition of their hearing, their training in experimental protocols, and something called "previous experience."[34]

"Previous experience" at first seems to be just that: prior experience with sound-reproduction equipment, familiarity with "live" sound in concert situations, playing an instrument, critical-listening practice, "and the general aptitude for detecting sonic differences in reproduced sound."[35] Yet this experience must be obtained, and it could be more or less easy to get depending upon who you are. To begin with, critical-listening practice, working with live sound or studios, and certain kinds of musical equipment are still heavily gendered. Engineering culture is still very male in both number and flavor, as are other areas of musical subcultural practices, from musical-instrument stores and record shops to music journalism.[36] Yet for all this, "previous experience" is an exceptionally desirable quality in a listening test subject. Bech notes that "there is a positive correlation between the degree of previous experience of a group of subjects and homogeneity of ratings within the group." People who have more experience with audio technology, who are more inculcated into the culture of working with it, talking about it, thinking about it, are more likely to produce similar results on tests. The culture's own values overcome differences in the listening-test scenario once it is well-enough established in the test subjects. This effect was pronounced enough that after finding it, Bang and Olufsen moved to establish a permanent listening team to test its equipment—essentially a panel of professional virtuoso testers.[37] Clearly, there are common cultural factors that lead people to enter a listening test with prior experience, as well as the subjects' greater amenability to the test scenario itself if they are already accustomed to talking about, working with, or thinking about sound. But even if common cultural factors don't bring them to the world of audio equipment, their time spent in that world may provide common culture enough.[38]

Aesthetic acclimatization may therefore be a more precise term than "experience" to describe the process that produces the ideal (or for that matter, nonideal) subject of a listening test. A still-cited study from 1956 of college students' loudspeaker preferences at Ohio State found that exposure to a particular sound system over time led students to have a preference for that system, that in general students preferred the sound of "low-fi" to "hi-fi" speaker systems, and that the type of music students listened to was correlated with their preferences for sonic characteristics of speakers. Although the author behaves like a good social scientist and cautions against overgeneralizing the results, the study has been cited repeatedly in subsequent literature on listening tests down to the present day. Taken together, these various implications of experience explain both the bias toward engineering values in the MPEG tests and the preferences among today's university students for 128k MP3 audio over other formats. Indeed, worries about the thirty-odd-year lifetime of formats indicated that the test designers considered experience as a problem in the future, and not just in the present.[39]

It is cliché to say that technologies reveal themselves and their limits at moments of failure, and best conceal themselves (or at least their technological character) at moments when they succeed. Yet "broken" or "unusable" is not an objective state of a piece of equipment, but rather a question of orientation and relationship. An object is not in itself unusable but unusable for somebody in particular. Sara Ahmed writes "a hammer might be too heavy for you to use but perfectly adequate for me. . . . What is at stake in moments of failure is not so much access to properties but attributions of properties, which become a matter of how we *approach* this object."[40] Even expert listeners who act like meters (to use Thibault's language) represent a broader potential field of users of the technology—something less than a universal subject. Here again we confront the question of representation: who listens for whom?

In interviews for this book, I posed the question of difference—in the selection of listening-test subjects and in the range of their experiences—to a number of people involved in the development of the MP3 format. I heard a wide range of responses, suggesting that the question is far from settled within the engineering field. Louis Thibault argued that no differences mattered except for the regularity of performance on the tests themselves. In other words, if people gave reliable and consistent results, that was good enough. If they didn't, it didn't matter who they were. JJ Johnston repeatedly asserted the importance of age and experience. As people grow older, they tend to lose some of their hearing but aging can be mitigated by experience with testing scenarios and audio-processing equipment. In fact, aesthetics may reemerge as an issue in the case of the very artifacts that test subjects are supposed to note and help eliminate:

> I still hear more coding artefacts than most people. But it helps to have 25 years of doing it. That's it in a nutshell. Now, I have observed a couple different kinds of listeners. Interesting enough they tend to rate quality the same on almost, but not all things. And this is where Karlheinz and I just diverge completely. He is hypersensitive to pre-echo; I barely hear it. He doesn't hear imaging artefacts; they drive me absolutely, stark-raving [mad]. If stuff starts moving around on the sound stage, I start to foam at the mouth. And he doesn't even notice it happening. But if he hears a pre-echo ... So, people actually hear the same thing, but they have very different intense dislikes.[41]

Johnston's account gels nicely with Ahmed's. One or another scheme for perceptual coding may fail for you but not for me depending on who we are, what we've heard in the past, and what we are doing.

Other people involved in MPEG also acknowledged that some kinds of differences could matter, even if there was a less systematic sense of how they might. Bernhard Grill and Marina Bosi noted that language was important for the speech samples, as nonnative speakers had more trouble separating signal from system (and speakers of tonal languages can more easily distinguish pitch). Familiarity might also matter for music in the same way. If listeners have not already formed an idea of how the music ought to sound, then they may not easily distinguish signal from system. Karlheinz

Brandenburg said that while they didn't consider issues of cultural difference in the tests, it is possible that it might have mattered. Bosi entertained the idea that women and men may hear best at different frequencies, a notion that has circulated for some time in audio-engineering lore.[42] Given the closed nature of the test, I hadn't expected such a wide range of opinion among people actually involved in developing the technology. Perhaps it is therefore better to consider that, via the category of "previous experience," the possibility of cultural difference is tacitly acknowledged in the design of the listening test, even as it is officially refused. The listening test seeks to overcome difference through its own guided particularist strategies. It attends to difference, but only by trying to eliminate it as a variable.

The psychological research on hearing acuity and difference is somewhat limited. Most of it deals with hearing tests (which use equipment to test subjects' hearing, rather than the subject's hearing to test equipment), and some of it is quite old. The clearest correlation is around age. As people get older, their hearing gets worse, especially at the high end. While there is a body of research on the relation between absolute pitch and tonal languages, the strong association between the two has been disputed by neuropsychologists and by Asian researchers, who argue that the findings are partly the result of Western researchers' lack of familiarity with and difficulty distinguishing registers of Asian tonal languages. A strand of hearing research has consistently found that women have more acuity in higher-frequency ranges than men, while men have greater acuity in lower-frequency ranges. But the explanations for this difference are also controversial. One argument involves the differences between the mean length of men's and women's cochleas. Other researchers, like Dennis McFadden and Edward Pasanen, have claimed that lesbians and bisexual women exhibit more similar otoacoustic emissions to men than to those of heterosexual women. While press accounts focused on their study as "proof" of biological differences between straight and queer women, the authors suggested that rather than being entirely biological, exposures to certain drugs and noise levels may be "a secondary consequence of the lifestyle adopted by the majority of these women and thus is not conceptually different from the hearing loss developed by rock musicians." Either explanation seems suspiciously heterosexist, unless the authors actually questioned the lesbians and bisexual women in their study about their lifestyles (they call for a longitudinal analysis which they themselves do not do), or measured their brains and androgen levels (instead they cite a single study published in 1993 which argues for enhanced prenatal exposure to androgens as the cause of homosexuality, which ignores that bisexual women are not lesbian). The scholarship on race and class is even spottier. Mostly, those differences do not appear at all as variables in hearing studies. Because there has been little interest shown among psychological researchers into these kinds of differences, we know very little about how race, ethnicity, nationality, and class mediate hearing. Some historical exceptions exist, like Bell Labs' tests at the world's fairs in New York and San Francisco in 1939, which had subjects check a box for "colored or white" and implicitly collapsed race and class, arguing that better economic status leads to better hearing acuity. A survey conducted in 2005 of hearing studies in the United States argued that children for whom English is a second language had more trouble with hearing tests (for the obvious reason that they were less familiar with the language), but also that African American and Hispanic children have higher rates of hearing impairment than white children.[43] Already the most basic hearing research assumes some familiarity with sound equipment. As it moves to studies of speech, music, and aesthetics—all fundamentally cultural materials—hearing research will need to develop better models for how to account for culture and difference in its tests.

What's on the MPEG Player? Genre and Style in the Tests

Particularism also affected which recordings were chosen for use in the tests. Engineers call *critical material* that which puts the most stress on the codec, since the purpose of the test is to be a worst-case scenario. Early on, this meant simply that the material was considered to be of high quality. Familiarity also was important. In his experiments conducted in 1979, Michael Krasner used digital copies of record releases from an audiophile label as his critical material. In 1985 JJ Johnston used a wider range of materials (see table 4). The article's presentation of the material tells us a lot. He aims to present a variety of styles and content, but the actual performers and recordings are presented in the article as though they are irrelevant to its argument. Just as the expert listener is supposed to be an anonymous listener who transcends context, so too critical material is supposed to anonymously stand in for all possible future source material in the test scenario and in its scholarly representation. This is why codes are preferable to names. Compared with other articles, Johnston's is striking for the amount of information it provides about the material used. Most

TABLE 4. Recordings used in Johnston's listening tests (1985)

Source material used in both the pe study and the pxfm listening test 15 kHz source material, sampled at 32 kHz

4 LETTER CODE	INSTRUMENT/STYLE
ATBR	Brass choir, baroque music
BJPV	Pop vocal '50s style
BPSP	Solo piano (soft passage)
DZFV	Female coloratura opera vocal
DZMV	Male operatic bass vocal
EHPV	Female pop vocal
JBHC	Complex harpsichord solo
JBOR	Baroque organ
LFSR	Solo recorder
LZBD	"Heavy metal" rock vocal/instrumental
PEAK	Whip-crack (surrounded by silence)
PERC	Percussion (African drums)
PHOR	Romantic style full organ
WHEP	Electronic pop/Ball synthesizer
WHSG	Solo guitar

Source: Johnston, "Transform Coding of Audio Signals Using Perceptual Noise Criteria," 315.

articles on perceptual coding from 1991 or earlier say little about the recordings used. Again, this is intended to deflect attention away from taste, but of course taste resurfaces here and there in the process. Johnston clearly had ideas at work about which material ought to be in a test. In an interview, he went through the list with me:

> Pop vocal was "The Longest Time" by Billy Joel. Solo piano was a portion of "Moonlight Sonata" from Rudolph Sorkin, played off an LP. Female Coloratura Opera Vocal was a piece of "Queen of the Night—Second Aria." Female pop vocal was not yet Suzanne Vega. I think it was Linda Ronstadt, but I'm not sure. Harpsichord was "Chromatic Fantasy in F"—a million notes. Baroque organ was either [sings a tune] or it might have been something out of "Fifth Organ Symphony." Solo recorder was [whistles tune] from *Beggars*. Heavy metal rock vocal was "Black Dog" from Led Zeppelin. The whip crack was a whip crack out of

the *Folies d'espagne*. The percussion was off of "Rhythm of the Saints," the Paul Simon album. Electric pop synthesizer . . . I don't remember what that was. Solo guitar was Norman Blake—I had to slip in some bluegrass.[44]

Johnston's account is an interesting contrast to the published article because he speaks of the music not for its universality but for its effect on him. His affection for some of this music—and music in general—came through clearly in his answer. This is one of the central contradictions of the listening test. It operates according to a logic and rhetoric of disinterest, yet it is equally susceptible to what Pierre Bourdieu called "the games of taste."[45] The idealized world of the listening test requires its participants to listen beyond their own interestedness, the immediate pleasures or gratifications the music brings, reaching instead toward a kind of pure audition. The listening test presents the technology as if it could work for anyone; but by virtue of playing music, its sonic address is always to and for someone in particular. Johnston's line that he "had to slip in some bluegrass," nicely encapsulates the contradiction between the questions of pleasure posed by listening and the depersonalizing rhetoric of the testing situation. The test scenario does not immediately obviate the love of music.

The rhetoric changes a little bit when we get to the MPEG tests. Because of the juridical nature of the tests—MPEG was setting a standard—exhaustive lists of recordings are given. Consider the recordings used for the MPEG test in July 1990 (table 5). Some of the recordings are tracks from commercially released albums: Tracy Chapman "Mountains o' Things" off *Tracy Chapman* (1988); Ornette Coleman "In All Languages" off *In All Languages* (1987); Suzanne Vega's "Tom's Diner"; a Haydn trumpet concerto. The other recordings, like the castanets, were made for the specific purpose of testing audio equipment. SQAM stands for "Sound Quality Assessment Material"—a CD that was released by the European Broadcasting Union for the purposes of having a uniform set of audio recordings with which to test different issues in audio equipment; the instructions that come with the CD even explain which recordings are best for which kind of test.[46]

From a certain aesthetic perspective, we might think of this list as balanced. There was classical music, radiophonic speech, free jazz, and acoustically inflected pop. When I asked Bernhard Grill about the musical selection, he said, "I think it was pretty balanced—maybe the latest pop music was missing. If you wanted to accuse the test of omitting something I don't

TABLE 5. Recordings used for MPEG "high-quality" listening tests (July 1990)

High-quality sessions (128 & 96 kb/s/channel)

1	Suzanne Vega	tr 1	0.22–0.42	A&M 395 136-2
2	Tracy Chapman	tr 6	0.36–0.57	Elektra 960 774-2
3	Glockenspiel	tr 35/1	0.00–0.16	EBU SOUAM 422-204-2
4	Fireworks	tr 1	0.00–0.30	Pierre Verany 788031
5	Ornette Coleman	tr 7	—	Dreams 008
6	Bass synth	—	—	RR recording (DAT)
7	Castanets	tr 27	0.00–0.20	EBU-SQAM 422-204-2
8	Male speech	tr 17/2	54.16–54.35	Japan Audio Society CD-3
9	Bass guitar (K B Enchill)	—	—	RR recording (DAT)
10	Haydn Trumpet Concert	tr 10	05.10–05.30	Philips 420 203-2

Intermediate quality assessments (64 kb/s)

2	Tracy Chapman	tr 6	0.36–0.57	Elektra 960 774-2
3	Glockenspiel	tr 35/1	0.00–0.16	EBU SOUAM 422-204-2
5	Ornette Coleman	tr 7	—	Dreams 008
8	Male speech	tr 17/2	54.16–54.35	Japan Audio Society CD-3
9	Bass guitar (K B Enchill)	—	—	RR recording (DAT)

Source: Bergman, Grewin, and Rydén, "MPEG/Audio Subjective Assessments Test Report," appendix 5.

think there was anything missing except for pop songs of the late 1980s." Of course, by other standards, this list of recordings isn't that balanced: there are no thumb pianos or distorted guitars in this bunch, no heavy backbeats or polyrhythms. Not even a gong. There is an evident middlebrowness to the selections, which should come as no surprise when we consider that they were chosen from the available holdings of a state radio network: Swedish radio. To follow Pierre Bourdieu, we could say that all of the music was either what he would call "legitimate music or music undergoing the process of legitimation"—that is, it was music sanctioned by state, educational, or other official cultural institutions. The recordings thus appealed to—or at least were meant not to offend—legitimate taste, a taste for works of art that have been canonized by academic institutions as "worthy of admiration in themselves," and inculcated through educational or domestic training, combined with a taste for works that are "in the process of legitimation," such as the singer-songwriter pop of Vega and Chapman, which in the early 1990s had moved from Top 40 to "good music." The use of legitimate music in a test scenario is not accidental. The music choices here are designed to not call negative attention to themselves, and the fact that the works have already attained (or are in the process of attaining) the status that marks them as worthy of aesthetic contemplation suggests aspirations toward isomorphism between the music and the system. Both were evaluated for quality, and any enjoyment that comes from listening to them should emerge in part as a result of the performance (of the musicians, of the equipment) and not in spite of it. Thus, musical aesthetics quietly takes its place in engineering discourse even as the tests are "not about the music."[47]

Or at least that was how it was supposed to work. Brandenburg recalled that some of the engineers in the tests found the trumpet playing on the Ornette Coleman record "disgusting," but aside from that example, the music was not the object of the tests. When I asked Brandenburg, Grill, and Johnston about cultural difference and the selection of expert listeners, they all replied that the most important thing was that the test subjects didn't dislike the music. Familiarity was important. Johnston put it succinctly: "First rule: don't annoy the listener."[48] Inasmuch as there was a good deal of class homogeny among the test subjects, the critical material also reflected a relatively coherent taste culture. Once again, the standards-testing scenario leans toward a cultural bias for knowledge of engineering. But while selection and genre play a role, we would miss the point if we made too much of the engineers' generic tastes in the MPEG tests or the attempt to draw conclusions about them based on their musical choices. The point is not how cultured or "hip" they were by whatever misshapen measure we might apply (or worse, for a reader to feel superior in his or her own aesthetic preferences). As Lawrence Grossberg argues, "Both 'taste' and 'pleasure' draw us, perhaps unwillingly, into individualizing and psychologizing discourses."[49] To do so is to derail the discussion into the "games of culture" derided by Bourdieu. The point is to understand how the sound of the music functions in the encounter between listening subject, codec, and objectified testing system. At the propositional level, the listening test is not oriented so much around the preferences of a listening subject as it is a sub-individual set of reactions to audible timbres.

The Sound of the Coded Sound of Music

The material's sonic signature is therefore crucial to the story. Almost all of the critical material used for testing followed a realist style of recording.[50] Realism, of course, is not the same thing as reality. Realism in cinema, for instance, can include dialogue edited in a shot-counter-shot fashion in which we see the face of one character and then another, yielding an impossible visual perspective. The same is true for auditory perspective in realist recording styles. The instrument is recorded alone in the studio, with a microphone that is relatively euphonic and placed at enough of a distance that some of the reverberant sound of the room is audible in the recording. The exceptions were the speech recordings, which were recorded with the speaker next to the microphone, with no audible room sound or echo, because that is the convention for presenting mediatic speech over telephone or radio. The result is an impossible acoustic perspective: someone speaking loudly, but whose voice has the reverberant characteristics of a person whispering in your ear. Conversely, a processed voice figures prominently in the singer-songwriter selections, since the human voice is so processed in most popular music that a recording with an unprocessed (or unusually processed) voice would sound unnatural to untrained listeners. All this is to say that at the level of style, the conventions of good recording and mixing, parsed out by genre, were perhaps more important than the genres of the music included in the tests.

In the years since the tests, one recording in particular has become synecdochically famous for the entire testing process: Susanne Vega's "Tom's Diner." By considering some of the sonic meanings of the recording we can elucidate how the aesthetics of recorded sound actually worked in the testing process. An account of Brandenburg's story has circulated widely on the internet and back into books (and now this one). My version comes from Suzanne Vega's own version of the story on a *New York Times* blog. It begins with Brandenburg speaking in a *Business 2.0* article:

> "I was ready to fine-tune my compression algorithm . . . somewhere down the corridor, a radio was playing 'Tom's Diner.' I was electrified. I knew it would be nearly impossible to compress this warm a capella voice." So Mr. Brandenburg gets a copy of the song, and puts it through the newly created MP3. But instead of the "warm human voice" there are monstrous distortions, as though the Exorcist has somehow gotten into the system,

shadowing every phrase. They spent months refining it, running "Tom's Diner" through the system over and over again with modifications, until it comes through clearly. "He wound up listening to the song thousands of times," the article, written by Hilmar Schmundt, continued, "and the result was a code that was heard around the world. When an MP3 player compresses music by anyone from Courtney Love to Kenny G, it is replicating the way that Brandenburg heard Suzanne Vega."[51]

The story presses all the romantic buttons on the machine that makes histories of technology. A lone genius inventor slaving away at his invention encounters a woman's enticing vocal art. The muse moves him and leads him to his discovery. We are now all the beneficiaries of this confluence of genius. Like all urban legends, the story has grown and mutated. For instance, one textbook on data compression called "Tom's Diner" "the first MP3, ever recorded . . . in 1993"[52] — a meaningless epithet since 1993 is two years after the tests were concluded and two years before the term *MP3* was coined and put into use. I asked Brandenburg about it and he qualified the story somewhat:

> Well, for OCF [the first functioning real-time codec discussed above] it was the last fine-tuning stage, but this was a predecessor to ASPEC, which was a predecessor to layer 3 . . . so in the later stages, when we worked with AT&T and then with Thomson and the others, that item was already there. In fact, the interesting thing is that on one hand, once we used better stereo coding techniques, the Suzanne Vega item became very easy to code using layer 3; within the test set it turned from a worst-case to a best-case item within six years. It was 1994 when we finally knew how to do it. But it's still critical enough that it survived into the current test set for MPEG-4 audio coding. We still use it as a test item today. And I once met Suzanne Vega and she knows about this story.[53]

For the official MPEG tests, the Vega recording was used in both high-quality and intermediate quality tests.[54] The Vega song was used, but not exclusively, solely, or finally, and Brandenburg did not perfect the algorithm on his own, nor has he publicly claimed to be the sole inventor of MP3. To the contrary, Brandenburg is careful to credit his many collaborators.

Even if it's not exactly true, the story certainly is powerful. The Vega story's considerable reach is tied not only to the romantic language in which invention is still often presented, but also to notions about the female voice, authenticity, and a cappella music. Many listeners certainly hear a cappella music as if it has some purity to it, but despite its being recorded in a realist style, the Vega recording is heavily processed. In addition to the engineer's careful choice of microphone for Vega's voice, the vocal recording is compressed to limit the dynamic range of her voice—a technique so standard in popular vocal recording that recorded voices sound strange without it. After the compression, her voice was bathed in artificial reverberation. If you listen to the recording in headphones or at a moderately high volume on speakers with a lot of definition, a massive echo is audible in the background at points in the song where her voice abruptly stops. In the first verse, listen for the sound after "diner," "corner," "argue," "somebody," and "in." In other words, if we take Vega's recording as having particularly heavy symbolic weight in the story of the MP3, it indexes the degree to which the MP3 format was designed to reproduce the aesthetics of "well-recorded," heavily stylized sound, especially music.[55]

All of the recordings for the tests are considered *good recordings* by generally accepted standards of audio engineering, especially broadcast engineering, and here they are far more homogenous than even the limited range of genres might suggest. Some of this is functional. Solo instruments and voices have a much weaker masking effect than ensembles playing together (or a wall of sound), so they are more likely to challenge the codec being tested. But some ensembles made it through. The one inviolable base line for critical material appears to be the aesthetics of the recording itself. Just as there are "legitimate" genres of music, we can also say that there are institutionally sanctioned styles of recording. These rules are far harder to break than the rules of style, even today with the explosion of music on the internet. To sound good to casual listeners, recordings must have certain types of compression, equalization, and mixtures of ambiance. The rules do vary by genre but not as much as other conventions might. The MP3 format is thus built on top of a long history of practices of making and listening to recorded sound. Its reference is not only CD-quality audio, but also what we might call the stylistics of quality audio. Hilmar Schmundt may have invoked a great-man history when he wrote that every MP3 replicates the way that Brandenburg heard Vega, but there is a different truth in the story. MP3s are tuned to the stylistic conventions through which radio listeners come to hear music as normal. They presuppose the circulation of sound recordings and the hierarchy of taste cultures that are still so central to medium aesthetics.

Particularism and Alternative Models of Expert Listening

In his study of audiophile controversies, Marc Perlman discusses debates around the ABX test, which is a close relative of the double-blind triple-stimulus tests MPEG used. The ABX test is designed to test different components playing the same material. A computer-controlled box switches between source A and source B, and then randomly selects one of the two as X. The listener must then identify X as either A or B. A pattern of correct guesses over time indicates that there is a significant audible difference between the two sources. A success rate around 50 percent, which is the same as guessing, indicates that no significant difference is audible. In audiophile culture, the ABX test is designed to account for subjective aspects of listening that aren't measurable by electronic instruments. In that sense, it is a contrived test to overcome the subject-object split in audio. If there are no good ways to measure certain aspects of auditory experience except through listeners' own responses, then the goal of testing ought to be to map a set of calibrated listener responses.[56]

As MPEG's engineers searched for just noticeable differences, they followed in the footsteps of psychoacousticians, even as they modified their findings. They turned qualities into measurable, verifiable, and repeatable quantities. Tests like ABX and double-blind triple-stimulus are meant to escape the prison of the adjective in the analysis of listening. Yet this kind of testing has come under considerable criticism from audiophiles who believe that "the ABX testing situation is so different from real-life listening conditions that its results cannot be projected into the real world. After all, we do not normally listen to recordings in an unfamiliar environment, in the company of strangers, switching every few minutes between one piece of audio equipment and another, feeling all the while that our self-respect is at stake." In tests like ABX, "we assume the device's behavior under test conditions can predict its real-life behavior. In other words, we *project* from the test case to the conditions of use."[57] That is why the worst-case scenario is so important, because it aims to turn the test into a context that transcends other contexts.

Yet the worst-case scenario can only go as far as engineers' imaginations. Today, even the fiercest defender of perceptual coders' sound quality will concede that perceptual coding should only be done at the moment right before final transmission or release to the consumer. This was not always the case. In 1992, when the ITU began testing MPEG and Dolby codecs for use in digital audio broadcast, they had hoped that audio could be perceptually coded at each stage of the broadcast chain: for instance, a report recorded in the field, again in the studio as the program was produced, and again as it went out to satellite for distribution to audiences. They quickly discovered that it was not possible to perceptually code audio more than once. Or rather, they learned that if audio was perceptually coded at one stage of the process and then run through the same codec again later in the process, it would become quite distorted. Marina Bosi recalled that "we certainly learned a lesson" from those tests. "If you think about it, we barely—just barely—finished the standardization, so we had a lot of things we were just learning at the time. So you learn through trial and error."[58]

Subsequent audiophile criticism of the tests also points out their limits, usually by refusing to accord the tests the transcendence they request. Suzanne Vega's story of her visit to the Fraunhofer Institute is worth quoting at length:

> The day I visited—"The Mother of the MP3 comes to the home of the MP3!" said the woman in charge of press (the slightly odd implication being that I would be meeting the various "fathers" of the MP3)—we had a press conference at which they played me the original version of "Tom's Diner," then the various distortions of the MP3 as it had been, which sounded monstrous and weird. Then, finally, the "clean" version of "Tom's Diner."
>
> The panel beamed at me. "See?" one man said. "Now the MP3 recreates it perfectly. Exactly the same!"
>
> "Actually, to my ears it sounds like there is a little more high end in the MP3 version? The MP3 doesn't sound as warm as the original, maybe a tiny bit of bottom end is lost?" I suggested.
>
> The man looked shocked. "No, Miss Vega, it is exactly the same."
>
> "Everybody knows that an MP3 compresses the sound and therefore loses some of the warmth," I persisted. "That's why some people collect vinyl . . ." I suddenly caught myself, realizing who I was speaking to in front of a roomful of German media.
>
> (Actually, I recently read an article that said the high end is distorted and the low end uncompromised, so I guess there is room for subjectivity in this argument.)
>
> "No, Miss Vega. Consider the Black Box theory!"
>
> I stared at him.

"The Black Box theory states that what goes into the Black Box remains unchanged! Whatever goes in comes out the same way! Nothing is left behind and nothing is added!"

I decided it was wiser at this point to back down.

"I see. O.K. I didn't realize."

They were happy again. Then they showed me a seven-point sensurround system, their latest project. But as I was a kid who grew up with transistor radios and lousy record players that were left at our house after our parents' parties, I kind of like a hard, tinny sound. The lo-fi approach works for me. Still, I appreciated their enthusiasm. It was a great day and I was very proud to have been a tiny part of history.[59]

The humor of the story comes from the question of whether we should doubt the engineers' ears, the songwriter's own ears, or both. She appeals to her experience but also to conventional wisdom ("everyone knows"); the engineer appeals to authority and precedents ("the black box theory"). Presented in story form, it seems absurd that anyone would doubt his or her own ears, especially a respected songwriter. And yet the double-blind triple-stimulus test is designed precisely on the principle that expectation shapes audition, and that too much knowledge might sway one's subjective evaluation of audio. The psychoacoustic theory on which the MP3 is based goes even further. We cannot trust our ears to perfectly reflect vibrations in the air or the measurements made by audio equipment, because their parts are also filters, shaping and modifying the vibrations in the air before they emerge as perceived sound in our minds' ears. Could Vega have heard something the engineers have missed? Certainly. We will never know for certain; Vega writes as if she won't either.

But there are those who are certain that the entire listening-test enterprise is a sham. Following exactly the subjectivist-objectivist pattern of debate that Perlman identifies in his article on audiophile culture, the journalist Robert Harley has attacked MPEG's listening tests as a failure on their own terms, using an apocryphal allegory:

> After announcing its decision, Swedish Radio sent a tape of music processed by the selected codec to the late Bart Locanthi, an expert in digital audio and chairman of an ad hoc committee formed to independently evaluate low-bit-rate codecs. Using the same non-blind observational listening techniques that audiophiles routinely use to evaluate sound quality, Locanthi instantly identified an artifact of the codec. After Lo-

canthi informed Swedish Radio of the artifact (an idle tone at 1.5 kHz), listeners at Swedish Radio also instantly heard the distortion.

How is it possible that a single listener, using non-blind observational listening techniques, was able to discover—in less than ten minutes—a distortion that escaped the scrutiny of 60 expert listeners, 20,000 trials conducted over a two-year period, and elaborate "double-blind, triple-stimulus, hidden-reference" methodology? The answer is that blind listening tests fundamentally distort the listening process.[60]

This account is interesting because it centers on privileged virtuoso listeners as the arbiters of sonic quality, but here authority is grounded in charismatic individuals (whether Harley or Locanthi), rather than statistics generated by an anonymous panel. In other writings, Harley celebrates technique, craft, and virtuosity as alternatives to the objectivism of the listening tests.[61]

One can easily hear the resonance between the critique of objectivism in twentieth-century philosophy of science and Harley's critiques of listening tests. Listening tests do not distinguish between practical and formal knowledge; they do not account for subjectivity and positionality; and their aspirations to universalism only universalize an anaesthetized, objectivist worldview. Yet Harley's conclusion, meant as an attack on objectivistic reasoning in audio, is strikingly resonant with it on at least one core issue: "Our common goal is this. When a faceless listener somewhere in the world sits down before his playback system with his favorite music, he experiences the greatest joy our technology can convey. Can audio engineering have a higher purpose?" In this turn of phrase he means to denounce the emphasis on measurement in the listening tests, even as he assents to its universalist aspirations (and the gendered language is both grammatical and indicative—audiophile culture is an even more heavily gendered space than professional engineering). As we have seen throughout this chapter, beauty of one sort or another was at the very center of what engineering practices sought in testing and tuning codecs, and appeals to their own love of music appear in the engineers' accounts of what they were doing— even as aesthetic language remained submerged in formal engineering discourse.

The battle between objectivism and relativism (or more accurately, subjectivism) is an old opposition in the philosophy of science, and it has been resolved repeatedly and to different ends. Writing more or less contemporaneously with the first tests of the codecs that would become MPEG, Donna Haraway rejected the choice altogether. "The alternative to relativism is not totalization and single vision, which is always finally the unmarked category whose power depends on systematic narrowing and obscuring.... The issue is politically engaged attacks on various empiricisms, reductionisms, or other version of scientific authority should not be relativism, but location."[62] If we believe all the critiques of objectivist epistemology that one can launch against the MPEG tests and the MP3 format, the startling thing wouldn't be that MP3s sometimes fail in their quest to be audibly indistinguishable from CD-quality sound. More startling would be how rarely this is the case and how infrequently its sonic qualities are noticed by listeners or made a tangible issue in their engagement with the audio it carries. I am not suggesting that listeners have been duped by lower-quality audio. The question of audio quality is too often asked apart from any actual context of lived sonic experience. The MP3 format works surprisingly often and well for something so limited and so imperfect by audiophile standards. Moreover, it works exactly according to the "higher purpose" that Harley sets out for audio engineering for more people in more situations than the ostensibly higher-quality audio equipment that both Harley and our MPEG engineers would no doubt themselves prefer.

In the cultural life of MP3s, the choice between objectivism and subjectivism is no choice at all. It leads repeatedly back into the political games of taste and status. Choosing objectivism or subjectivism requires us to act as arbiters who appeal to either the authority of an objective science as a context that transcends all context, or the authority of charismatic experience as a judgment that sits atop all others. Perhaps making fun of himself, Bourdieu wrote that it is impossible to outline the "games of culture" without participating in them.[63] Harley aims to discredit and deflate the idealist strategy of the listening tests but only to assert the centrality of a different set of arbiters of sonic taste, namely refined listeners such as himself.

Although it also cannot avoid the games of sound culture, my own approach in this chapter has been somewhat different. I have chronicled the strategies for universalism within the listening tests while noting their cultural biases, internal contradictions, and political implications. The listening tests were at once technical, political, practical, and aesthetic. They aimed to extract perceptual capital from existing digital audio technologies, so that they could invest it in new ones. Golden-eared experts sitting in a room passing judgment could certainly have told us which audio they preferred and why, and it is possible that the testimony of such experts might have produced a codec with completely different sonic artifacts, one that would no doubt sound better to them. But they could not have performed the many other functions that the listening tests did.

Like good pragmatists, engineers and the communication systems they build do not ask whether something is universally true, but simply if it works. This is the logic of perceptual technics. To locate the MP3 and the listening subject it addresses and inscribes, we move away from the test to the mediatic world of sound that it inhabits. The MP3 reflects not only the habitus of its relatively elite and subcultural test subject (if one can consider engineering as subcultural). Recall that the listening test is, in engineering parlance, a worst-case scenario: "difficult" material that is timbrally or temporally complex for the coder, "difficult" listeners and a listening environment to which few people have access. One might then ask what a better or best-case scenario for the MP3 would be—and the answer is listeners who are less hung up on sound quality, material less difficult to reproduce, and a sound system and listening environment that is full of irregularities. The more distracted the listening subject, the worse the speakers, headphones, or room, and the denser the recording, the better an MP3 will work. The MP3 format is designed for casual users, to be heard in earphones on trains or on the tiny speakers of a computer desktop, to be aggregated in blogs, to be carried around in large volume on portable audio players, to be sent in e-mails and IMs and through file-sharing programs. This is not all music for all time, but it speaks to the condition of music in contemporary urban life in many places around the globe.

To put it another way, the MP3's shadow of the listening subject may fail tests for scientific universality, but it achieves a kind of social ubiquity because of the musical and sonic contexts through which MP3s circulate. To the extent that MPEG succeeded in its ambitions for interoperability and transnationality (goals one also finds in the practice of professional recordists, broadcast engineers, and mastering engineers), a relatively limited set of sonic aesthetics obtain for an incredibly large and diverse array of recorded media. This is the case not just across musical genres, but across other types of content like broadcast speech. The sonic referent of the MP3 is a massive, polymorphous, interlaced global network of technologies, practices, and institutions. Although the system is neither closed nor total, its reach cannot be underestimated. MP3s may confront an almost infinite and unmeasurable multiplicity of listeners, but they do so within a surprisingly limited set of contexts and aesthetics of "good sound." A standardization of sonic aesthetics may suggest a standardization of musical or sonic subjectivity. But that is not necessarily the case—one can have an infinite multiplicity of listeners even if there are only a handful of important mastering engineers in the world. The MP3 codec's standardized listening subject may serve to modulate real instances of listening, but it is just that—a modulation, an adjustment, and not strictly a determination. The subject inside the codec is real and works, but it does not need to be true or even representative in order to reach out toward ubiquity.

Listening tests turn historical relationships among ears, technologies, and music in upon themselves, as objects of manipulation in the service of generating perceptual capital, but also, as I have shown here, as objects worthy of contemplation. The MP3 reflects several ongoing histories of the ear and listening, and the ways in which those histories—and the sedimented sets of practices that descend from them—are now the raw material out of which the contemporary mediality of music and sound is fashioned. The psychoacoustic models used and refined from the listening tests covered in this chapter no longer inhabit many modern MP3 encoders. They have been replaced by new models. But the tests from 1990 and 1991 live on, both in the standards for listening tests they begat, and as the sonic base line from which the MP3 format has developed. No one single group of listeners has determined the sound of the MP3 format—the engineers listened, then their expert listeners, then the MPEG panels, and since them countless other engineers and expert listeners, as well as massive and distributed audiences. Every time a coder makes a file, it references those auditory histories. As engineers try new coding schemes and select or modify psychoacoustic models, as users install or update a coder, they simultaneously extend and mutate a history of listening in the service of perceptual technics.

6. IS MUSIC A THING?

If we look back over the past quarter century, it would appear that the commodity form of music has undergone a massive transformation. Twenty-five years ago, it was dominated by recordings on physical media: compact discs, tapes, and (though in decline) LP records. Sheet music made up a portion of the market as well, as did mechanical rights, reproduction rights, and payment for live performance. Today, the world's largest music store sells digital files. Sales of recordings on CD have plummeted precipitously, and though vinyl is enjoying a resurgence, it is not going to pick up the financial slack. Before the financial crisis in 2008 and resulting recession, the industry had already lost about a sixth of its mass, a shrinkage of over

$6 billion. Labels and artists alike have sought out new business models, and it is generally acknowledged that the recording industry is in crisis. The blame, more often than not, is laid at the foot of the phenomenon of mass file-sharing.[1]

We should not be too quick to accept this simple explanation. The record industry is prone to crisis. In the late 1970s, sagging profits were blamed on the failed promise of disco and on lost profits from home taping. Between roughly 1990 and 2000, record-industry profits were artificially elevated by format changes and resale and repackaging of back catalogues. Once LP collections were replaced by CDs this market dried up and, with it, a substantial portion of industry profits. Some have argued that the failure to agree on a high-definition audio standard to supersede the compact disc must also be considered a part of the crisis. Experimental high-definition formats like HDCD and DVD-A found no commercial success, so no new avenue exists for the back-catalogue business that propped up sales figures in the 1990s.[2] Digital files or not, a sizable dimension of the market for physical recordings has dried up.

However we read the current conjuncture, it is true that worldwide, more recordings now circulate through channels that do not carry the official sanction of recording industries or states. The iTunes store may be the world's largest music retailer, but an extended web of Gnutella and BitTorrent sites and the users who frequent them make up an even larger, transnational swap meet for recordings. For the industry, this state of affairs poses the important economic question of what is to be done. Perhaps the music industry will undergo yet another of the massive reorganizations that have characterized the last five hundred years. But we should not view the current crisis as purely a practical question for those who make money from recorded music, or a matter of whether or not industries should be preserved in their present form. The debate over the explosion of file-sharing and the MP3's role therein also opens out into a long-standing argument about the nature of music in contemporary culture — as process, as practice, as thing.

The nature of music is worth revisiting for reasons pointed out by Walter Pater, of all people. Pater is famous for his overused dictum that "all art constantly aspires to the condition of music," but let us consider why. In the very next sentence, Pater wrote "while in all other works of art it is possible to distinguish the matter from the form, and the understanding can always make this distinction, yet it is the constant effort of art to obliterate it."[3] In some ways, Pater's aesthetic proposition was quite forward-thinking. Form and matter were inseparable for him in music, and other arts did indeed strive toward the same condition. Consider what happened to all the arts in the modernist moment that emerged some years after the publication of Pater's book in 1873 and in the avant-garde practices that continue down to the present day. In much of this work, its form is part of its expression. Certainly, Pater offers a hierarchy of the arts common to nineteenth-century thought and unfashionable for that of the twenty-first. And certainly he idealizes musical performance and listening as, in Alan Durant's words, a "condition without conditions, a condition at all times and in all places identical, even as surrounding conditions shift and vary."[4] But in the intervening years, both artists and scholars of art and music have expanded their concept of form from a formal concern to one of cultural, economic, or physical form. In that sense, the form and matter of music remain closely bound today, especially if we consider the ways in which recordings move across formats and milieus.

As Bill Brown writes, "If the topic of things has attained a new urgency..., this may have been a response to the digitization of our world."[5] Journalists and humanists have worried as much about the purported dematerialization of tangible musical objects as have label executives and policymakers, and not for entirely dissimilar reasons. Depending on whom you ask, analog and early digital recording media either led us to hold music in our hands or to think that we did. A similar dichotomy exists now. Either music has dematerialized, or its materiality now exists on a different scale. When ads can talk of effortlessly holding 25,000 songs in your hand, recorded music moves more freely and into more places than ever before.

Is music a thing? If it was, is it still? The MP3 format's widespread success demands that we reconsider these basic questions. This chapter begins by tracing the contours of a long-standing debate among scholars and critics as to what music is, framed against the changing status of recordings over the last twenty-five years. Having set out the range of positions on music's thingness, it offers an account of how the MP3 format became the most common form of recording audio in the world. For most of this book, I have been concerned with the conditions of possibility for creating perceptually coded audio. But the establishment of the MPEG standard in no way guaranteed that its third layer of audio would become commercially successful. Thus, I consider both the traditional market strategies used by Fraunhofer—the owners of important MP3 patents—alongside piracy as forces in the format's promotion. First, the software to make and play back

MP3 files was itself cracked and distributed through unauthorized channels, then promoted through large-scale unauthorized music sharing online and the repackaging and resale of music on city streets around the world.

The term *piracy* implies the unauthorized distribution of copyrighted material but is more complicated than it initially appears. As a figure of speech, the term accomplishes a ridiculous conflation. *Piracy* collapses people who make mix CDs for their friends with kidnappers who operate off the coast of Somalia, among other places. It suggests lawlessness and seems to authorize military or police vigilance against its spread. As Patricia Loughlan points out, the term stretches beyond a legalistic definition to "instances where, despite the action of the 'pirate' not being against the law, it is contrary to what the writer thinks ought to be the law, or contrary to the writer's view of what the 'natural rights' of the intellectual property owner in question are."[6] Adrian Johns mentions file-sharing and the digital copying of music just a few times in a six-hundred-odd page book on piracy, arguing that those who reduce piracy to intellectual property evacuate the term of its history, which predates the invention of intellectual property as a concept. "What is piracy? It is not entirely clear that we agree on the answer. An official study for the European Union once defined it rather impishly as whatever the knowledge industries said they needed protection from.... In the end it may even be the most adequate definition we can get; but it will scarcely do as a starting point." Johns goes on to discuss examples of piracy that appear to have nothing to do with intellectual property, such as British "pirate buses" in the 1850s, arguing that piracy ought to be considered those activities characterized by contemporaries as "piratical" while acknowledging that "we cannot simply take such characterizations at face value. Those who were called pirates . . . repudiated the label as inaccurate and unjust. The point is that when they did so, they often triggered debates that threw light on major structural issues and had major consequences as a result."[7]

Even within the realm of copyrighted digital audio files, piracy conceals the diversity of practices it names, ranging from copying practices that recording industries simply don't like, to unauthorized file-sharing online, to the unauthorized sale of MP3 CDs on city streets, to unauthorized webcasting, and on and on. For the purposes of this chapter, I will refer to a specific activity such as peer-to-peer file-sharing when there is one, and use *piracy* to denote a range of unauthorized copying and distribution practices considered collectively (rather than issuing a long, drawn-out list each time). The key attribute here is *unauthorized*—the activity may be legal, moral, or appropriate to its context or it may not, but that is a separate matter. Even here, there is an ambiguity. The recording industry may treat file-sharing as unauthorized, but as we will see, the consumer electronics and broadband industries clearly profit from it and de facto authorize it. Piracy may not itself always be an economic activity, but it enables all kinds of other market activities.

Viewed from the litigious wing of the recording industry, piracy is a dangerous, antimarket force. But as I will argue, the mass piracy of music was actually quite productive as an economic force. Record companies may view mass copying as a threat to capitalism, but copying generates all sorts of value for other industries like consumer electronics, broadband, and even other kinds of intellectual property, like the patents on MP3s. Piracy also reveals and calls into question the social organization of music. "In many parts of the world, media piracy is not a pathology of the circulation of media forms but its prerequisite," writes Brian Larkin in his study of film in Nigeria.

> Piracy and the wider infrastructure of reproduction it has generated reveals to us the organization of contemporary Nigerian society. They show how the parallel economy has migrated onto center stage, overlapping and interpenetrating with the official economy, mixing legal and illegal regimes, uniting social actors, and organizing common networks. This full flowering media, and the infrastructure it relies on, presents a stark contrast to the state sponsorship of media in the colonial and early postcolonial era. Now political control exercised through the governmental, pastoral care of developmental media has been replaced by an economy shorn of its political objectives.[8]

Larkin's point about Nigeria could be generalized to the global record industry with just a little substitution: international intellectual property regimes and trade agreements in place of a developmental media, and a corporate regime that stood outside that transnational regulatory sphere. The worldwide proliferation of MP3 files announces the end of the artificial scarcity of recorded music, but it does not guarantee a more just or democratic organization of music. It simply reopens the organization of music—and the infrastructure that supports it—as a social question. The traffic in MP3s thus brought to elite economies a set of questions that had been more commonly asked in the developing world.

Some Different Kinds of Musical Things

The argument over whether music is a thing and what that might mean has a long history in the scholarly study of music, though it is often found at the margins of what we now call *music studies*. The debate most often wells up in response to industrial and technological transformations. One tradition considers music as a social practice and process that may produce artifacts but is not itself something that can be objectified as a thing, except as a kind of reduction. Meanwhile, writers who refer to music as a thing may refer to its technologized forms, its status as a commodity, or its essence as a work independent of any particular performance. Against these definitions, I consider music as a bundle of affordances, thus borrowing some of the process language and some of the thing language. To clarify my argument here, let's consider some of the other, better-known positions on whether music is a thing.

TECHNOLOGY

Unsurprisingly, writers who focus on music's technologization tend to tell their story as a series of landmarks, indicated by—for instance—the invention of notation in the ninth century, the invention of movable type for musical printing in the sixteenth century, the invention of sound recording and radio in the nineteenth century, and the digitization of music at the end of the twentieth. In such a scheme, the "condition of music" is a technological condition. Technological change remains an important historical signpost for marking major shifts in the form and matter of music, even if it doesn't simply cause historical change in music. Scholars often use each stage of music's thingness as shorthand for different social relations of and around music. Chironomic notation—the diacritical marks one finds, for example, in Hebraic chants—and the notation that arises in the ninth century clearly embody and enable different orientations toward music-making. The former is a mnemonic device, the latter is not. The changes to Western music that we associate with these kinds of transformations were not immediate or causal—they happened over time. The modern sheet music industry did not appear out of thin air the moment it became possible to print sheet music on presses using movable type at the beginning of the sixteenth century. It slowly formed over the next three centuries alongside a new kind of domestic amateurism and new forms of patronage and profitability for composers.[9]

COMMODITY

Recording also follows this pattern. In the chapter of *The Recording Angel* from which I have borrowed my chapter title, Evan Eisenberg argues that music became a thing not with the invention of sound recording but with the development of the recording industry and prestige records like Victor's Red Seal line in the early twentieth century. "Whereas dances, recitals and soirées remained slippery ground for the bourgeois, records brought music to his home turf, which was acquisition."[10] For Eisenberg, music is a thing because it is a commodity; the technological fact of sound recording is necessary but in itself insufficient to objectify music. Later in the chapter he writes, "When I buy a record, the musician is eclipsed by the disk. And I am eclipsed by my money—not only from the musician's view but from my own. When a ten-dollar bill leaves my right hand and a bagged record enters my left, it is the climax. The shudder and ring of the register is the true music; later I will play the record, but that will be redundant. My money has already heard it." Ten years ago Eisenberg's $10 for a record would be charmingly dated. As I write, many albums (if not records) are again available for $10, which speaks to a host of changing economic and cultural formations. But Eisenberg's point remains salient. Relations that once existed between musicians and audiences were transformed into relations among cash and records. In other words, relations among people became relations among things.[11] The current recording economy breaks music into units that can be owned, sold, or loaned.

There are many possible implications of commodification. Some writers of a certain anthropological (or at least ethnographic) bent may understand and lament music as a process that is compromised in its objectified, recorded form. For instance, Charlie Keil writes, "I have nurtured a deep ambivalence, at times masking outright hostility, toward all media for many years. I treat records badly; they aren't real music." Although Keil treated media with more care and sophistication in his scholarship than in his record and tape collection, the question of what to do with recordings has been an ongoing source of anxiety since the first vexed encounters between sound recording and the study of music.[12] Keil's problem with notation, recording, and industrialization is quite simply the containment, fixing, and instrumentalization of this process. Writers like Keil and Christopher Small have argued that we are so deep into this system of objects and objectification that we have forgotten how to think about music as a vital force in life, one driven by involvement and participation, and this forgetting has limited the possibilities for ourselves and for a more just and egalitarian world.[13]

PROPERTY

Keil and Small's argument can be extended to a critique of copyright as such. While property talk has suffused most discussions of the products and processes of creativity, intellectual property isn't really property. Simon Frith writes, "In the music industry itself, a song—the basic musical property—represents 'a bundle of rights'; income from the song comes from the exploitation of those rights," and the relevant rights in the bundle can change over time. There are the rights to sell the music, but also secondary rights like licensing fees from other users, for instance, when a song is used in a film soundtrack.[14] These rights, which are really rights of circulation, are the basis of the value of the commodity. A copyright, patent, or trademark may appear to refer to a concrete thing, but this analogy quickly disintegrates. As Majid Yar argues, intellectual property talk enacts a "myth of equivalence between tangibles and intangibles." It is more accurate to describe intellectual property as a temporary trade monopoly guaranteed by a state. Copyright is not outright ownership to begin with—it expires. The notion of the song as intellectual property converts a trade relationship—the temporary but exclusive right to copy—into a quality of the thing, where no such quality (and possibly no such thing) exists. From this perspective, mass file-sharing threatens not just specific properties, but the very legitimacy of the recording industry's monopoly of distribution. In a way, the property argument is a ruse, since supporters of the existing intellectual property regime often seek to extend the domain of property rights, to make this equivalent to "a natural, unquestionable right to control, use and decide upon the dispersion of property." In the world of tangible property, a vendor cannot assert any control over my use of a product once I purchase it. Mattresses sold with tags that say "it is illegal to remove this tag" are the subject of countless bad comedy skits, and most customers would balk at chairs that came with user agreements regarding where they could be sat upon, or musical instruments that came with a list of allowable genres and venues for the musicians to use them in. When it comes to digital files, the name intellectual property conceals a wide number of strategies for restricting use and circulation through contracts or trade monopolies. To use Siva Vaidhyanathan's terms, we would be better off talking about intellectual "policy" than intellectual "property."[15]

Some industries have attempted to convert their desire for control over circulation from an ideological project to an engineering project. Digital rights management (DRM) schemes neatly illustrate how intellectual property is really an attempt to enforce a trade monopoly. Digital rights management most often takes the form of encryption, so that a digital file won't operate in a given hardware or software system unless it meets certain criteria, for instance, that the system is officially sanctioned, or that a user code has been entered to show that the owner paid for the file. It may also work as an attempt to prevent copying, for example, by scrambling the output of an audio or audiovisual device when hooked up to a recorder. Although the term DRM is new, the practice is old. When the global cassette industry was threatened by mass unlicensed duplication, it attempted to create media that only sanctioned dubbing houses could write onto (basically re-creating the artificial scarcity that existed with LPs). Digital rights management is about asserting control over an economy by force when law and custom are not enough. But DRM has been both a technological and a cultural failure. Its technologies are routinely circumvented, and DRM undermines legitimate rights that users may have since it cannot, for instance, distinguish between fair use and other kinds of uses.[16] It is therefore not properly understood as a conservative or status quo technology, but a reactionary one, since it erodes users' legitimate rights in the service of extending effective trade monopolies.

IDEALIZED WORK

The "it's not really a thing" line of reasoning is not the only ground for the critique of music's technological or juridical objectification. Adorno also had a profound ambivalence toward media in the form of recording, radio, and scores, but he believed that music existed as a different kind of thing. In his unfinished manuscript, the *Theory of Musical Reproduction*, Adorno considered notation to be "the notation of something objective, a notation that is necessarily fragmentary." Musical reproduction presupposed "the existence of works that are fixed through writing and print, and thus independent precisely from empirical music-making. The end of improvisational practice, the work's attainment of independence and its separation from interpretation at once instigate its self-sufficiency. The work can only be rendered once it is estranged. Interpretation, as an autonomous form, is necessarily confronted with its contradiction, the autonomous musical construction."[17] This quote is notable precisely because the tone is so strikingly different from Adorno's more famous laments about radio and recorded popular music. Here, he takes for granted that there is something like a work of music that exists independently of its performance. It might be tempting to view him as the product of a particular industrial-technological regime: symphonic music, concert performance, sheet music, and patronage of composers by states or elites. But we should keep in mind that the main ingredients of Adorno's notion of music were already present in Pythagoras, Plato, and Aristoxenus. In this tradition, music may have an existence independent of its notation, capture, and storage in a physical object, or its insertion into a money economy.

BUNDLE OF AFFORDANCES

Although Adorno would find awkward company in the figure of Martin Heidegger (as do I), there is an affinity between them around the status of things. Heidegger argued in his essay on "the thing" that things are things because they offer people affordances by virtue of their presence.[18] A thing lets you do something that you could not otherwise. Heidegger's key example is a jug. A jug can hold liquid, which can then be poured out at the user's will, so long as the jug is ready-to-hand. In this arrangement, a picture of a jug on television won't do if you've got a garden to water.[19] On the other hand, a digital recording will do just fine in many cases when you've got music to hear.

Considered outside the realm of technologization and commodity exchange, the thingness of music has also been framed in terms of its affordances, which are largely about the well-being of subjects and societies. Plato wrote that "a change to a new type of music is something to beware of as a hazard of all our fortunes. For the modes of music are never disturbed without unsettling of the modern fundamental political and social conventions." The medieval church's prohibition of certain modes and Martin Luther's adage that "music is a fair and glorious gift of God" are opposite inflections of the same proposition that music is a divine gift meant for worship. Adorno's derision of the hit song was based on his belief that the massive symphonic work presented its listening subject with an opportunity to apprehend the totality that lay behind the fragments that capitalism presented to the senses.[20] We need not accept Platonic, Christian, Heideggerian, or Marxist moral precepts to grasp their great insight into music. If music is a thing, then it is *for* something, and that affordance is both the substance and the object of critique.

DIGITAL THINGS

Digital data have a materiality even though they are not available to unaided senses. A digital song takes up space on the platter of a hard drive or in the channel of a DSL connection. In a review of Traktor, a software DJ program, Philip Sherburne pauses to note that digital audio formats and their manipulation on computers reflect "the ongoing dematerialization of music (or perhaps a better term would be 'micromaterialization' since even MP3s live in silicon, invisible as they may seem)."[21] So MP3s remain things but in a special way. In his jug example, Heidegger notes that the jug does its holding through the void created by its walls and base: "the empty space, this nothing of the jug, is what the jug is as holding vessel."[22] Like the jug, the MP3 is defined by the interior space that it creates. The MP3 is a container technology.[23] All communication technologies can be considered as container technologies, but the MP3 is a special kind. For most of this book, I have considered the space inside the MP3 as an attempt to fabricate a medium adequate to the spaces left by the gaps in the hearing subject. In its conception, the MP3 holds only the sound that can be heard; it discards the rest and attempts to hide its own excesses. But the MP3 is also a container for recordings—records, CDs, and so forth. It is a container for *containers for* sound, and it codes the space within itself. To borrow a phrase from Lisa Gitelman, new machinery is constantly interposed into recording.[24]

Lewis Mumford first wrote in 1966 that technology scholars' emphasis on tools over containers overlooked containers' equally vital role.[25] He postulated that one reason why container technologies are often neglected in the history and philosophy of technology is that they are usually coded as feminine. While the gender coding may be a bit dated, Mumford did have a point about activity and passivity. More recently, the feminist scholar Zoe Sofia has picked up Mumford's thread. She qualifies Mumford's argument, positing that even though the container category may be considered feminine in some cases, container technologies may in fact be as connected with men as with women.[26] Sofia argues that the misogyny story is only part of the explanation for the neglect of container technologies: "to keep utensils, apparatus, and utilities in mind is difficult because these kinds of technological objects are designed to be unobtrusive and . . . make their presence felt, but not noticed."[27] An MP3 is useful but does not call attention to itself in practice. It takes up less space than other kinds of digital recordings, and when it is listened to, it is experienced as audio and not as a file format. Mumford and Sofia both use the term "apparatus" to describe a container that transforms as it holds.²⁸ The MP3 clearly belongs to this category, but it differs because it holds other containers. Like an oven that holds a casserole and transforms its contents, the MP3 is a holder for sound recordings. It is a media technology designed to make use of other media technologies.

When an encoder makes an MP3, two things happen. It shapes the audio in the file in relation to a model of experience based on the history and culture of sound recording, as we have already seen. But it also places the audio in a definite economy, both within the file and in relation to other MP3 files. Like the jug, the space inside the encoder makes the audio ready-to-hand for users. Less like the jug, it manages the conditions under which that readiness-to-hand occurs and requires an ensemble of other technologies. At a formal level, this is totally unremarkable. The MP3 is yet another in a long line of storage formats that, with the aid of other equipment, allow recordings to be managed and played back. The difference is that the MP3 allows for audio to circulate in ways that it otherwise couldn't. It takes the temporalities of recording and subjects them to a new set of possible paths of circulation.

For all this, we would be wrong to see this as a quality inherent in the technology alone. Users can only experience music's readiness-to-hand in use, and they may only become conscious of it when it breaks down. Only when MP3s actually store music, circulate, accumulate on hard (or flash) drives, or play back through headphones or speakers do they make music ready-to-hand.²⁹ Only when the playback technology is not up to its task do users take note of its contours, and MP3s fail a lot less often and in many fewer contexts than records, CDs, or tapes. The MP3 *player* may run out of batteries, but the format itself is much more robust in everyday use than its larger cousins. In use, whether on computer networks or in some portable form, the format much more effectively makes music more ready-to-hand than even the relatively durable and portable cassette. As Michael Bull writes, "The carrying of large slices of one's musical library in a small piece of portable technology liberates users from the contingency of mood, place and time, making redundant the contingencies of future moods and circumstances."³⁰ Again, it is not the technology but the act of carrying that "liberates."

Even though this aspect of MP3 use belongs more to the realm of practical reason than formal reason, we can understand a bit of its contours by considering how the MPEG protocol presupposes certain kinds of relationships, and allows for others, as well as the ways in which the actual historical development of the MP3 as a technology of circulation superseded this code. The code carried in MP3s and the historical trajectory they have followed appear to have increased the number of things that "go with" music—and this is one of the key nonsonic aspects of the MP3's thingness.[31] Now the music lives in its recorded container, which is enfolded into another container. Every MP3 is divided into frames, and every frame conforms to an arrangement of bits specified in the MPEG standard. Each frame is divided into a *body* that contains information correlated to the sound waves the playback software and hardware will reproduce. The audio data in each frame represents 8 ms of audio (at a 48 kHz sampling rate). Each frame also contains a *header* that is information about the information in the body. Although one might assume the intent behind such basic coding is purely technical, it was clearly ideological in its attempts to represent existing and hoped-for delivery formats and industrial structures in every frame of every MPEG audio file. Just as we can analyze a film or song for the worldviews it contains, so too can we analyze a header. As Wendy Chun puts it, "In our so-called postideological society, software sustains and depoliticizes notions of ideology and ideology critique. People may deny ideology, but they don't deny software—and they attribute to software, metaphorically, greater powers than have been attributed to ideology." The code just *is*. It speaks to the machine in ways we can't and in ways to which we do not attend.[32]

In figure 28, each block represents one of 32 bits that make up the header. The first eleven are used for syncing the frame with all the other frames in the stream. Since one of the original purposes of the MPEG standard was to render audio suitable for digital broadcast, it required that a receiver could at any point tune into the data stream and "find" its place in the playback. The header also identifies which version of the MPEG standard the file conforms with; the bitrate of the file; the sampling rate at which the audio should be played back; whether the audio is coded according to layer 1, layer 2, or layer 3 (MPEG-1 only); and so forth. But three of the bits in the header provide clues as to the world for which the MPEG format family was designed: the private bit, copyright, and original/copy. The private bit was reserved for third-party applications, for instance, if the frame in the file would trigger some kind of event. The copyright bit does not hold any copyright information but simply indicates whether it is permissible to copy the

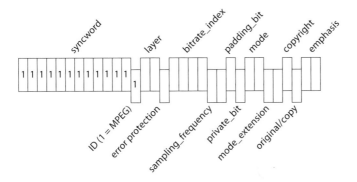

28. MPEG header syntax. Source: Pan, "A Tutorial on MPEG/Audio Compression," 68.

file, according to the people who created the file: set to zero, the audio is copyrighted; set to one, the audio is not. The original/copy bit set to zero means "copy of original media" and set to one means "original media."[33]

The existing and hoped-for technological forms of music in 1991 became the model for the MPEG standard.[34] The commodity forms of music—as recordings for sale, as something provided by a central broadcaster—were written into the code. As Karlheinz Brandenburg explained: "You will find two bits in the header which duplicate the interior copy management scheme introduced for DAT (digital audio tape) and CD recording. Two bits saying whether it is copy protected or not, and if it is whether it's the original or the first copy. The idea is that then equipment wouldn't allow an additional copy to be made of something that had already been copied." But no software ever looks at these bits, and even if it did, the scheme would be easy for users to circumvent.[35] There was no internet in 1991 as we currently understand it. The MPEG audio standard was conceived in terms of emerging delivery systems of the time, which were all freestanding hardware devices or devices that were part of a closed network like the relays between remote reporters, a central radio studio, and a broadcast tower. If an MPEG audio player was going to be like a radio receiver or a CD player or a digital audio tape player but with a hard drive, then it would be possible for such simple measures to be effective deterrents against users making copies that copyright owners did not want to be made. In fact, it would have been more difficult to copy MPEG audio than it was to make a tape of a radio program or a CD, for instance. Like Heidegger's jug, the inside of an MPEG file is an affording space. In coding and tasking the divisions within its void, the

IS MUSIC A THING? 197

MP3 file shapes both the sound of the material within it and the form that material takes. It recognizes the music's status as a commodity and tries to preserve it, however daintily.

How MP3s Became Ubiquitous

As the header story illustrates, MPEG may have intended to preserve and promote one kind of media system, but its formats succeeded in very different ones. The MP3's rise to global preeminence was a product of contingency, accident, and opportunity. Its affordances harmonized with other, broader cultural, technological, and political forces in the 1990s. The MP3 would proliferate thanks to file-sharing, thanks to a motivated computer and consumer electronics industry, and thanks to a series of happy accidents that put the format in the right place at the right time. The standard had to be set, then sold, then used. It was not the copyright bit that determined the MP3's historical significance, but its amenity to being copied and shared. MPEG initially split its audio into layers to broker an industrial compromise among competing interests. Layer 2 was backed by major companies like Philips and Panasonic. Layer 3 was the result of collaboration among Fraunhofer, Thomson, and AT&T. Once the MPEG standard was set, competition that had been somewhat mediated by the organization spilled entirely outside it. Given the much greater power and influence of layer 2's backers, we might have expected layer 2 to win the day in the marketplace. By some measures, it was a wonder that layer 3 became a successful commercial format at all, much less a household name.

Shortly after the establishment of the MPEG standard, layer 2's backers scored two quick victories. The format was chosen as the specification for the two key emerging applications of digital audio that were known at the time: audio on video compact discs and stereo digital audio broadcast, which included satellite radio, terrestrial digital radio, and satellite TV. From these two adoptions alone, there are hundreds of millions of layer-2 systems in use to this day. Each early adoption of the layer-2 standard made it increasingly difficult for layer 3 to get into the business. The Fraunhofer engineers grew progressively more worried. Here's Karlheinz Brandenburg: "In 1992–1994, the main focus was to find companies who would really use this MPEG audio layer 3 and with the exception of some professional applications in the first year, layer 3 was out of luck. Everybody else decided to go with layer 2." Harald Popp added: "We were very afraid that it would wind up a paper tiger. In the video market there were much better video systems and in the end they did not succeed and VHS made it, so we very often heard . . . 'you did a great academic job with a highly complex algorithm—yes, it works well in the labs but in real-world applications it's not really useful.'"[36]

A phenomenon called *path dependency* helps to explain the importance of early victories in the development of media formats and standards. Once manufacturers and users adopt a system built around a certain standard, the standard becomes a self-reinforcing phenomenon. Both manufacturers and users have interests in the persistence of the standard (or "path"), since a change in standard means a transformation in manufacturing equipment and sometimes major purchases for users. So the potential advantages of a new standard have to outweigh the cost for either manufacturers or users. If manufacturers decide to eliminate an old standard, like the floppy drives on computers or vinyl records, they can push a large swath of users into line over time. As we will see, users can exercise a similar power over manufacturers in some cases.[37]

Path dependency can also define the limits of a technological system. If companies build a successful radio network or millions of video compact disc players that all work according to the layer-2 standard, it is quite likely that the standard will persist for the life of the format and that devices conforming to the standard will be more appealing to users, even if a "better" technology comes along. In fact, this is exactly what happened with digital audio broadcasting. Layer 2 was initially adopted in 1993. In 1995, when major players in European Digital Audio Broadcasting met to reevaluate their decision, they decided to stay with layer 2, mainly because they had already been working with it for three years. Arguments about efficiency and similar sound quality at lower bitrates were less relevant than the hassle of retooling entire technological systems. Harald Popp called it "one of those battles we lost at the time."[38]

While Philips was securing the markets for traditional broadcast and audiovisual applications, Fraunhofer sought out other opportunities to commercialize layer 3. Popp and Brandenburg developed a slide presentation that they took on tour to electronics shows, to corporate boardrooms, and to meetings of organizations interested in the development and promotion of digital audio. Layer 3 did manage some sales. Fraunhofer built a few ASPEC-branded coder boxes to help sell the concept (see figure 29). The self-proclaimed first commercial application for MPEG audio was by Telos Systems in 1993, a Cleveland company that manufactured a box called the

29. The ASPEC audio codec. This version was built after the MPEG standard was set as an early attempt by Fraunhofer to market MPEG layer-3 audio. Image copyright © Fraunhofer IIS. Used with permission.

Zephyr. The Zephyr realized one of the original goals for MPEG audio: it digitally transmitted voices in real time over ISDN lines. The Zephyr and its descendants are used for remote telephonic connections in radio broadcasting—for instance, for call-in shows or shows from temporary remote locations. Telos also teamed up with Macromedia to develop an alternative to RealAudio for streaming web audio.[39]

Layer 3-technology was also used for a short-lived music-on-demand-via-ISDN service by Deutsche Telekom and a slightly longer-lived one in Britain. Cerberus Sound and Vision, the British service, first appeared in the news in 1994, and in some ways prefigures more modern online schemes for selling music. It consisted of proprietary coding and decoding software, complete with a copy protection scheme. Customers would dial in to a server through their modems and download songs that they could play back on their computers. Although originally scheduled for launch in 1994, Cerberus could not come to an agreement with major labels, and the service finally went live in 1995 with a catalogue of entirely independent-label music. By 1997 the service had signed deals with EMI and the Harry Fox Agency, which handles music licensing in the United States. That same year, they began a "virtual pressing plant" initiative, through which labels could transmit music digitally to kiosks in record stores, where people could audition music and have it burned onto a CD on demand.

Fraunhofer had other success as well. It signed a contract with a German jukebox company, NSM Music. WinPlay3, a Windows application that could play MP3 files, made it easier for home users to try out the technology on their own PCs. In 1995 Fraunhofer reached an agreement with Microsoft to include the MP3 specification in applications that came bundled with its operating system; in 1996 it first appeared in "Netshow," a predecessor of Windows Media Player. In 1997 Fraunhofer signed a deal with Worldspace, a satellite-broadcasting initiative. Apple's QuickTime had MP3 compatibility by 1999. But apart from the software deals (and then only a few years down the road), none of these deals were of an industry-changing nature. They were business plans that operated within the parameters of existing schemes for making money from sound technology such as broadcast, software playback, auditioning in record stores, jukeboxes, and the direct purchase of recordings. Had layer 3 remained in these niches, it would simply have continued to tread water in the crowded pool of digital audio formats. It would have had little industrial or broader cultural importance.[40]

During this period, Fraunhofer devoted a lot of effort to marketing. Harald Popp, who had come to Fraunhofer as an engineer, switched over to marketing and authored a FAQ on MPEG layer 3. The version of the FAQ from December 1996 is mostly concerned with describing possible applications. It lists music links via ISDN, digital satellite broadcasting, audio-on-demand, audio-on-the-internet, recording studio equipment, sound on CD-ROMs, and sound on solid-state memory (like RAM) that could then be put in portable players. Where it did not have commercial takers or enough resources on its own, Fraunhofer engaged in proof-of-concept applications. It built a solid-state MP3 player that could play back one minute of audio. Although one minute isn't useful for much, Fraunhofer assumed that increasing computing power and storage capacity would eventually allow such a player to be bigger and more powerful. Another proof of concept was a pair of command-line programs called L3Enc and L3Dec, released in July 1994 and priced at $250 (this was before WinPlay3 became available). Fraunhofer also released a shareware (initially free) version of the programs that would allow users to encode and decode twenty seconds of stereo audio, as a sort of test-drive for the more expensive version. An Australian hacker acquired L3Enc using a stolen credit card. The hacker then reverse-engineered the software, wrote a new user interface, and redistributed it for free, naming it "thank you Fraunhofer."[41]

But in 1994, there was no such thing as an MP3, and there were only small communities online dedicated to file-sharing. The Internet Underground Music Archive, founded in 1993, was dedicated to allowing unsigned bands to post their music and download other bands' music. Today, IUMA is hailed as the first major player in online music distribution. It was later purchased by eMusic, a company in California that was one of the first online stores that successfully sold MP3s via download. The IUMA started out using the layer-2 specification, but switched to layer 3 after "thank you Fraunhofer" and WinPlay3 became available in 1995.[42] For the subculture of hacking and file-sharing developing in the 1990s, layer 3 had one significant advantage over layer 2 because it provided higher quality at lower bitrates and lower file sizes. In 1994 most internet users were still using dial-up connections, and variations of even a few kilobytes made a big difference in terms of how long it took to upload or download a file. In 1993 and 1994, it would probably have taken about a half hour to download a single song encoded as a layer-2 file, while a DSL, cable, or 3G connection allows the same download to occur in seconds today. (Given computers' storage capacities at the time, file size was also a major concern for early users.) With the "thank you Fraunhofer" hack, layer 3 was also free to use, which meant that users who had some technical facility could simply acquire the software and start converting their digital audio files into layer-3 recordings that they could then upload to sites set up for the exchange of files.

There was also the matter of the name. "MPEG-1 audio layer 3" does not roll off the tongue, and as Fraunhofer got more involved in marketing its scheme to users as well as to companies, it realized that it needed a different name for its file format. After considering generic names like ".bit" and ".son," a vote among people at the institute resulted in the adaptation of "MP3" as the name for the file format. An e-mail from Jürgen Zeller to the rest of the organization on 14 July 1995 announced the result and urged *"for WWW pages, shareware, demos, and so on, the .bit extension is not to be used anymore. There is a reason for that, believe me."*[43] The name served several functions at once. It promoted Fraunhofer's technology; it made uses and applications easier to track; and it concretized the format in the minds of users. An MP3 was a thing, like a .doc or a .pdf. Naming the format helped demystify and make banal digital audio for users: your word processor documents are .docs, your spreadsheets are .xlss, and your music files are MP3s.

The relative absence of innovation in the mainstream recording industry is crucial to the MP3 story. Most labels ignored the potential of online distribution, and as late as 1998 many companies did not have any staff assigned to monitor the internet. The retired police officers who made up the RIAA's antipiracy enforcers were more likely to troll flea markets for cassettes than websites or newsgroups for files. Apart from a few scattered and tentative forays into online marketing, it appears that the Recording Industry Association of America's first major engagement with online music distribution was a lawsuit in 1997 against illegal FTP sites. A year later the RIAA and complainants filed suit against Diamond Multimedia, whose Rio device was one of the first commercially viable portable MP3 players. The RIAA won the FTP case (which was in many ways a dry run for the more famous Napster case a few years later) but lost the suit against Diamond.[44] In other words, its first action regarding the online distribution and digital copying of music was to try to make it stop. The recording industry's inaction until 1997 is a key part of the story because it allowed other industries to develop and organize the online music environment according to their needs. The recording industry may or may not have been able to shape online distribution to its liking in the 1990s. But while it did experiment with online sales and distribution, the lack of a coherent strategy and its litigious attitude toward file-sharing guaranteed that other industries would shape the environment.

MP3 music is thus a classic case of transectorial innovation, a process whereby "new technologies are no longer confined to a single application, to a single sector; they are disseminating and interpenetrating the whole economy." Technical innovations in one industry may have massive effects in another because of the vast interconnections among technologies and standards. The phrase *transectoral innovation* comes from economics and social studies of technology, to provide a language for discussing the ways in which innovations in one area may cut across another. In studies of sound technologies, the phenomenon was first noted by Paul Théberge, because digitization led to the subordination of the music instrument industry to the computer industry, a phenomenon that perseveres today. Théberge was interested in synthesizers, but the phenomenon he noted has now suffused almost every kind of music-making or editing technology (apart from high-end markets in several fields that specifically distinguish themselves as "analog," "boutique," or "hand-wired"), especially if we include their manufacture in our considerations.[45]

In its formative years, online music was not the province of the recording industry, which had hitherto done a fairly good job of controlling its distribution channels. Online music—which was at its core a mode of distribution, a relation to infrastructure—was instead the province of companies like Fraunhofer and Philips, Microsoft and RealNetworks. It was dominated by manufacturers of computers and consumer electronics, providers of broadband internet, software companies, and startups dedicated to some aspect of online distribution.[46] As the internet grew in size and scope, this posed more of an economic and social problem for the recording industry. The industry had already lost control over distribution in some parts of the developing world during the heyday of cassette recordings, but now the multinationals were losing ground on their home turf. For generations who grew up after prestige records became the norm in the first decades of the twentieth century, there were relatively few channels of musical distribution: labels, record stores, radio, television, the mail, hand-to-hand. With the exception of the mail and hand-to-hand, users had little chance to distribute music themselves and benefit from that distribution. Because it presented an alternative, especially in these early years, file-sharing betrayed the social character of musical exchange to its users, putting the recording industry's privileged position directly into question. The internet was a space of circulation where the record industry did not assert its dominance, and in that moment, file-sharing served synecdochically to call into question the industry's dominance tout court. If Eisenberg's exchange of money for records hid the social relations of music from him, the absence of that particular economic relationship, the opportunity to acquire recordings in some other fashion, could illuminate music's social character.

The term MP3 entered into wide journalistic use in reference to sound recordings for the first time in 1997 and really took off in 1998. Before that, a few scattered articles mentioned other computer software and hardware schemes that used the name MP3, a multipurpose plow that hitches to horses, and a United States Air Force space and missile systems center which had an "MP3"—a "Manufacturing Problem Prevention Program" (let us hope this last one was effective).[47] The MP3's popularity as a new topic roughly coincided with a sea change in online file-sharing between 1995 and 1997. The MP3 became a popular format for putting music on hard drives and sharing on the internet thanks to the crack of L3Enc and its later inclusion into applications that came bundled with PC and Apple operating systems. College campuses in the wealthier parts of the world (especially the United States and Canada) began to wire their dormitories for high-speed internet connections, bringing many more students online, and cable and telephone companies started to offer high-speed broadband internet as part of their consumer services in some communities. Broadband became increasingly common at large institutions like universities (which were hubs for file-sharing in the late 1990s) and eventually began to expand in the consumer market, though it did not really take off as a product for home users until the 2000s.[48] CD burners also began to drop precipitously in price (from initially over $10,000 to under $1,000), and by 1997 were an option on consumer-grade computers. It was possible for anyone with a few hundred dollars to purchase a CD burner, and by converting CD files to MP3s, a user could place over one hundred songs on a single compact disc. Unauthorized CDs containing MP3s started to appear for sale on the streets of major cities around the world. But the rate of estimated music piracy in the United States actually *decreased* between 1993 and 1997, despite the availability of new digital tools. This was likely because most internet users did not yet view it as a medium for the distribution of music. At the time, a 2-megabyte MP3 was still considered a relatively large file, both in terms of the hard-drive space it took up and the bandwidth it required (and therefore the time it took to upload and download on a slow dial-up connection).[49]

When I interviewed Brandenburg, he cited 1997 as the turning point at which MP3 became a mass phenomenon online:

> In 1997 the elements were forming. We already heard earlier that in the USA students at colleges and universities played around with MP3 and put their music on the university servers using MP3; we had the licensing contract with Microsoft in place who wanted to put it into the Media Player attached to the operating system; we had the first consumer electronics application in view with the WorldSpace application.
>
> In mid-1997 there was the first legal action of the RIAA against illegal FTP sites at universities and there was an article about that in *USA Today*—think that was clearly a moment when everybody knew about it and everybody wants to use it and this can't be undone. It was the same year that we tried to get into closer contact with the music industry to

make them aware of the format saying, "Look—you have to be careful what you do next because there is something rolling that is running in the way people want it."⁵⁰

At the same time as these developments, a number of commercial and unauthorized ventures combined to help promote the MP3 format. In November 1997, a Hungarian fan with a prerelease promotional tape made clips from U2's *Pop* album available online prior to its release. By January 1998, music by Van Halen, Metallica, Eric Clapton, and Madonna appeared on the internet before its official release. The site MP3.com was launched in October 1997. It began as a clearinghouse for information on MP3 and internet music and quickly morphed into a massive distribution service for artists who signed a nonexclusive agreement with the site, meaning that they retained many more rights than artists who signed recording and distribution contracts with major labels. The site's revenue was generated from banner ads and sales of consumer data to clearinghouses like DoubleClick. Until a lawsuit stopped them, MP3.com also provided an online streaming service for its users' collections. Users would deposit MP3s in online "lockers" and then be able to access their music anywhere they were online. The service did not actually allow for downloads, and users had to "prove" they owned the music by inserting the relevant CD in the CD-drive of their home computers. Meanwhile, eMusic (formerly "Goodnoise Corporation") experimented with a subscription-based service where users would pay a monthly fee to download tracks in MP3 format from the eMusic.com site. As 1998 opened, MP3 was the second most popular search term on the internet, second only to pornography. By 1999 MP3 was the most popular search term on the internet. August of that year saw the public launch of the file-sharing service Napster.⁵¹

By the time Napster appeared on the scene, online file-sharing was well established. Its main innovation was the interface, which made it easier and faster to share more music with more people. Napster was a single, massive, synthetic database of recordings that could be searched by keyword. Users who downloaded the client software and logged in would make the music on their hard drives available to other users through the database. They could also search the database and potentially download music from the millions of other users logged in at the same time. Napster claimed that over 28 million people downloaded its software, though other estimates place the number of users closer to 7 million. Either way, the company became incredibly valuable as a property despite its lack of revenue generation. Napster didn't sell anything: not ads, not user information, not subscriptions. Its value was generated by its user base, upon which investors assumed they could cash in at a later date. The RIAA saw Napster as a threat and sued to shut it down in April 2000. After a string of cases and appeals, along with a dwindling user base (and with it dwindling investor interest), the company filed for bankruptcy in 2002. Other less traceable software protocols emerged in the wake of the Napster shutdown. The people who built Aimster, Gnutella, and BitTorrent all learned from Napster's mistake and skipped the centralized database part of the program, thereby creating a massive swarm of users who shared files directly and anonymously with one another. Individual IPs could be traced, but the whole system was difficult to shut down. Today, sanctioned search engines like Google are almost as effective for finding pirated material as specialized torrent sites like the Pirate Bay.[52]

Although there exists a long-standing debate regarding the actual cost of file-sharing to the recording industry, it is clear that the massive boom in file-sharing between 1998 and 2001 is the moment when MP3 became the dominant sound format online and for pirated music. This feeds into the usual story that is told about the format's dominance: the MP3 was destined to come to dominance; it only needed the right helping technologies in place—broadband internet, CD burning, sound cards, and portable audio players. In an interview, Fraunhofer's Harald Popp characterized it as almost epiphenomenal, as though MP3 was swept up in an unstoppable technological revolution:

> [Piracy] made MP3 the breakthrough format on the internet. There's no question about it. Sometimes I think people in the music industry had some nasty thoughts about MP3 but we were never approached by the industry with criticisms about killing the music industry—those were headlines from the press but not the music industry itself. The real challenge is not MP3 but the internet—and it still is the internet. This is the thing that traditional businesses, media businesses, music businesses and video/movies as well have to think about—the availability of the internet and computers in every household. This is the challenge and the opportunity. It's not MP3—that was always our clear answer: "MP3 was just a catalyst. If it was an ISO standard it could [just as easily] have been a proprietary scheme." The technology is out there and you cannot

avoid some audio compression—we were very happy that the appearance of MP3, when it took off, that people took MP3 to code their music and use it. Of course, in such [cases] you are sometimes violating the law, and that's not good, but that is just an aspect of a technical revolution.[53]

The Fraunhofer patents now bring in over 100 million euros a year in revenue—the MP3 patents are the company's main revenue stream. Their value derives from the MP3 format's ubiquity. But this ubiquity is the result of a confluence of sanctioned and unsanctioned markets. Even though the MP3's growing proliferation depended on the emerging infrastructural preeminence of the internet, the growth of broadband was necessary but not sufficient for MP3 to become the most common format for recorded audio. Fraunhofer did what it could to establish the possibility of path dependency in computer operating systems and consumer electronics.

Fissures between Industries: The Political Economy of Piracy

Piracy was also a central catalyst in the MP3's rise to preeminence and the growing value of MP3 patents in this period. As we saw, the format was first widely disseminated through the "thank you Fraunhofer" hack and later through its inclusion in Microsoft and Apple software that was free to users. And while there were concerted efforts to sell MP3 music online from 1994 on, the vast majority of music that circulated online did so through unauthorized channels—FTP, IRC, and file-sharing services. Although the pirates paid for their computers and their internet connections, they experienced the recordings as something they acquired for free (even though strictly speaking, they were still operating inside a money economy).

Nowhere is this clearer than in the behavior of a conglomerate like Sony, which held both record labels and consumer electronics interests. When Sony Electronics introduced a CD player that would play MP3s burned onto a CD in 2001, Sony Music was enraged, but the electronics division felt it had no choice: it had an opportunity to cash in on the explosion of file-sharing.[54] Multisectorial innovation and path dependency worked together to produce the value of the MP3 format at precisely the moment that recordings could be obtained for free. While Sony Music sought to stop the growing swarm of MP3s, Sony Electronics either had to capitalize on it or be left out of the market. The conflict inside Sony encapsulated the conflict across industries. Pirated music was very good for some people's business even if it was bad for that of others. In subsequent years, this has continued to be the case. A study published in 2006 estimated that 70 percent of all internet traffic was peer-to-peer file-sharing, most of it pirated movies, TV shows, games, software, pornography, and music.⁵⁵ That statistic suggests that ISPs built a large portion of their business on unauthorized traffic.

As already intimated, the MP3 was not the recording industry's first crisis. Two other tales of piracy set the current declarations of emergency in relief: Britain's pirate-radio adventure in the 1960s and the international wave of cassette piracy in the 1980s. Although various forms of unlicensed radio broadcasting had existed in Britain since the government began to regulate broadcasting, pirate radio was central to British popular culture of the 1960s. At the time, the British Broadcasting Corporation was limited by law to playing only a few hours of recorded music per week, and much of that time was devoted to classical music. Into this void stepped a group of pirate broadcasters. If radio audiences wanted to hear the latest recorded pop hits, they needed to tune in to a pirate station. Unlike the BBC, which received its income from license fees charged to listeners, pirate stations sold time to advertisers (as in the model of broadcasting in the United States), providing them with a steady source of income. When violence and scandal in the pirate industry eventually led to its shutdown, the British government recognized that it needed to provide an alternative where audiences could hear the latest popular records. Born in October 1967, BBC Radio One filled this void.⁵⁶

The pirates' challenge to British broadcasting was not only institutional and commercial but also resolutely ideological. Coincident with the rise of the pirate stations of the 1960s was the emergence of the Institute for Economic Affairs (IEA), which published tracts like *TV: From Monopoly to Competition* (1962), *Competition in Radio* (1965), and *Copyright and the Creative Artist* (1967). The IEA was part of a tradition of antimonopoly writing in England, which had previously found its strongest statement in Ronald Coase's *British Broadcasting: A Study in Monopoly*. Coase attacked the idea that the BBC's monopoly was natural or desirable and called into question its value as a source of cultural uplift. The IEA tracts "tried to teach a lesson about the social role of so-called 'pirates' in general." The radio book argued, "Pirate broadcasters had a pivotal role to play in the development of a new politics of communication and public culture. Pirate broadcasters were examples of a broader type. They represented a form of commercial life that recurred frequently, but that the state and existing institutions always regarded as *immoral*." In contrast, the IEA argued that piracy was a "business force" and that resistance to it was "a reflex reaction by established interests to unwelcome and adventurous competition." Yet the IEA arguments offer a very different genealogy for today's anticorporate and anticopyright crusaders. The IEA and its allies lost the battle in the 1960s with the ban on pirate stations and the opening of BBC Radio One, but they would win the war as free market ideology and deregulation would come to dominance in British telecommunications policy in the 1970s and 1980s, and eventually around the world. Anticopyright politics can be libertarian or pro-corporate just as easily as they can be anticapitalist.[57]

The British case is also instructive in its terminology. *Legitimate* broadcasting was broadcasting that had official state sanction, alongside the section of the broadcasting industry that was sanctioned by the state. *Pirate* operations were not anticapitalist or anticommercial. They simply operated outside of the bounds of legitimacy as defined by the state and state-sanctioned industry. As the pirate industry gained influence and purchase, it facilitated the transformation of the state's policies and industries, breaking old oligopolies but leading to the establishment of new ones in the process. Thus, piracy was a business force not only in terms of breaking open markets that were previously closed, but also in creating the next generation of institutions that would themselves come to dominance in a market. Despite the degree to which terms like *legitimate* and *pirate* carry normative weight, they are strictly relative terms and exert a certain ironic force in their invocation (or at least they should), since their application always begs the question of who legitimizes and under what conditions transgression occurs. The government treated the BBC as if it spoke for the British listening public, even as audiences of the 1960s wandered away to find the music they sought elsewhere. Radio pirates made it their major ideological mission to undermine the BBC's legitimacy as well as that of the government as cultural arbiter.[58]

Similar stories can be told about cassette piracy. Two decades after Britain's pirate-radio crisis, the worldwide recording industry was in an uproar around the mass piracy of music through cassette duplication. The most famous effort during the 1970s and 1980s was the British phonographic industry's PR campaign slogan "home taping is killing music." But home taping was dwarfed by massive pirate cassette-duplication operations, which sometimes even used the logos of legitimate labels (or crude imitations thereof). These cassettes sold for a fraction of the cost of their authorized counterparts and weren't even treated as illegal in many countries, because in the 1980s many national copyright codes hadn't been updated since the early twentieth century.

Piracy thus generated huge profits for two kinds of enterprise: locally and nationally based cassette-duplication operations and multinationals that manufactured blank tape, such as BASF (a subsidiary of Siemens) and Philips, which went on to be a major player in MPEG. Both the duplication houses and blank-tape manufacturers benefited from low overhead and huge profit margins on the sale of their products. But the political valences of their interventions are somewhat different. As with the example of Sony's record label and consumer electronics divisions working against each other, many conglomerates owned both record labels and tape manufacturers (for instance, Siemens owned half of PolyGram, as well as BASF). So it is possible to see the issue as at once a conflict between the interests of two different industries and, at the same time, large-scale multinationals essentially maintaining diversified portfolios to hedge their bets. Meanwhile, the duplication houses in developing countries were often the first institutions to dislodge the hegemony of foreign multinational conglomerates that had previously dominated national and regional recording industries.[59]

Consider the example of T-Series, an Indian company. Today, T-Series is a major and diversified music conglomerate at the center of the legitimate music industry in India. But in its early days, T-Series functioned as a label and distributor for new music, as well as a pirate duplication house that churned out copies of EMI releases, whose market dominance and near-monopoly were leftovers from British colonialism. T-Series employees tell stories of HMV artists (HMV was an EMI label) coming to them begging for pirate releases, as HMV could not keep up with demand. These are not apocryphal stories. In several cases during the 1980s, legitimate labels turned to pirate duplication houses to meet consumer demand for recordings. T-Series thus offered an Indian-owned alternative to a multinational that had dominated India's recording market for decades, an opportunity for artists to reach larger audiences than they might otherwise, and a new outlet to release music that would compete with a previously locked-down national record industry. One important implication of this arrangement is that the pirate cassette business predated the legitimate cassette business in India.[60]

The current uproar around file-sharing follows a similar pattern to the cases of British radio and the transnational cassette industry, although it clearly operates on a different scale. The distinction between home taping and mass duplication in factories has morphed into the difference between the so-called sneakernet ripping of CDs and exchange of hard drives and the massive online swap meets of Gnutella and BitTorrent. A legitimate online record industry appeared via a wave of independents, followed by the iTunes Music Store and now competitors like Amazon. And yet a sense of crisis persists, largely because of the sheer scale and apparent impossibility of stopping the file-sharing tide. So while it shares some important structural features with the earlier examples, it may also be the case that the MP3-enabled file-sharing breaks some of the conceptual machinery of both classical and Marxist explanations of capitalist markets. To use the language of the former, a vast body of recorded music has become demonetized. To use the language of the latter, that body of recorded music has lost its exchange value while retaining its use value.

This phenomenon is normally explained through variations on a story about the internet as a gift economy and its radical potential for changing the way we think about exchange. Examples drawn from the free-software movement support this proposition. Yochai Benkler's *The Wealth of Networks* also argues that the free exchange of labor and goods online can generate economic value in new ways. The argument is that sharing online can lead to new, more humane modes of economic exchange and social collaboration, and may pose a threat to existing economic hierarchies.[61] Yet there are at least two problems with this formulation. If we return to Marcel Mauss's classic work on the gift, he is quite clear that a gift logic already obtains in large industrial societies and that it is part of the capitalist economy. Contemplating various forms of social insurance produced by states, corporations, and unions, ranging from socialized medicine to employers' family benefits to union-organized unemployment insurance in France and Britain, he calls these forms of collective gifts not "an upheaval, but a return to law." These techniques attach people to their employers and impose social obligations on them, just as the gifts in smaller societies impose reciprocity and escalation. In a footnote clarifying his position, Mauss writes that "of course we do not imply any destruction; the legal principles of the market, of buying and selling, which are the indispensable condition for the formation of capital, can and must exist beside other new and old principles."[62]

There are also some key ways in which "gift" model doesn't quite work for online file-sharing of music. Gift economies are normally distinguished from commodity economies in at least three ways. Gifts come with obligations bestowed upon the receiver that may themselves be coercive. To receive a gift is to receive an obligation; to give a gift is to create one. Gifts are inalienable because to transfer the gift to a third party would undermine the relationship between giver and receiver—gifts "carry traces of the gift giver." Commodities are by design alienable. It matters that my friend painted the painting he gave me, it matters not whose hands touched the movie poster on its way to the store display where I purchase it. In traditional gift economies gift exchange is also characterized by a time delay. As Andrew Leyshon points out, most forms of online exchange of recordings do not follow these patterns. In file-sharing networks, MP3s are alienable; it does not matter who made them or ripped them. They carry no mark of the individual who passed them on, and although some groups enforce norms of reciprocity (like a 1:1 download-upload ratio on a BitTorrent site), only some specialized networks actually enforce this reciprocity, which in any event is considerably more abstract. The temporality of file-sharing is not serialized as in a traditional gift economy.[63] File-sharing also differs from the open-source movement in that the former is not, strictly speaking, any kind of self-conscious movement. Some elements, like Fraunhofer and Napster, quite explicitly sought profits for their work. Others, like the Australian hacker who cracked L3Enc or the millions of students who started sending files back and forth on networks in 1997 and 1998, had other, more idiosyncratic motivations and may or may not have seen themselves as part of a massive collective enterprise.

The idea of online collaboration and sharing as a gift economy has already been debunked in the scholarship of open source. For instance, Steven Weber argues that the logic of the gift is wholly insufficient for understanding free software, and that network externalities better explain why people contribute to a nonrivalrous, nonexcludable resource like open-source software. The larger the scale of a technological form, the easier it is to use, access, and make available.[64] Chris Kelty goes further to argue that

> coordination is important because it collapses and resolves the distinction between technical and social forms into a meaningful whole for participants. . . . Such coordination would be unexceptional, essentially mimicking long-familiar corporate practices of engineering, except for

one key fact: it has no goals. Coordination in Free Software privileges *adaptability* over *planning*. . . . Geeks not only give expressive form to some set of concerns (e.g., that software should be free or that intellectual property rights are too expansive) but also give concrete *infrastructural* form to the means of expression itself.⁶⁵

The collectives that move music online are neither as self-conscious nor as coordinated as the recursive publics of the free-software culture. Certainly protocols exist, and something like BitTorrent can be read as an infrastructural politics that is a direct response to the RIAA's shutdown of Napster. But there is no MP3 community, no self-defined or self-conscious public of file-sharing. There is a much more atomized, clustered, rhizomatic series of collectives, some of which share interests and empathy, and many of which don't. If anything, browsing a single search engine, like TorrentBox or Pirate Bay, gives an illusion of a cohesive public where there is considerably less collective consciousness to be found beneath its well-designed interface.

The MP3 that travels over a file-sharing network partakes of both commodity and something else. If Evan Eisenberg's money listened for him in the cash exchange at the record store, it may seem as if file-sharing obliterates this relationship, since file-sharers do not directly pay for the files they download. And yet there is a sense in which the concept of the collection persists, along with the bourgeois sense of ownership that subtends it. Users may be able to handle MP3s quite differently than the recordings they possess in larger physical forms like records or CDs, but they still talk about MP3s as things—things that are owned, and which offer affordances to their users. Consider this review of the iPod and iTunes Music Store: "You'll even find that you listen to music in new ways. Recently the Talking Heads' sublime 'Heaven' popped up on my jukebox in random play mode; I'd owned the CD for years but hadn't played it much and never noticed this amazing song. That kind of discovery happens all the time now that my music collection has been liberated from shiny plastic disks."⁶⁶ Listeners may refer to the dematerialization of music in discussing their practices of use, but they insist on treating the music as a thing when they discuss it in terms of possession. Music micromaterialized is still recorded music. "For a collector," wrote Walter Benjamin, "ownership is the most intimate relationship that one can have to objects."⁶⁷ This appears to be the case even when people do not pay for the objects they collect.

For both the classical and the Marxist economist, ownership without payment presents something of a contradiction. MP3s appear to walk and talk like the regular products of capitalism. They act *as if* they had been received in exchange for money—even the modes of exchange follow the model of the money economy more closely than that of the gift economy—and yet in most cases, they were not in any direct sense acquired for a price. By definition, a thing is only a commodity when "its exchangeability (past, present, or future) for some other thing is its socially relevant feature."[68] Sure, there were markets. There were markets for hardware, software, and bandwidth. There were markets for software licenses and blank media (and what is a portable MP3 player if not a very sophisticated blank medium?). But because of the recording industry's absence in the formative stages of the online music economy, music itself was *demonetized*, to use the term in current business parlance. Along with videos, websites, news, e-mail, social networking, and pornography, it helped to sell the internet and all of the commodities that went with it. It was not itself sold. Thus, we need to think differently about markets and capitalism. A pair of feminist scholars who write under the pseudonym J. K. Gibson-Graham offer an interesting way out by applying the insights of poststructuralist feminism to the theory of capitalist markets. Gibson-Graham reasoned that if it was the case that a range of configurations of gender and power could flourish "on the ground" in a society characterized by a prevailing condition of patriarchy—an argument routinely advanced by feminists in other fields—a similar dynamic might be at work under prevailing conditions of capitalism and market economies. "The market, which has existed across time and over vast geographies, can hardly be invoked in any but the most general economic characterization. If we pull back this blanket term, it would not be surprising to see a variety of things wriggling beneath it. The question then becomes not whether 'the market' obscures differences but how we want to characterize the differences under the blanket."[69] Beneath the veneer of capitalism and the market, we might come to see "feudalisms, primitive communisms, socialisms, independent commodity production, slaveries, and of course capitalisms, as well as hitherto unspecified forms of exploitation."[70]

The music market may be capitalist in a general sense, but not in its totality. Even in the most fully monetized situation, commodities can move in and out of market exchange over the course of their lifetimes. If relations between people can become relations between things, changing relations among people may lead to changing relations among things.[71] In this sense, the music contained inside an MP3 partakes of a well-established pattern. Although the vast majority of MP3s in the world today were not purchased, they are copies of recordings that were produced in the money economy. The most popular MP3s in circulation at any given moment are all popular commercial recordings, most of which were released by major labels through the usual industry channels, then pirated and recirculated. Although the correspondence is not exact between the most popular MP3s and the biggest hits (country music, for instance, remains quite underrepresented in file-sharing networks), the phenomenon is striking.

In MP3 form, recorded music's behavior is deeply related to its commodity status even though buying and selling represents only a minority of the occasions upon which an MP3 circulates. Although Gibson-Graham's work focuses on self-conscious, progressive alternative economies, the case of the MP3 extends their argument in another direction. Alternative, nonmarket economies within capitalism may not themselves be anticapitalist. It may appear that file-sharing and sampling challenge particular market economies, but that does not necessarily mean that they challenge the broader capitalist condition of music. There remain music-related markets for software licensing, blank media, and infrastructure (and access to bandwidth), not to mention vast economies of licensing for other media (such as film and video games). This may also be true for music, musicians, and audiences.

DJ Danger Mouse's career is an instructive example of how piracy can generate value, even for artists. In 2004 Danger Mouse released the *Grey Album*, which was a mashup that combined material from the Beatles' *White Album* with Jay-Z's *Black Album (Acapella)*. As a mashup, the sampling rights were impossible to obtain. Danger Mouse released his recording in a limited-pressing vinyl edition and on the internet for free, for which he received a cease-and-desist letter. Scholars published articles celebrating Danger Mouse's challenge to the corporate hegemony of the record industry, analyzing the political dimensions of mashup culture through the recording techniques it used, or using the Danger Mouse case as another example of the need for alternatives to copyright law that effectively render sample-based music illegal. These were all valid and important arguments. But Danger Mouse himself moved on to a successful career as a producer for bands like Gorillaz, the Rapture, and Sparklehorse, and eventually wound up as half of the very successful duo Gnarls Barkley. Danger Mouse's career thus shows a possible connection between piracy and the promotional culture of the official record industry. His unauthorized copying leveraged future market-based activity—his successful production and recording career. In many ways, Danger Mouse's move was a classic piece of independent musician entrepreneurship: the internet equivalent of the local DJ or band that plasters posters all over a city in the hope of attracting an audience.[72]

Both the industry lawyers who decry file-sharing and the culture critics who celebrate it mix up their forests and their trees (see figures 30 and 31). To threaten an industry's incumbents is not to threaten the economy itself, despite incumbents' protestations to the contrary and their critics' glee at the prospect.[73] Tiziana Terranova sums up the challenge: "The question is not so much whether to love or hate technology, but an attempt to understand whether the internet embodies a continuation of capital or a break with it. . . . It does neither. It is rather a mutation that is totally immanent to late capitalism, not so much a break as an intensification, and therefore a mutation, of a widespread cultural and economic logic."[74] File-sharing is not automatically or necessarily more progressive or egalitarian than the economy it challenges. It could just as easily go the way of British radio or the Indian cassette industry.

Politically, this leaves us with an ambiguous situation. Most positions in the file-sharing debate are unable to articulate a strong ethical position beyond self-interest. The interests in the recording industry speak for themselves as copyright holders, and not for the musicians they claim to represent or for music itself. But the same must be said of the people who enable or participate in file-sharing. Often enough, they too have themselves in mind, either as industries who affect music transectorially and find profit as recordings slip out from inside the money economy, or as users who simply get something for free because they do not want to pay for it and do not have to (though they pay for the tools and network access that enables file-sharing). There are many mechanisms for supporting music-making and listening in societies, and there is no divine decree or moral precept that a hundred-year-old recording industry must be preserved in perpetuity. The continued existence of music schools and community-funded bands and orchestras suggests that many cultures are already comfortable with alternatives to market-driven music-making and listening. In countries with culture ministries, the ethos extends to rock and hip-hop as well as

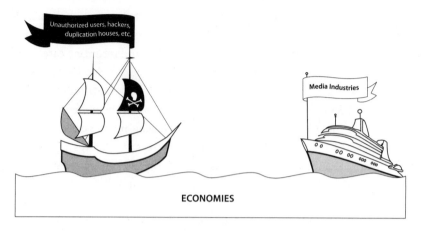

30. Piracy: how it's usually represented. Content owners represent piracy as a force external to the legitimate media industries. Image by Liz Springate.

art music. A real social critique of music markets must extend beyond the *usually named* music industries, to broadband, consumer electronics, computers and software, and a range of other related fields.

The Right to Music, and Other Things

The ambiguous economic status of the file-shared MP3—not quite a gift, but not entirely a commodity—marks a historical conjuncture full of possibility. It does not fully conceal social relations in the way that Evan Eisenberg's trade of cash for records might. But a file-shared MP3 also lacks the revelation of social relationships that would occur with a gift. To borrow a term from Marx, MP3s are part of "social intercourse," but of what kind? Commentators, even those sympathetic to the file-sharing phenomenon, routinely refer to mass file-sharing as "promiscuous," and the normative overtones of the sexual analogy are instructive.[75] In a classic gift economy, the gift requires a certain level of intimacy between two definite individuals, a relationship of obligation. Because it is largely aggregated, file-sharing has none of that structuring obligation written into it—but neither is it promiscuous in a strict sense. The normative implication behind the epithet *promiscuity* is that more regulated transactions—transactions that more closely follow the gift model or the commodity model—are somehow

218 CHAPTER SIX

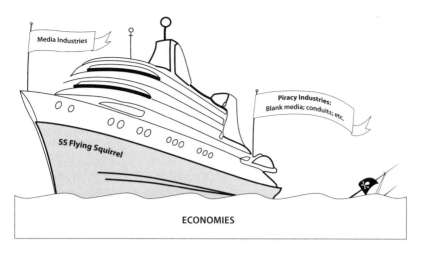

31. Piracy: more like how it is. A subset of media industries have an economic interest in piracy. The piracy industries are media industries that sell blank media, conduits, and connectivity. Image by Liz Springate.

less deviant. But from what norm do they deviate? As we have already seen, the unsanctioned economy of file-sharing has been immensely valuable and value-generating for several legitimate software and hardware industries. British pirate radio extended the reach of the for-profit radio industry and a sector of the recording industry. Pirate activities always occur in particular historical and political conjunctures, which themselves change over time. The end of recordings' artificial scarcity has no inherent moral or political valence written into it, but it does have potentials.

One potential is clearly greater industry control. When there are fewer shiny plastic discs, some sectors of the industry imagine moving from a product to a utility model. Patrick Burkart and Tom McCourt refer to this model as the "celestial jukebox," where content providers could charge users for each time they listen to music or use it in some other way. In other words, when people legally purchase MP3s, they are not purchasing commodities (as was the case with records or CDs) but instead buying a license, as is the case with software. The license can restrict, curtail, or order use, and it provides the vendor with an opportunity to maintain a relationship with the buyer that would have concluded at the moment of purchase in the old commodity regime. Accessibility and universality, the "anytime, anywhere" advantages of networked digital culture, may entice users. According to Burkart and McCourt,

The Jukebox may promise more innovative music, more communities of interest for consumers, and lower prices for music; in fact, however, it gives us less (music in partial or damaged or disappearing files) and takes from us more (our privacy and our fair-use rights) than the old system.... Rather than a garden of abundance, the Celestial Jukebox offers a metered rationing of access to tiered levels of information, knowledge and culture, based on the ability to pay repeatedly for goods that formerly could be purchased outright or copied for free.[76]

In other words, mass piracy provided training for users to abandon the commodity model of music, but that does not mean they will leave the money economy forever. They may reenter on other terms. As Andrew Leyshon writes, the "hi-tech gift economy . . . is seen by some of its influential participants as a precursor to a more distributed, more efficient market economy, with a strong libertarian edge."[77] Or to put it another way, the short-term loss of compact discs as property can be overcome by industry players who seek ownership and control of the means of distribution as the main model for profit generation. In this way, music companies would become more like phone, electrical, cable, or satellite companies. The service, not the recording, becomes the commodity form. Inasmuch as users already pay for their broadband connections or their electricity, they are already accustomed to this order of things.

In the MP3, there is a potentially explosive commingling of property and thing. Clearly, the MP3 has the potential, as do all recordings, to exist as property in the conventional sense. Forget for a minute about the "intellectual property" epithet and consider an MP3 as a very, very small miniature that is analogous to a compact disc or LP record. Its power comes from its holders' abilities to acquire it by expending their own energies, and from its holders' abilities to put it to work. It represents a stockpile of accumulated and alienated labor which the user or listener can put to work. But the corollary to these kinds of property rights, if an MP3 is a commodity to be bought and sold, are the rights to "buy, sell, produce, and trade, without monopoly, arbitrary regulation, or arbitrary taxation."[78] When recordings move from a good purchased to something subject to licenses and contracts, DRM schemes, or monthly service agreements that impose artificial economies of scarcity on recordings, those corollary property rights become easier to abridge. In fact, a radical reading of the noncircumvention clauses popping up around the world in telecommunications and intellectual property law—which make it illegal for end-users to circumvent whatever copy protection vendors wish to put in their hardware and software, no matter how heinous—suggests that politicians, lawyers, and industry lobbyists have done more to erode the status of music as a form of property than all the pirates in the world.

But music is not only a thing because it is a commodity. In MP3 form it is ready-to-hand because it affords greater storage, greater portability, more opportunities for playback, for sound, for pleasure. In this way, the MP3 is part of a long history of transmission and proliferation that has spanned centuries. And in this way, the condition of digital reproduction is not that different from mechanical reproduction as described by Walter Benjamin. It offers large numbers of people an opportunity to bring the arts nearer to them. The distance between art and everyday life was only ever a social distance. It may be a necessary distance inasmuch as a division of labor is a constitutive feature of complex societies, but the social character of the distance between music and its listener is now an open question for the first time since prestige records began to erode elite and popular disdain for recorded music.

Peter Szendy has put forward the provocative notion of the *right to music*, which for him includes the rights of appropriation, the rights of listeners as well as authors (and authors as listeners, as in the case of the arranger, the DJ, the person who shares a recording). "*Who has the right to music?* This question can also be reformulated thus: What can I make of music? What can I do *with* it? But also: what can I do *to it*, what can I do *to* music? What do I have the right to make *of*, do *with* or *to* music?" Szendy locates this problem at the historical intersection of copyright, artistic, and technological reproduction (first the arrangement of musical works based on their scores, then sound recording), and changing listening practices. If listening changes over time, if it is plastic, then "we must think of the conditions and limits of the *right of the listener*."[79] To Szendy's liberal language of rights, we might also add a republican language of responsibilities: the responsibility to music, and the responsibility to listen, as well as the limits of those rights and responsibilities.

To fully understand the circulation of MP3s, we need to look beyond market economies as our models of circulation to other confluences of infrastructure and circulation. File-sharing has a deliberate and considered dimension to it, but it is not only exchange. It also depends on seriality and networked sociability with people one doesn't know. For decades, one strand of social theory approached this aspect of modern life as the basis of alienation and destabilization—a position most famously stated in Georg Simmel's work. Simmel argued that the money economy, combined with the intensely varied experience of urban life at the turn of the last century, produced a "blasé attitude" whereby people cut off their inner emotional lives simply to get by in city cultures characterized by large and impersonal institutions, the division of labor, and bureaucracies. Simmel's people compensated for this by turning inward and sensation-seeking in other ways, most notably by searching out "disposable" emotional experiences, a point that has been picked up more recently as an explanation for the increasing saturation of contemporary life by media in all their different forms.[80] Certainly one can think of many examples where this holds true in bureaucratic and economic interactions, where both parties basically move through scripts in the transaction in order to negotiate the fact that they do not otherwise know one another. But while this way of life struck Simmel as new and different, a century later there is no necessary reason why we need to maintain his particular classification of what is authentic or alienated. A simple ride on any major city's metro system demonstrates the folly of attempting to have a full, meaningful intersubjective interaction with everyone we encounter on a daily basis. If modern transit requires alienation, a nonalienated condition would be psychologically unbearable. In fact, we could just as easily look to transport to tell a very different story about circulation in the current moment.

Modern life is full of spaces like the metro where we interact with strangers peacefully and cooperatively—or at least there is a tendency to order even if order is never fully achieved. These interactions are not necessarily economic. Just as often, they are about simply moving through life, and some tacit cooperation is necessary in order for the system to work. The same could be said of a vast file-sharing network. We normally think of "non-places" as spaces of physical transition—the inside of a rail car or airplane, a waiting room, a freeway, a line outside a ticket booth, the hallway of an enclosed mall—but the file-sharing network operates in a similar way. A non-place refers to "two complementary but completely distinct realities: spaces formed in relation to certain ends (transport, transit, commerce, leisure), and the relations that individuals have with these spaces."[81] It is, strictly speaking, impossible to do an ethnography of a non-place because the subjects are not consciously relating to one another. "Try to imagine a Durkheimian analysis of a transit lounge at Roissy!" jokes Marc Augé.[82] The same is true of a file-sharing protocol like BitTorrent, because it is precisely not a community. There are, of course, file-sharing communities, and even customs and norms for using the protocol built into software interfaces, but those self-formed and identified communities are exceptions. In these social non-places, one can imagine the overall shape of interactions is perceivable—at least in its outlines—even if it would be impossible to do a traditional ethnography of a BitTorrent swarm.

File-sharing is guided by a whole range of protocols that actually leave users relatively little freedom (the genius of the BitTorrent protocol is that it effectively forces users to share content they acquire through the network). To borrow Augé's language if not exactly his point, we could say that even if every MP3 has its life to live in circulation, that life is not lived in total freedom, not simply because no freedom could ever be lived in society at large, but more precisely because there is an ordered and coded character to traffic.[83] Against the backdrop of these larger flows, the individual acrobatics of users or files disappear—deliberately—into the massive traffic of bits, discs, and sounds in the world. The rules governing their movement do not exactly conform to an economic theory of gifts or commodities, but they do follow the injunctions and affordances of pull-down menus in interfaces, as well as protocols and specifications for passage through the infrastructure. Those protocols in turn govern what is recognizable as a coherent entity—establishing the insides and outsides of files by applying rules for assembly and disassembly. Interfaces' constructions confront users with instructions and indications: "30523 seeders," "60991 leechers," "ratio: 1.95," the available space or "number of songs" in a library, or the gradations of color on the bottom of a CD-R that indicate whether it has been written upon. All of these apparently banal characteristics underscore the collective and ruled character of the traffic in MP3s. Again, the metaphorical extensions of *piracy* mislead. File-sharing is intensely governed, not lawless.

Inasmuch as intellectual habit has sedimented into tradition, the argument about MP3s and the future of music breaks down into a few familiar positions. There are those who believe that we are moving toward an even more restrictive moment in the history of cultural industries, with Celestial Jukeboxes and the creation of massive, for-profit archives of the world's knowledge by Google and its competitors. There are those who believe that the MP3 signals an age where the conduit industries rule the content industries and not the other way around. There are those who believe that through its sonic characteristics and its peculiar mode of exchange, the MP3 format has helped to take the sheen off the products of the culture industry for its users, who will hereafter no longer treat its products as exalted. But all of these positions begin from the assumption that the rules of economy—gift, commodity, or others—govern the circulation of MP3s. Certainly they play a role, but they are clearly not the entire story. The circulation of MP3s is also very much a conditioned interaction among people through heavily regulated protocols. It is a kind of social circulation.

The MP3 story could well herald the future of all media, and not simply because sound files take up less bandwidth than video files. Although less tangible than recordings on LP or CD, MP3s continued to act like commodities even when they weren't exchanged like them. If there is a distinctive "thingness" to the MP3—and to music in its current digital form—it is that it occupies an ambiguous position that is both inside and outside market economies. As was the case with records, the MP3's particular thingness did not emerge sui generis along with the initial publication of the standard. It was only with the emergence of vast online networks and the creation of powerful path dependency through a confluence of licensing agreements, software cracks, and user networks dedicated to music piracy that the present form came into existence over the period of a decade.

Music remains a thing in all the senses I have outlined in this chapter. It retains many of the trappings of the commodity form, but it also retains a purpose beyond commodity exchange so long as it is ready-to-hand. Music also exceeds any definition of a thing as it always has—it contains irreducible dimensions of social practice. If anything, the so-called MP3 revolution has made more music more ready-to-hand for more people than at any other time in human history. If copyright law has served to create an economy of artificial scarcity for recorded music, file-sharing has announced an age of musical abundance.[84]

The age of artificial scarcity privileged a commodity form for music that resided in liberal notions of property, alienable labor, and ownership. It is easy to rail against the ways in which the industry exploited artists and audiences through that system—and we should not diminish the ways in which those forms of exploitation followed existing patterns of power across race, class, gender, and nation. But it is also easy to forget that aspects of that system also gave form and meaning to the rights of users and artists that are now once again under contest. The end of the artificial scarcity of recording is a moment of great potential. Its political outcome is still very much in question, but its political meaning should not be.

Anahid Kassabian has recently described a new type of musical subjectivity, "ubiquitous listening," which has emerged from the extensive availability of music in daily life in many parts of the world: "Those of us living in industrialized settings have developed, from the omnipresence of music in our daily lives, a mode of listening dissociated from specific generic characteristics of the music. In this mode we listen 'alongside' or simultaneous with other activities. It is one vigorous example of the non-linearity of contemporary life."[85] As a phenomenon, ubiquitous MP3s are symbiotic with ubiquitous listening and ubiquitous media. The global circulation of MP3s resonates with ubiquitous listening in its present form, although clearly ubiquitous listening (and the desire for ubiquitous media with which Kassabian connects it) predates MP3s by decades, maybe centuries depending on how it is periodized. The wide availability of MP3s is a testament to an enduring collective passion for recorded music, but Kassabian's thesis suggests that it may be a particular kind of passion. The desire to have music present and ready-to-hand—available for listening but not requiring engagement—lies just behind the gigabytes of music files that accumulate on hard drives and the "complete collections" that reside on CD-Rs sold in city streets.

MP3s are particularly striking because they are so small but circulate on such a massive scale. Whatever the fate of existing recording industries, we need not worry about the future of music as a vital component of human cultures around the world. If the MP3 story demonstrates anything, it is that the desire for music and to be with music is greater than anything the recording industry has been able to imagine or provide. This is the main point for those who seek out music in one form or another. The politics of the MP3 mixes issues we normally keep separate through artificial, analytical divisions between culture and technology, the economy, and everything else. Under the umbrella of culture, we ask after how music should and does enter into people's lives. Under the umbrella of technology, we ask after the means of moving recordings. But the means incorporate their own answers to the questions of how music should and will enter into people's lives. This is readily apparent in existing copyright law, which is why debates about intellectual property are so close at the mention of the word *MP3*. It is readily apparent to activists interested in net neutrality or free public broadband, though these are still often wrongly dismissed as more specialized or technical issues. We cannot assume that a new system of circulation, even one that was to totally destroy the music property system as we know it, is an automatically progressive development for musicians and audiences. The questions we now face go far beyond intellectual property or even the will to cooperate or collaborate among individual users. We now face an age where the very technological basis for culture is available for discussion, for the first time in well over a generation. However slight it may be, there is an opportunity to build a cultural infrastructure based on values other than buying and selling, even if they have some connection with commerce.[86] If we do not begin our discussions of the future of music by considering what music is for—and by extension, what culture is for—the answers will be decided for us.

THE END OF MP3

Measured by the pace of internet fashion and the product cycles of consumer electronics, the MP3 is already old technology. So long as it brings in revenue, its developers look after it. But ask them and they will freely recommend successor standards, like Advanced Audio Coding (AAC, represented in file names like .m4a or .mp4).[1] Formats like AAC give higher definition at lower bitrates and have fewer audible traces. Other alternatives, like .ogg are nonproprietary, open standards. Yet MP3s remain the most common form in which audio circulates today. "Ostensibly 'old' and discredited technical forms may actually represent the majority use pattern, be considered more appropriate to the context of their use, and undergo innovation of their own; they may co-exist indefinitely with their 'newer' rivals, and may even supplant them."[2] Compressed file formats persist even when there are opportunities to move to higher-definition formats. Part of this has to do with the simple mathematics of effort and expense. Successful "standards have significant inertia and are difficult to change."[3] If MP3 files remain compatible with the largest number of audio playback scenarios, then there is a strong institutional inertia behind them. If your MP3 files were acquired for free or next to nothing and you are happy with them, there is little incentive to pay for an upgrade to higher-quality (but larger) audio files. You may have little incentive to pay anyway—record-industry executives often lament that there is now a generation of consumers who are not accustomed to paying for recorded music. If your MP3 files were purchased, the need to repurchase music in a higher-quality format is not necessarily an obstacle, but it could be. If the MP3 files were ripped from users' own CD collections, then they would have to spend a good deal of time re-ripping the same CDs to achieve the improvement in quality. Only those who care a great deal about audio quality are likely to put forth the effort to reconvert or repurchase their collection—or to purchase it for the first time.

For all this inertia, it is still unlikely that MP3s will remain the most common audio format in the world forever. To succeed, the MP3's eventual replacement will require its own combination of technical processes, multi-industrial and transnational regulatory formations, user practices, and opportunities. Whatever it will be, we know that simple technical improvements or new business models are never enough. If they were, MP3s would already be forgotten bits of history, like RealAudio files from the mid-1990s. They once littered the internet but are now very difficult to come by. As a medium or format fades over time, two processes of forgetting occur. First there is the loss of particular documents or examples, as when I no longer remember to whom I lent a record or CD. But over time, a second process occurs, a forgetting of forgetting, where we forget there is even something to search for. Most of history works this way. The past is lost to the present, and bits and pieces are gathered together in archives. In a certain sense, digital files are no more ephemeral than paper, which also easily disintegrates over time if not given proper care. But digital files also require a playback mechanism. In the absence of hardware and software to play back the audio—an interface and someone who can make sense of it—digital audio files simply cease to exist as audio. This issue is especially clear in the preservation of early digital games. Melanie Swalwell describes the problems facing the curation of "Malzek," an arcade game released in 1981. "This game still cannot be played as it was intended: no one has seen it working for 20 years, no one knows the correct colours, collisions are not working, and there is no sound. Anyone can download a copy of this (sort of) mass-produced digital work, but in this case redundancy does not ensure the survival of the game."[4] The same conditions apply to digital audio. Not only will metadata be lost, but many aspects of the files themselves may be too. Archival specialists also expect that preserving digital sound recordings will require more in resources than their analog counterparts. The added expenses come from all the things that come with digital storage: duplication and backup, the need to maintain proper equipment and expertise for "reading" the digital files in whatever format. Any digital rights management scheme will only make it harder for the file to be heard.

There are other issues as well. Consider the case of an unauthorized recording genre like mashups.[5] A mashup is made by combining two or more recordings and beat-matching them so that they work together as a new kind of song. Because mashups are made without any kind of permission or sample clearance, they have an ambiguous status in relation to many regimes of intellectual property. Many of them are anonymous and circulate through file-sharing services that are of themselves of controversial legality in some countries. Although such recordings are available in abundance and for free, I know of no major archival institution that has begun the process of collecting them, despite the fact that many music libraries and sound archives—including national archives—now understand the importance of preserving popular music. Current selection and collection policies could actively prevent archival institutions like the Library of Congress from collecting and cataloguing mashups. Thus, an important popular cultural formation of the current decade will remain largely undocumented. Eventually, many of the currently popular mashups will move out of circulation and perhaps even disappear from most of their owners' collections if they are not cared for and backed up. A few dedicated collectors will no doubt keep meticulously organized collections and perhaps, a few decades hence, one such collection will find its way to a major archival institution that exists in a world of more enlightened intellectual property laws. This person's idiosyncratic collection will thus become an important historical resource for anyone interested in what mashups might tell them about the first decade of the 2000s.

The story may sound speculative, but the condescension of legitimate archival institutions toward popular culture in the early part of the twentieth century meant that they collected nothing for decades. Highly idiosyncratic collections, like the Warshaw Collection at the Smithsonian National Museum of American History in Washington, D.C., have since come to play important roles in current historiography, despite the fact that the collections themselves had no clear logic of acquisition beyond the collectors' tastes. Contemporary reactions to the abundance of digital sound recording replay attitudes that emerged a century ago in the earliest ages of recording. People hail the possibility of keeping, cataloguing, and making available all of the world's music, all of the world's recorded sound, at the same time that they lament the passing of time and the passage of the available material into obscurity. These laments often go hand in hand with practices that actually hasten the disappearance of the music itself.[6]

There may one day exist an MP3-based music archive, but for now the best hope that anyone's collection will be audible in ten, twenty, or fifty years is that for whatever succeeds the MP3, backward-compatible hardware and software will continue to be available. The collections themselves will also disappear over time from neglect, since hard drives fail easily if they are not used for a few years, and everyday users are not nearly as careful as archivists in their backup practices. Compact discs, records, and tapes that haven't been played for twenty years sit quietly on shelves waiting to be played again. What physical medium would hold MP3s for the same period of waiting? Put a hard drive with your entire digital music collection on it next to those CDs, records, and tapes—or burn a stack of writable CDs and DVDs—and wait a few years. Your digital music collection will fade away.

The Persistence of Compression

Whatever happens to MP3 as a format, compression has a long future ahead. Every time a potential expansion of definition, storage, and bandwidth occurs, it is matched by a set of new compression schemes that attempt to wring more efficiency out of the system. Compression also allows media content to proliferate in new directions, where it might not otherwise have gone. Innovation in digital technology tends toward finding new sites, contexts, and uses for file compression, rather than eliminating them. Consider the case of mobile phones. Since the mid-1990s, mobile phones have been able to play music of one sort or another. The first form of mobile phone music was monophonic ringtones, which played melodies and were programmed in simple text languages. They required very little computing power and very little memory (by contemporary standards). By 2000 mobile phone companies were marketing phones that could play polyphonic ringtones, which used a basic synthesizer chip and MIDI (Musical Instrument Digital Interface) language to program more complex music that could include multiple parts playing at once. In the same year, Samsung and Siemens produced the first truly MP3-capable mobile phones, though both had an extremely limited range of function. In fact, MP3s required more bandwidth, storage capacity, and processing power than most phones had at the time. Audio format innovations, like AMR-WB (Adaptive Multi Rate Wideband Extension) allowed for the use of MP3 files as ringtones, and by 2003 MP3 ringtones were commercially available. Increasing storage capacity and computing power also made the use of a phone as an MP3 player more feasible. By the time that Motorola joined with Apple to release an iTunes music phone in 2005, MP3s were a growing part of the mobile phone landscape.[7]

Audio playback is now a standard part of smartphones and even many basic phones can play MP3s. As computing power, storage capability, and bandwidth efficiency increased over the span of about five years, MP3 capability on phones went from a marginal capacity to a more widely available feature. Even though many mobile phones now have the power and space to play uncompressed audio files, the standard is still compressed audio. Compressed formats like MP3s allow users the maximum possible mobility and flexibility for their audio files. This resonates well with the more distributed form that music collections now take. Where people once had a collection that moved around in pieces, now larger subsets or even the whole thing has become mobile. People who want audio files to be compatible across their home stereos, computers, portable MP3 players, and mobile phones still tend to use compressed file formats, or use multiple file formats for their music depending on the intended playback device.

Compression itself will continue to develop in multiple directions with multiple ends. Further increases in computing power and bandwidth may be used for higher definition, but they will also be used for more elaborate compression schemes. There will be no post-compression age. There is a recurring pattern in sound history, where people attempt to design technologies by modeling the characteristics and functioning of the voice and then gradually shift to building sound technologies around models of the ear.[8] But that is only one tendency. A broader view of innovation patterns would reveal a set of crisscrossing, shuttling paths. The shift from vocoders to perceptual coding is only one part of the story in perceptual technics. Another story would highlight the resurgence of research done for adaptive predictive coding in the mobile phone codecs widely in use today, and there are still others. Like normal science, technological research tends to work out the unfolding questions of a paradigm or model until such time as it reaches a question it cannot answer. At that point, researchers either give up the problem or switch perspectives. In the process, they transform the questions they ask and the problems they pose.[9] The domestication of noise is one such case. As noise moved from a problem to be eliminated to something that could be managed, new attitudes toward masking began to make sense, and perceptual coding could emerge. Engineers' move to redo basic research into critical bands with music and speech (instead of sinusoids) is another example: the existing research assumed a circumscribed laboratory context and used perceptual problems from telephony as a privileged formation of media. As the formations changed, the actual results of basic research into hearing changed as well.

There are a wide variety of patterns and shifting models in sound history—from mouth to ear, yes; but also ear to mouth and countless others. For instance, one new approach to digital coding would put a massive dictionary of basic waveforms on each device (called a *big dictionary*). Using a vocoder-like principle, devices would then only need to transmit or store a small stream of numbers that referred to cells and combinations of cells in the big dictionary. A phone would decompile my speech into cells in the big dictionary, transmit a numerical sequence that instructed your phone on how to use its big dictionary, and your phone could then recompile my speech on the other end. This could also theoretically work for music files. Other compression techniques may find additional ways to reduce redundancy in a signal. Besides perceptual coding, the MP3 made use of new mathematical operations that greatly reduced the intensity of computation required, like the Fast Fourier Transform and the Modified Discrete Cosine Transform. Other mathematical innovations may introduce other kinds of efficiencies into signal processing.

As compression has proliferated, so too has perceptual technics. For instance, the ratings company Arbitron's Portable People Meter (PPM) system uses masking effects to allow media systems to talk to one another without people—or animals—hearing. As more and more ambient media suffuse everyday life, it becomes harder for traditional audience measurements to work because they rely on stationary media (like a television in a living room), interviews, or media-use diaries. The PPM is a small, mobile phone-sized device that people carry around with them as they go through their day. At night, they plug it into a portable charger, which also uploads its data to an Arbitron facility for compilation and analysis. The PPM is essentially a listening technology, where machines listen to one another. It picks up codes embedded in the soundtracks of broadcast content via a patented encoding device. This device uses auditory masking to hide digital data from listeners. The *New York Times* reporter Jon Gertner describes a demonstration by Ron Kolessar, Arbitron's top engineer: "Kolessar led me into Arbitron's sound studio to listen to what his team came up with after a decade. He cranked a recording ... through a pair of $20,000 speakers and switched back and forth between the coded and uncoded versions. 'You can't tell the difference,' he said, more in a statement than a question. I agreed."[10]

In the PPM scheme, masking hides the noise of a data signal from listeners, rather than saving space in a channel. That signal is meant to be ubiquitous but unheard by people or animals—ever present in operation and ever absent in perception. (In fact, the masked signal could not be too high in pitch or it would bother dogs and cats.) Beyond standard broadcast, Arbitron has aspirations to track media in bars, airports, health clubs, hotel rooms, or any space through which its wearer moves, so long as the audio portion of the media content has Arbitron's masked signal. Gertner also lists other potential nonbroadcast applications: DVDs, video games, Muzak, MP3 listening (with a special headphone attachment), and supplemental technologies like GPS (to track the wearer's movements) and RFID (to track print documents the wearer reads). For its initial trial in Houston, Texas, Arbitron convinced nearly all radio and television stations in the area to run their material through the device. Today, there are PPM measurements for more than fifty cities.

Masking allows a whole level of interaction among media devices that is not perceptible. In itself, this is not a new thing since the electromagnetic spectrum is absent to the unaided senses. But apart from noisy tooth fillings, human beings have never been able to perceive what happens in the electromagnetic spectrum unaided. Masking allows this level of imperceptible interaction among media to work in a space normally within the horizon of perception. It builds audio media around the measurable absences of perception, or it uses the gaps of perception as channels for machines to talk with one another without distracting (or alerting) humans to the interaction. Perceptual technics designs media for functional complementarity with the senses. Masked media establish sympathy with the senses to move with them, around them, and between them. Like the MP3's perceptual-coding scheme, PPM takes perceptual technics a step further by operating within the horizon of the perceptible without venturing into the space of the perceived. The grooves of perception can be notched into technologies for a whole range of purposes. The gaps and absences of perception can themselves become conditions of possibility for transmission, for connection, for mobility, and for many other actions.

Diluted Media, Distributed Subjects

The sound of a 128k MP3 has settled over time. Already, some artists and avant-garde musicians have experimented with using the artifacts of "bad" or low-bandwidth MP3 encoding, and it is too early to tell whether some genres of music will be especially attuned to the sound of the MP3. (The telltale signs of MP3 encoding are called *artifacts* because they emerge from the encoding process itself, not the "missing frequencies" removed by perceptual coding, as some might imagine.) But many other sonic artifacts have become central aspects of the sounds normally associated with particular technologies or genres of music. Originally created by overloading an analog device such as a speaker, an amplifier, or oversaturating an audio tape, distortion is now a central component of a host of music genres—the overdriven guitar amplifiers of so many rock subgenres, the telephone vocal effects in R&B songs, the boom-bap of hip-hop, the moaning bass of dub, and the distorted vocal overtones in popular music from Bollywood films are but a few examples.[11] Psychoacoustic principles have also been used for similar purposes, and avant-garde musicians have experimented widely with incorporating psychoacoustic theories into their work.[12] Audiences and musicians may grow accustomed to various kinds of harmonic distortion in their music and eventually come to find its absence disconcerting. Whatever standards engineers may bring to the music are often lost in the context of reproduction, either through playback systems that are not themselves high definition or playback in noisy contexts, where even rapt attention will not yield finely grained sonic detail.

As artists, engineers, and others remix and reuse, they transform the original material.[13] The same could be said for the sonic artifacts of the MP3 format. MPEG's engineers originally hoped that it would be possible to encode audio multiple times. In other words, they hoped that users would be able to make an MP3, edit it, and then run the audio through an MP3 encoder a second time. But it turns out that if you send MP3-coded audio into an MP3 coder, the artifacts of the encoding process are hypertrophied. As a result, the official line is that MPEG audio is an "end-use" format, a position reiterated by the people who worked on perceptual coding in the 1980s and early '90s. As JJ Johnston put it in a presentation on the subject, "There is a one-word solution to the problem of multiple encodings: DON'T!"[14] But of course MP3s are a consumer technology. People routinely recode and recirculate MP3s.

Just as a vinyl record melds static, dust, and scratches into the sound of music over time, MP3s bear the mark of their use, even if those marks arrive differently. It is too early to tell whether the artifacts of MP3 encoding will retrospectively become part of the defining sound of a genre or set of genres. But we do know from examples in America, Jamaica, India, and Europe that such things do occasionally happen. The format's heightened artifacts are sometimes audibly present in mashups. Artifacts can also be heard in many live DJ sets as performers dump MP3s into performance-oriented software programs like Ableton Live and Native Instruments Traktor and then twist and manipulate them on stage. The format also breaks down differently than other kinds of analog and digital audio. A DJ's hand slowing down a vinyl record will produce one kind of deep low end as the sound slows down and decomposes; a .wav file decomposes differently, and slowing down an MP3 file sounds audibly different again. In the context of contemporary visual art, Rosalind Krauss has argued that we have entered a "post-medium condition." She uses the phrase *technical support* "as a way of warding off the unwanted positivism of the term 'medium' which, in most readers' minds, refers to the specific material support for a traditional aesthetic genre." Krauss's argument cannot be directly transported to creative audio such as music and sound art, as the stakes are somewhat different. We would do better to consider media as diluted forms, rather than as entirely superseded ones. My emphasis on formats here echoes Krauss's concern that the relevant materiality for creative work exists beneath what is connoted in a word like *medium*. Formats like vinyl or tape or MP3 or .wav audio impart particular characteristics to sound and respond to a musician's or artist's touch in particular ways. A medium as such—like *sound recording*—is much too general to describe this phenomenon.[15]

The exaggerated artifacts that come from particular uses and itineraries of MP3s are synecdochic of the format's larger place in mediatic culture. Aesthetically, compression technologies exist in relief against at least two other formations: the shapes of infrastructures and formats at larger scales; and the more mundane and material world of everyday use at smaller scales. Consider the paperback book, or even the newspaper before it. The point of having portable printed material is that it is relatively lightweight (compared to a hardcover book, anyway), possibly smaller, and therefore easier to take from place to place. While such portability might come at the cost of durability and fine production values, those values are subordinate to others, like reading on the move (as opposed to reading in the hallowed halls of the monastery or the library, or in the formal spaces of the bourgeois home). Reading in transit has long been considered a way of managing alienation and placelessness.[16] We don't ask after the immersiveness of someone's experience of reading on a train or outdoors. The valued experience is precisely the combination of the mediatic and place-specific experience.

Michael Bull's study of iPod users suggests that something similar can happen with portable audio. It interacts with the physical space through which the listener moves. As in so many contemporary contexts of listening, iPod users' experience of music listening is diffuse, distracted, and interconnected. Music becomes a soundtrack to other activities, which can mean many different things. It can be an attempt to overcome alienation and boredom; it can render the strange familiar and the familiar strange; it can be a tool of solipsism and mood management; it can help pass the time or shape the rhythms of movement through town; it can be a form of self-assertion against an indifferent world; or it can be a form of self-dissolution in a world that presses down upon the subject. Portable listening occupies a place in between social registers, just as the listener moves through a series of spaces in between others. For Bull and his interviewees, iPod listening folds into the range of subjective states and politics of everyday life. Music contributes to the warmth or chill of a situation; it aids its listener in various subjective acts, but it is not itself an object of direct contemplation.

Listening is not an end in itself. It does not aspire to a single aesthetic standard or political project.

Without denying the historical specificity of the iPod or its users, we can make this point more generally about music as well. It is part of an affective or emotional economy that "exists at a level of materiality that need not be consciously experienced nor represented as such."[17] The same could be said for anything that can exist in multiple formats, travel across multiple sites or scenes, and be experienced in multiple levels of definition — from early broadsheets and paperbacks to e-books and portable video games. Bull's iPod study doesn't address sound quality, immersion, and definition of audio because his subjects were generally unconcerned with those features. In the few cases where they expressed concern, it was based as much upon what they'd heard about data compression as it was based on their actual impressions of its sound. For instance, one interviewee commented, "Because the music downloaded has some compression scheme on it I hesitate to download classical music. I am afraid I will lose some of the fidelity because of the compression scheme used."[18] There is at least as much ideology here as technological savvy, for if listeners are concerned about the iPod's sound quality, they may wish to start by using better earbuds or headphones, which will make a considerably larger sonic difference than the encoding scheme.

Just as university professors care about good grammar and proper citation, audio engineers and sound professionals care about matters of sound quality, as well they should. But it is one thing to maintain a professional interest in something that is the stock of a trade; it is another to elevate that value to a philosophical or aesthetic concern above all others, or worse, to confuse it with a political value. While a full life may have some transcendent aesthetic and immersive experiences, it is also characterized by the attempt to inject beauty and meaning back into alienating, difficult, or simply mundane parts of everyday life. Like the noise it hides under masking sounds, the MP3 moves in the gaps and disjointed social spaces of the modern subject, rather than attempting to suffuse them. For all its resonance with ideas of mobility, MP3 history also offers us another occasion to reflect on the opposite scenario: stationary media and stationary listeners.

In 1958 Mitch Miller, the director of pop A&R at Columbia Records, addressed a convention of disc jockeys in Kansas City. Noting that the total dollars spent by the record-buying public went up but that the numbers of singles purchased went down, Miller lambasted the emerging Top-40 format and called for more originality in radio programming. Miller's speech is remembered today as a tirade against youth audiences and rock-and-roll, but there is another part worth remembering. Miller noted that the "75 percent of the nation that is over 14 years old" was "buying hi-fi record players in unprecedented numbers, setting them up in the living room, shutting off the radio, and creating their own home made programming departments."[19] Like today's critics of compressed sound, Miller read changes in musical taste and practice from changes in consumer electronics.

In the decades after the Second World War, "hi-fi" audio systems proliferated in American living rooms. They made use of the newly popular LP format for records, which allowed for higher definition and longer musical selections. Advertisers and audiences connected hi-fi listening with experiences of immersion and transcendence. As Keir Keightley has elegantly shown, these ideals were very closely tied to anxieties about gender in the middle-class household. "The conception of home audio as a masculine technology that permits a virtual escape from domestic space is a significant development in the history of sound recording. Before World War II, the phonograph and recorded music were not especially associated with men. By the 1960s, however, home audio sound reproduction equipment had hardened into masculinist technologies *par excellence*." Periodicals of the time trumpeted the hi-fi boom as a rejection of the mass, feminized tastes embodied by television; in these articles, "high fidelity is cast as high, masculine, individualistic art, and television is portrayed as low, feminine, mass entertainment." Magazines and advertisements presented hi-fi as cultivated, sophisticated, and edifying. A hi-fi system was said to promise access to the extremes of experience and an escape from the world of middle-brow taste and the leveling effects of mass culture. It offered opportunities for immersion and transcendence through contemplative listening.[20] The stereo system would eventually replace the hi-fi, but stereo listening operated within the same cultural logics of middle-class domestic space, and male desires to escape it.[21]

Today, that rhetoric still exists, but it operates in a nostalgic voice, rather than as the voice of the future. One can hear echoes of hi-fi listening's challenge to mass culture in published critiques of compressed sound. "Analog's demise means the end of high fidelity. Digitally recorded, produced, and distributed music suffers sonic degradation at every step, meaning the new wave you listened to in 1981 might actually have sounded better than the nü-metal of today," declared a *Wired* magazine article in 2005. "Most iTunes users never fiddle with their encoder settings, which means that they're ripping 1,411-Kbps CD audio into 128-Kbps MP3s—a shadow of a shadow of the music's original self."[22] Every year, dozens of newspaper and magazine articles advance the same argument. Though it is connected to a more sophisticated commodity aesthetics, today's vinyl resurgence has a similarly tinged connection to class distinctions. If the middle-class living room was the site of a pitched battle between record players and televisions in the 1950s, then sometime in the last twenty years, the television finally won. To be fair, the TV didn't exactly win—its screen became a general-purpose display for an ever-growing field of audiovisual formats. A specialized stereo system is totally inappropriate to this mode of consumption, and indeed stereo receivers gave way to more generalized "AV" (audiovisual) receivers in the 1990s. A general purpose set of "multimedia" speakers is designed for the range of media experiences now afforded in middle-class homes—live sports, serial drama, internet videos, movies, video games, and countless other uses. In 2010 the *Wall Street Journal* reported that MP3-player docking systems were one of a very few growth areas in home audio.[23] Transcendent musical experience may or may not be part of the plan. Today, when audiences sit in their living rooms paying rapt attention, they are more likely to lose themselves in movies, games, or sports than the latest record release. Consumer electronics now tend to move across domains that we used to imagine belonged to separate media.

Music playback has folded into other media. In 2010 more record players were sold than CD players. This makes perfect sense, since so many different devices can now play a compact disc but you need a record player to play a record (for now). The closest many people may come to listening to music and doing nothing else is in transit, either on a portable audio player or in a car. Car audio may be the last place outside a specialized hi-fi market where high definition—specifically for music—is still part of the sales pitch.[24] The real transformation here is in the middle-class living room— both its reality and how it is imagined. It is quite likely that the contemplative music listener who sat, listened to the hi-fi or stereo, and reveled in the experience was more common as an ideal than a practice. The ideal was promoted alongside the equipment to perform it in the 1940s, '50s, and '60s. Today it is in decline, and it is certainly getting no help from the advertisements and lifestyle journalism that once sold it as a masculine escape from mass culture (except possibly in cars). Other musical ideals—portability, modularity, malleability, access—have replaced contemplation.

The history of sound may well contain clues to the future of all communication. To this end, the decline of the stereo in the middle-class living room offers a rich allegory. The living room stereo disappeared into the home theater, and where we once had a few privileged spaces for music, we now have a much more distributed social field of music. Other iconic media of the twentieth century—newspapers, magazines, telephones, phonographs, radio, television, cinema, and now even computers, games, and wristwatches—are also melting into multiuse devices. The technological history of communication has always happened at multiple registers, but writers have traditionally rendered communication history legible through media machinery. Today, that is a much more difficult and much less rewarding proposition. Future turning points in communication technology, at least in the near- to midterm, are more likely to occur at the levels of infrastructures that shape the possibilities for media; or at the level of formats, standards, platforms, and protocols that shape their sensual qualities and machinic compatibilities.

Media remain on the scene, but they are diluted. Whatever historical forces were once crystallized within them have weakened some, or migrated elsewhere. In analysis and in politics, we should be careful about which hardware we fetishize, if any. For if the future development of communication technology is moving away from what media history has traditionally called *media*, then the legal and governmental legacies of broadcast policy may well not be the important battlegrounds for shaping tomorrow's communication systems. Future confrontations over democratic media systems and the right to communicate will be held over infrastructures, protocols, formats, portals, and platforms. Legal debates may need to happen on the transnational, rather than the national level.[25]

Plural Ontologies of Audition

Reflection on the diluted state of media today also challenges existing habits of psychoacoustics and information theory. Both fields were built around particular, and particularly dense, conceptions of media. Both fields now must let go of those conceptions. There was a reason that engineers around the world could not simply import tables of critical bands into their perceptual coders. Those tables were developed with a particular set of presuppositions: they were made in laboratory-like conditions; they used sine waves as their basis rather than speech or music; and they drew the majority of their epistemological orientation from a set of problems defined by the phone company. Psychacoustic researchers were noting the limits of this paradigm already in the 1970s, but it has come under additional criticism in recent years. In the *Intelligent Ear*, the psychoacoustician Reinier Plomp identified a host of problematic tendencies in psychoacoustic research: the dominance of sinusoidal tones, the preference for a "microscopic" approach to sound perception, the privileging of psychophysical explanations over less mechanistic and reductionistic alternatives, and a hostility to context (with its concomitant desire to abstract from "dirty" everyday conditions to "clean" laboratory conditions). Against these tendencies, Plomp argues for more naturalistic research methods that study the perception of multiple, complex sounds at once, and the perception of speech as it actually happens in everyday life. In essence, he argues for softening the firm distinction between message and content upon which twentieth-century psychoacoustics was built. He also suggests replacing technical models of perception with human-centered models: "There are no compelling reasons why our brains might function similarly to a technical tool working at a much lower level of sophistication. . . . There are many ways in which machines have been made to mimic human performance, such as moving and writing, but the performance goals are reached by quite different means in mechanical as opposed to biological systems."[26] Psychoacousticians had abstracted up from the telephone and isolated sounds into component elements that could be measured in laboratory settings. Plomp is arguing for a plurality of sounds, listeners, and contexts as the basis for new research into hearing.

If twentieth-century psychoacoustics tended to draw too tight a line around its listening subject and its laboratory environment, twentieth-century information theory drew too wide a circle, moving out from the phone system to suck all of human endeavor into its intellectual orbit. Already in 1956, Claude Shannon published an editorial titled "The Bandwagon" warning against expanding information theory into a theory of everything. Reflecting on the expansion of concepts like *information*, *entropy*, and *redundancy* into fields like biology, psychology, physics, economics, and organization theory, Shannon wrote that information theory

> has perhaps ballooned to an importance beyond its actual accomplishments. . . . Although this wave of popularity is certainly pleasant and exciting for those of us working in the field, it carries at the same time

an element of danger. While we feel that information theory is indeed a valuable tool in providing fundamental insights into the nature of communication problems and will continue to grow in importance, it is certainly no panacea for the communication engineer or, *a fortiori* for anyone else. Seldom do more than a few of nature's secrets give way at one time.[27]

Although Shannon's concepts are doubtless more useful than he originally imagined, they are in no way universal concepts. They emerged from specific problems of communication systems in the second quarter of the twentieth century, and the models in *The Mathematical Theory of Communication* reflect the problems of particular media formations that may no longer obtain, at least with the same intensity. To elevate information theory as a general science of humanity and machinery is to elevate the will of the managers and engineers of communication systems into a kind of divine force.[28]

Plomp's critique of psychoacoustics resonates with a long-standing humanist critique of artificial intelligence. So too might Shannon's warning about information theory. The issue, as traditionally cast by computer scientists and humanists alike, is one of surface and depth: is listening a form of information processing or is it about deeper meaning and subjectivity? Marvin Minsky, a mid-twentieth-century proponent of artificial intelligence, put it thus: "It may be so with *man*, as with *machine*, that, when we understand finally the structure and program, the feeling of mystery (and self-approbation) will weaken."[29] In a way, this attitude simply extends from hundreds of years of scientific inquiry that aims to demystify the mind and body. Here is J. David Bolter in a book published in 1984, from an earlier stage of the contemporary new media boom, warning us not to accept an informational model of the human subject:

> The goal of artificial intelligence is to demonstrate that man is all surface, that there is nothing dark or mysterious in the human condition, nothing that cannot be lit by the even light of operational analysis. . . . Electronic man creates convenient hierarchies of action by dividing tasks into subtasks, routines into subroutines. The end is reached when the "subproblems" become trivial manipulations of data that are clear at a glance. In a way, all complexity is drained from a problem, and mystery and depth vanish, defined out of existence by the programmer's operational cast of thought.[30]

Bolter's position is a classic: machines dehumanize people by objectifying them. So it was in action, so it will be in thought, he worries. And certainly, there are many situations in which we do not wish to be reduced to numbers. At the same time, large-scale social life demands some abstraction simply to exist, and pretty much every form of inquiry today must objectify and abstract what it studies to some extent simply to understand it. The challenge is to assess our abstractions and models in more robust normative terms that account for both humanity and the massive scales on which humans operate collectively. Bolter's words are especially powerful when set against Shannon's and Plomp's. When we reduce some aspect of humanity to a single model, we elevate one set of problems and interests above all others. We take industrial and technical formations like the AT&T monopoly out of context and elevate them to the transcendent ground for discovering human truth.[31]

We could begin by simply naming those supposedly transcendent grounds for what they are. Perceptual technics increasingly shapes the sounds, sights, tastes, smells, and surfaces we experience. Mediality increasingly subtends everything we claim to know about a sensing subject in the state of nature. But mediality, that web of reference among communication technologies, itself changes over time. Psychoacoustics and information theory privilege particular configurations and goals of communication technology. Today, the accelerating product cycles of consumer electronics, competing standards unregulated by any governmental authority, and evolving infrastructures dilute the iconic media of the twentieth century. If all knowledge of hearing comes from the interaction between ears and sound media, and if the sound media are diluted and reconfigured, then it makes perfect sense that a plurality of contexts emerges in place of a single, golden ground for theories of hearing and for models of interaction between subjects and technologies. Both Plomp and Shannon gesture toward multiplicity as the starting point for even the most basic scientific inquiry and model-building.

In her book *For More Than One Voice,* the philosopher Adriana Cavarero advocates a "vocal ontology of uniqueness" as an alternative to universalism. Criticizing the philosophical tradition for ignoring "uniqueness as such," she writes that in the universalist tradition "speech is separated from speakers and finds its home in thought . . . a mental signified of which speech itself, in its sonorous materiality, would be the expression—its acoustic, audible sign. The voice thus gets thematized as the voice in general, a sonorous emission that neglects the vocal uniqueness of the one who emits it."[32] As it is for speech, so it is for hearing; as it has been for psychoacousticians and information theorists, so has it been for humanists and philosophers. We need not pursue an auditory ontology of uniqueness—plurality will do. The grounds that oriented hearing research and sound technology in the twentieth century have shifted. Their legacies remain with us, but going forward, our challenge is to produce a more humane sound culture. Psychoacoustics and information theory contain within them pregiven answers to the normative questions that animate communication—why communicate, to what end—and they devise how best to go about it from their answers to those basic queries. An auditory ontology of plurality—or for that matter an auditory pragmatics of plurality—reopens those normative questions for technology in the human sciences and in the arts. For what is the purpose of critique if we don't also imagine alternative worlds?

Sonic thought across science, technology, medicine, the arts, and culture will need to develop a set of models that attend to a variety of situations and values. Lilian Radovac's study of noise ordinances in New York reveals the political fallout that occurs when a sonic monoculture is imposed via urban planning and zoning ordinances. Bob Ostertag complains that computer music embodies the narrow avant-garde values that obtained at the moment it emerged: algorithmic composition and timbral innovation. He argues for greater openness and flexibility to lead the field out of its aesthetic cul-de-sac. As Mara Mills has shown, cochlear implants were developed around the values of telephonic communication and a medical model of disability: they embody "the privileging of speech over music, direct speech over telecommunication, non-tonal over tonal languages, and black-boxed over user-customizable technology." Mills suggests that they should be developed from a different set of values that would take sonic diversity and the needs of Deaf and hard-of-hearing culture as their basis.[33] Like all sensation, hearing is at its core an engagement with alterity—that which lies outside the subject. If science, engineering, design, policy, and art managed to contain the range of that alterity for most of the twentieth century, they will need to develop better ways of handling plurality and multiplicity in this century.

Between the psychoacoustic effects on which it depends and the distraction it expects, the lowly MP3 may tell us more about the aesthetic experience of media and sound than many works of high art. The history of the MP3 points to a crucial circuit in the relations between communication technology and the human subject. If it is true that the MP3 is structured by scientific knowledge of hearing and the dominant forces of the sound culture from which it emerged, it is also true that all prior scientific knowledge of hearing is touched by the cultural field in which ears and sound technologies interact. The MP3 collapses the distinction between media of transmission and recording, a distinction that still structures much contemporary thought about communication. Its history pushes us to move beyond assessing the so-called impact of digital technologies, because from the perspective of sound history, divisions between analog and digital were never that clear. It points to the changing nature of possession and collection, and music's movement from a restricted set of specialized practices, technologies, and situations to a ubiquitous phenomenon.

The MP3 exists as part of a century-long project of perceptual technics that shows no sign of abating. By following the MP3 across technical and cultural domains, we are led to name and question a small group of sonic monocultures from which it developed—just as issues raised by perceptual coding unsettled and eroded the worlds of information theorists, psychoacousticians, communication engineers, media industry professionals, musicians, artists, and everyday listeners. Our challenge is to continue and extend this project of refraction. Within, above, and beneath the surfaces of sonic monocultures, there are many subjects, many epistemologies, many technologies, many histories, and many economies.

NOTES

FORMAT THEORY

1. The sentence was originally drafted in 2005 and revised in 2011. During that time, overall online file-sharing traffic appears to have stabilized, even as its characteristics have changed. Although overall online sales of recordings have increased, legal action appears to have had no long-term effects on file-sharing traffic. For instance, a recent court decision to shut down the popular LimeWire service led to a temporary 18 percent dip in Gnutella traffic until users discovered other sites. The vast majority of online file-sharing is difficult to measure: one-click sites like RapidShare (which may well one day host file-shared versions of this very text) are impossible to track, and while it is possible to track BitTorrent users, as yet nobody has come up with a robust measure of its traffic in files (nor has anyone discerned the proportion of copyrighted audio files to television, film, pornography—and legally shared material). Author interviews with Joel Fleischer of BigChampagne

.com. See also Boudreau, "Illegal File Sharing Showing No Letup"; Schulze and Mochalski, "Ipoque Internet Study 2008/2009"; Arrango, "Judge Tells Limewire, the File-Trading Service, to Disable Its Software." In 2001, Andrew Leyshon traced a single MP3 file on Audiogalaxy and found that it was simultaneously available to 15,560 users in at least sixty-four countries. Leyshon, "Scary Monsters?" 539–40.

2. Jones, "Music That Moves"; McLeod, *Freedom of Expression*; McLeod, "MP3s Are Killing Home Taping"; Taylor, *Strange Sounds*, 17–18. For a textbook statement of the "circuit of culture" model, see du Gay, Hall, Negus, Mackay, and Janes, *Doing Cultural Studies*.

3. Here I tip my hat to Rudolf Arnheim, who made the same point about radio in 1936. He begins by noting radio's feats of transmission but quickly turns to its cultural significance: radio "has constituted a new experience for the artist, his audience and the theoretician." Arnheim, *Radio*, 14, 15. Subsequent radio histories have confirmed his position. See, e.g., Hilmes, *Radio Voices*; Hilmes and Loviglio, *The Radio Reader*; Douglas, *Inventing American Broadcasting*; Douglas, *Listening In*; Newman, *Radio Active*.

4. For instance, a three-minute stereo CD file takes up about 30 megabytes of disk space; a three-minute MP3 of average quality takes up 3–4 megabytes of disk space.

5. For a technical description of perceptual coding, see Hacker, *MP3*; Pohlmann, *Principles of Digital Audio*.

6. The implied user is, of course, a widely accepted general feature of technologies and part of a well-established practice of their critical analysis. Akrich, "The De-Scription of Technical Objects."

7. When they consider its aesthetic characteristics, many existing histories of digital culture tend to emphasize its visual and visible dimensions. See, e.g., Mirzoeff, "Introduction to Part Three"; Manovich, *The Language of New Media*; Friedberg, *The Virtual Window*; Rodowick, *The Virtual Life of Film*. Manovich and Rodowick in particular overemphasize the historical significance of cinema. While there are obvious connections between cinematic modes of representation and new media forms, cinema is only one character in a much bigger story.

8. Dyson continues: Sound technologies have "laid the groundwork for notions of immersion and embodiment: the primary figures that characterize new media." Dyson, *Sounding New Media*, 3. Although I focus on telephony here, Dyson rightly looks across audio technologies for the seeds of contemporary new media. Film sound, artificial reverberation, and radio created physically immersive and semi-immersive spaces of virtual and mixed reality long before it was possible to do so with visual or haptic media. The distraction and multitasking so often noted in contemporary discussions of social media probably have one of their origins in everyday radio listening practices and

telephony. Sterne, "What's Digital in Digital Music?"; Russo, *Points on the Dial*; Hillis, *Online a Lot of the Time*; Verma, *Theater of the Mind*.

9. Mills, "On Disability and Cybernetics."
10. The historical importance of the telephone system (alongside military and academic institutions) to information theory and computer culture is well established among writers in science and technology studies. See, e.g., Edwards, *The Closed World*; Nebeker, *Signal Processing*.
11. Radio and telephony traded many technologies and had some interdependence over the course of the twentieth century, but so far as I can tell, radio research was not nearly as crucial to the development of theories of hearing or theories of information that would lead to the digital technologies that now surround us. Instead, radio defined broadcasting, advertising, and common national cultures, and offered a new kind of experience of social time for listeners that would proliferate into a wide range of mass media. To my knowledge, no writing exists yet on the historical links between radio and contemporary social media, but the morphological similarity—as ambiance, as punctuation, as timeliness, as collectivity—begs for a more thorough historical investigation. Loviglio, *Radio's Intimate Public*; Goodman, "Distracted Listening"; Russo, *Points on the Dial*.
12. Denegri-Knott and Tadajewski, "The Emergence of MP3 Technology." Denegri-Knott and Tadajewski make a similar point to mine regarding telephone research, though they erroneously cite PCM audio as a compression technology and cite quantization as the primary innovation of Bell. (As we will see, the quantification of sound and hearing happened earlier.)
13. Following Michel Chion, I use the term *definition* to describe the available bandwidth or storage capacity of a medium in terms of how much of its content can be presented to an end-user at any given moment. See Chion, *Audio-Vision*, 98–99.
14. Most of the journalistic discussions of MP3 sound take some version of the position that greater technological potential and less "perfect" files are something of a paradox. This is also a question I heard frequently after delivering talks on parts of the book. For articulations of this position, see for instance, Atkinson, "MP3 and the Marginalization of High End Audio"; Bemis, "The Digital Devolution"; Milner, *Perfecting Sound Forever*.
15. Fukuyama, "All Hail . . . Analog?" The academic work on the sound of MP3s is considerably more limited. John Shiga spends a few pages discussing the political dimensions of MP3 technology aside from its distribution; Aden Evens argues that diminished sound quality blurs the line between musical form and commodity. Evens, *Sound Ideas*; Shiga, "Translations." More common is a brief description of the compression technology behind the MP3 on the way to discussion of it as a technology of distribution, as in Katz, *Capturing Sound*, 160–61.

16. Chion, *Audio-Vision*, 98–99. To argue that the primary purpose of media is to reproduce experience as it unfolds in life is to miss the point on two grounds. This position proposes that consciousness prior to technology is somehow "unmediated" (as if language and culture did not exist) and it reduces all large-scale social activity to some kind of an aberration of face-to-face communication. I outline these positions in *The Audible Past* (especially in chapters 5 and 6, where I question the idea of fidelity), and I will only briefly cover them here. The critique of realist aesthetics is also a well-established line of argument in cinema studies and alongside Chion, James Lastra's *Sound Technology and American Cinema* advances an elegant analysis of the problem with judging sound in terms of realism. Lastra, *Sound Technology and American Cinema*. Rodowick also has an interesting discussion of the aesthetics of "photographic credibility and perceptual realism" in the digital register in *The Virtual Life of Film*, 99–107.

17. Mowitt, "The Sound of Music in the Era of Its Electronic Reproducibility"; Rose, *Black Noise*; Toop, "Replicant"; Waksman, *Instruments of Desire*; Walser, *Running with the Devil*. Manuel, *Cassette Culture*; Poss, "Distortion Is Truth"; McLuhan, *Understanding Media*; Goffman, *Frame Analysis*; Hall, *Beyond Culture*; Marvin, *When Old Technologies Were New*.

18. Ethnographies of music listening in everyday contexts illustrate this to be the case for their subjects, and new work in the psychology of music has also moved to a model that privileges distraction over attention. Crafts, Cavacci, Keil, and the Music in Daily Life Project, *My Music*; Wolpert, "Attention to Key in a Nondirected Music Listening Task"; Bull, *Sounding Out the City*; Denora, *Music in Everyday Life*; Hargreaves and North, *The Social Psychology of Music*; Konecni, "Social Interaction and Musical Preference"; Sloboda, O'Neill, and Ivaldi, "Functions of Music in Everyday Life." How much this has changed from earlier periods, however, remains to be seen. While there is much anecdotal evidence for people sitting in their living rooms and listening to records or radio, there aren't good data to determine whether this was ever the dominant form of listening, since distracted listening is already a category in early radio research. See, e.g., Adorno, "A Social Critique of Radio Music"; Cantril and Allport, *The Psychology of Radio*.

19. Katz raises the issue in the context of a discussion of a different kind of compression—compression of dynamic range. But as the quote shows, he sees this as eminently related to the kinds of data compression in an MP3. For further discussion, see Katz, *Mastering Audio*, 187–88, quote at 88; Sterne, "The Death and Life of Digital Audio."

20. This is not to say I simply want to replace a grand narrative of ever-increasing fidelity with a grand narrative of ever-increasing compression. I am merely proposing compression as one possible basis for inquiry into the history of communication technology—in the same sense that *representation* has

served. We need to describe and debate long-term histories, as most of the currently available long-term histories of communication (whether directly cited or merely implied or assumed in contemporary work) have not taken on board the insights of recent decades, especially those drawn from work on globalization as well as postcolonial and poststructuralist thought. Such new histories would not need to function as teleology, nor need to approach universality, but ambition and breadth certainly seem appropriate in our moment.

21. See, e.g., Hillis, *Digital Sensations*; Grau, *Virtual Art*.
22. Pardon the pun, but this is a point I hope to develop elsewhere. There is also time-space compression as a feature of modernity, although *that* compression is more phenomenological than technical (though the two are related).
23. Kirschenbaum, "Extreme Inscription." While they are invisible, digital data do have tactility: "The drive resides within the machine's external case and is further isolated inside of a specially sealed chamber to keep out dust, hair, and other contaminants. When a drive is opened for repair or data recovery the work is done in a clean room, similar to those used to print microprocessors" (94). See also Kirschenbaum, *Mechanisms*.
24. Even as a legal and policy term, *new media* is old: Sandra Braman traces the phrase to a United States Supreme Court case in 1948 that discussed a bullhorn. Novelty is a routine occurrence in media history. Braman, *Change of State*, 2; Pingree and Gitelman, "Introduction"; Marvin, *When Old Technologies Were New*; Stabile, "Introduction." Taylor, *Strange Sounds*, 7.
25. See, for instance, The Representation and Rendering Project, "Survey and Assessment of Sources of Information on File Formats and Software Documentation."
26. Braman and Lynch, "Advantage ISP"; Lessig, *Code Version 2.0*.
27. A report by the Radio Manufacturers' Association Television Committee in 1936 first suggested the 4:3 aspect ratio, which was then set in federal policy by the National Television Standards Committee in 1941. See Boddy, *Fifties*, 34–35. Douglas Gomery discusses Hollywood experiments with widescreen in Gomery, *Shared Pleasures*, 238–46. See pages 259–60 for a discussion of editing films for TV.
28. Thanks to Charles Acland for this point regarding film study.
29. In English, the term is a recent coinage, since *medial* refers to a mathematical phenomenon, but one can find that it is already in common use among scholars, artists, and media professionals to refer in some fashion to qualities of media or medium-like qualities (though definitions appear to vary widely). I prefer *mediality* to two other, similar words, *mediation* and *mediatization*. *Mediation* implies historical sequence, where media come to the thing after the fact. "Mediatization" is a term used by Jean Baudrillard in an almost entirely pejorative sense, and like *mediation*, it implies the falseness of media in con-

trast to the reality of bare life. To be fair, Philip Auslander uses "mediatization" as a more ambiguous term in *Liveness*, 10–60. For its original usage, see Baudrillard, *For a Critique of the Political Economy of the Sign*, 175–76.
30. McLuhan, *Understanding Media*, 8. The Bolter and Grusin quote comes from Bolter and Grusin, *Remediation*, 45.
31. Quoted in Williams, *Marxism and Literature*, 98. The original quote is from a translation in *Working Papers in Cultural Studies*, Vol 2. In this section, Williams argues that "all active relations between different kinds of being and consciousness are inevitably mediated, and this process is not a separable agency—a 'medium'—but intrinsic to the properties of the related kinds." Lawrence Grossberg has more recently expanded this line of thought to argue for an understanding of mediation as "the movement of events or bodies from one set of relations to another as they are constantly becoming something other than what they are. [Mediation] is the space between the virtual and the actual." Grossberg, *Cultural Studies in the Future Tense*, 191.
32. The "literariness" analogy is taken from Winthrop-Young and Wutz, "Translator's Preface" in Kittler, *Gramophone-Film-Typewriter*, xiv. I don't, however, follow their reasoning that identifying the phenomenon of mediality requires that we "not resort to the usual suspects—history, sociology, philosophy, anthropology, and literary and cultural studies." I would rather have us bring out these suspects—and several others—to analyze different dimensions of mediality.
33. David Wellerby, "Foreword" in Kittler, *Discourse Networks, 1800/1900*, xiii–viv. In a literary context, John Johnston has defined mediality as "the technological conditions that make specific media possible within a delimited historical epoch and therefore to the cultural and communicational setting within which literature can appear and assume a specific shape and function." Johnston, *Information Multiplicity*, 268 n. 9.
34. Koschorke, *Körperströme und Schriftverkehr*, 11. Many thanks to Andrew Piper for translating the relevant quote for me: "A media theory that seeks to make such self-revolutionary processes explainable as completely as possible must develop a methodology to understand the *interdependence of technological mediality and semiosis*, the narrow overlap of the 'form' and 'content' of such signifying events" (emphasis in the original).
35. Media are "socially realized structures of communication, where structures include both technological forms and their associated protocols, and where communication is a cultural practice, a ritualized collocation of different people on the same mental map, sharing or engaged with popular ontologies of representation." Gitelman, *Always Already New*, 7.
36. My claim here rests not on the state of media theory in 1974, but rather on the state of media in 1974: the very idea of a medium that exists independently of hardware is easier to imagine today because of (for lack of a better

way to put it) the everyday hardware flexibility of media practices, even though such flexibility has existed for well over a century.

37. On articulation and technology, see Slack and Wise, *Culture + Technology*,
38. As of January 2010, Apple announced that it has sold more than 250 million iPods since introducing its MP3 player. Add in competing MP3 players and the proportion of the world's 4.6 billion mobile phones and smartphones that play MP3 files, and the number of portable MP3-capable devices is probably in the billions. International Telecommunications Union press release, "ITU Sees 5 Billion Mobile Subscription Globally in 2010" (15 February 2010), http://www.itu.int. Connie Guglielmo, "Apple's Jobs Unveils $499 Tablet Device Named iPad" (27 January 2010), Bloomberg News, http://www.bloomberg.com.
39. Douglas, *Listening In*, 226–27. See also Schiffer, *The Portable Radio in American Life*.
40. Douglas, *Listening In*, 221. See also Rothenbuhler and McCourt, "Radio Redefines Itself, 1947–1962," 378–79.
41. Hosokawa, "The Walkman Effect," 165.
42. "That with which our everyday dealings proximally dwell is not the tools themselves. On the contrary, that with which we concern ourselves primarily is the work—that which is to be produced at the time. . . . The work bears with it that referential totality within which the equipment is encountered." Heidegger, *Being and Time*, 99.
43. Gitelman, *Always Already New*, 7–8. In part, her point reads as a critique of certain strands of German media theory (and perhaps residual strands of the so-called Toronto school) that aim to study the "ontology" of particular media.
44. The idea that the waveform as inscribed on a record corresponds to sound's status in nature is usually advanced to critique the way in which the CD standard codes audio. See Rothenbuhler and Peters, "Defining Phonography"; Evens, *Sound Ideas*.
45. Adorno, "Opera and the Long Playing Record," 65.
46. Ibid., 63.
47. Whether the fate of opera really did shift after 1969, when Adorno wrote, remains an open question.
48. Immink, "The Compact Disc Story," 460. The more common story centers on Ohga (who was not yet president of the company when Sony and Philips were working out the standard). For an example of the Sony story, see, e.g., Schiesel, "Ideas Unlimited, Built to Order."
49. Immink, "The Compact Disc Story," 458–59.
50. Ibid., 460. For Philips's version with the Beethoven story intact, see "Philips Celebrates 25th Anniversary of the Compact Disc" (16 August 2007), http://www.newscenter.philips.com. The variable length of classical performances

on CD also suggests the story doesn't work—Beethoven's Ninth is sometimes longer or shorter than seventy-four minutes.

51. Gould, *The Glenn Gould Reader*; Hecker, "Glenn Gould, the Vanishing Performer and the Ambivalence of the Studio." See also Symes, *Setting the Record Straight*.
52. Gomery, *The Coming of Sound*, 79.
53. Author interview with Harald Popp. See also Pohlmann, *Principles of Digital Audio*, 592–93. Pohlmann points out that 64 kbps per channel (for two channels totaling 128 kbps) was standard for what was called "basic rate ISDN." The first common carrier digital transmission system, Bell's T1 system (put into service in 1962), had a bandwidth of 64 kbps. This was the first commercial communication system to use Pulse Code Modulation (or PCM), a digital system where an analog signal is sampled regularly and then translated into a digital code that can later be used to reconstruct the analog signal. This technique was first developed by the phone company and then was later applied to digital audio in general, most notably the compact disc. O'Neill, *A History of Engineering and Science in the Bell System*, 538–65. Nebeker, *Signal Processing*, 58–60.
54. Daniel Levitin, personal conversation with author; Ahmed, "Young Music Fans Deaf to iPod's Limitations." On familiarity, see Kirk, "Learning, a Major Factor Influencing Preferences for High-Fidelity Reproducing Systems."
55. On radio and television design see, e.g., Barnett, "Furniture Music"; Hartley, *Tele-Ology*; Schwartz Shapiro, "Modernism for the Masses."
56. Sobchack, "Nostalgia for a Digital Object," 317.
57. Bourdieu, "Men and Machines," 305.
58. "Infrastructure studies," as it is sometimes now called, is an emerging area of interest among scholars working in various corners of media studies. It has also in recent years become the subject of a good deal of political work around the internet given current debates surrounding "net neutrality"— the idea of the end-to-end model of the internet. For recent work of interest, see, Sandvig, "The Structural Problems of the Internet for Cultural Policy"; Parks, *Cultures in Orbit*; Parks, "Air Raids"; Noam, *Interconnecting the Network of Networks*; Jackson, Edwards, Bowker, and Knobel, "Understanding Infrastructure"; Barry, *Political Machines*; Star, "Infrastructure and Ethnographic Practice"; Bowker and Star, *Sorting Things Out*; Mattelart, *The Invention of Communication*; Mattelart, *Networking the World*. Of course, infrastructure is not a new area of interest in the study of technology—since it is a theme in classic works by Lewis Mumford, Harold Innis, James Carey, Thomas Hughes, James Beniger, and many others—it is only the interdisciplinary self-consciousness that is new.
59. Elmer, "The Vertical (Layered) Net," 163–64.
60. Bowker and Star, *Sorting Things Out*, 33–34. Their definition of infrastructure

also points to its textured dimensions: embeddedness in other structures; transparency of operation; reach or scope that spans multiple sites and practices; taken-for-grantedness; links with conventions of practice; and embodiment of standards. It is also built on an installed base; visible upon breakdown; and fixed in modular increments, not all at once or globally (35).

61. The internet is often contrasted with the phone system, which was built around the idea of "dumb" terminals—the relatively limited and simple phone at each end of the line—and "smart" switches that managed the flow of traffic from place to place. On the end-to-end ideal and "smart" and "dumb" components as they relate to the politics of media systems, see Sandvig, "Shaping Infrastructure and Innovation on the Internet." As Sandvig notes (and his argument is actually a variation of this), the end-to-end ideal has come under a lot of criticism recently: the actual workings of the network don't follow this pattern exactly, and of course there are those who would like to change it. Business and totalitarian governments want to route users to and away from certain sites and content, privileging some traffic over others. Reformers want the system to better facilitate democratic and participatory communication.

62. At one level, this is simply to extract out a processual turn in media studies, especially as video game theorists attempt at once to construct a historical context for understanding video games and at the same time to differentiate between video games and other forms of media. See, e.g., Bogost, *Unit Operations*; Frasca, "Simulation versus Narrative"; Galloway, *Gaming*; Wark, *Gamer Theory*. These kinds of approaches give us routes back into longstanding questions about the temporal and perceptual dimensions of mediatic experience, as in Deleuze, *Cinema 1*; Hansen, *New Philosophy for a New Media*; Turino, "Signs of Imagination, Identity and Experience"; Williams, *Television*.

63. The MP3 isn't historically exceptional; I am simply arguing that the format's history allows us to pay attention to these three different registers in the history of sound at once. Braudel, "History and Social Science."

64. Though the term *psycho-acoustic* was first applied in 1885 to the part of a dog's brain responsible for sound perception, its use in studies of humans slowly gained traction in the first half of the twentieth century, with a key definitional moment being the founding of Harvard's Psycho-Acoustic Laboratory in 1940 to conduct speech, hearing, and communication research during the Second World War. *Oxford English Dictionary*, s.v. "Psychoacoustic, Psychoacoustics." The OED's dates are a bit off. They chart the first reference to Harvard's laboratory to 1946, though according to Harvard University's archives the laboratory had been in existence since 1940 (Harvard University Archives, Records of the Psycho-Acoustic Laboratory, UAV 713.9, accessed online at http://oasis.harvard.edu:1080/oasis/deliver/deepLink?_collection=oasis&uniqueId=hua08005). This, in turn suggests that the term was in com-

mon use among researchers interested in auditory perception from sometime in the 1930s.

65. "In short, a proposition must fulfill some onerous and complex conditions before it can be admitted within a discipline; before it can be pronounced true or false it must be, as Monsieur Canguilhem might say, 'within the true.'" Foucault, *The Archaeology of Knowledge and the Discourse on Language*, 224. Throughout this book, I try to consider psychoacoustics as at once a science that actually works in many cases, and as a cultural product that promotes a particular worldview and set of social relations. As we will see, the field has not been entirely successful in this endeavor, but its habits and dispositions shaped discussions of sound in the engineering world for the better part of the twentieth century.

66. Writers interested in the aesthetic dimensions of music, meanwhile, still often refuse the distinction between percepton and meaning and are therefore more likely to use terms and concepts derived from psychoanalytic theory. This is true for classic as well as contemporary studies. See, e.g., Chanan, *Musica Practica*; Langer, *Philosophy in a New Key*; Meyer, *Emotion and Meaning in Music*; Silverman, *The Acoustic Mirror*; Schwartz, *Listening Subjects*. If each approach to psychological questions has its own model of the human subject, it could safely be said that the psychoanalytic subject and the psychoacoustic subject are two very different people. If we want to understand the MP3, we must acquaint ourselves with the subject of psychoacoustics—both the human subject and the academic subject—a little better. And to unravel the implications of the psychoacoustics built into the MP3, we must understand some of the basics of the science.

67. Shannon and Weaver, *The Mathematical Theory of Communication*.

68. Peters, *Speaking into the Air*, 23.

69. Johnston, "Transform Coding of Audio Signals Using Perceptual Noise Criteria," 314–15. "Rather than base the coding algorithm on source models, the new coder, called the Perceptual Transform Coder, or PXFM for short, uses a human auditory model to derive a short-term spectral masking curve that is directly implemented in a transform coder." The term had an earlier conceptual life to describe the information-processing routines of the senses and some usages in the 1980s seem to suggest a hybrid version of the concept. See, e.g., Carterette and Friedman, *Handbook of Perception*.

70. The term *masking* was borrowed from visual research. Schubert, "History of Research on Hearing," 63.

71. Digital rights management schemes undo this cross-compatibility.

72. MacKenzie, "Introduction," 6. Such "births" only matter retroactively, after an invention or standard comes to market dominance and becomes a recognizable technology through subsequent use, innovation, and negotiation. See, e.g., Bijker, *Of Bicycles, Bakelites, and Bulbs*, 84–88. Although Bijker's de-

scription of "closure" and "stabilization" has come under criticism in recent years, the point is well taken in that there is a social threshold where a technology takes on a broadly recognized form, and such a moment occurs sometime *after* the so-called invention of the technology.

73. Fuller, *Media Ecologies*, 95–96.
74. Despite my inattention to other paths, I do not mean to suggest that MP3 was a layer of destiny, or that the people involved in its invention were more visionary or original than their competitors. This book is a history of perceptual coding and perceptual technics, and not a history of scientific or technological controversies. Methodologically, my approach is therefore closer to the cultural study of technology than the social construction of technology, though I certainly also build on insights from the latter. Pinch and Bijker, "The Social Construction of Facts and Artefacts"; Bijker, *Of Bicycles, Bakelites, and Bulbs*; Slack and Wise, *Culture + Technology*; Winner, "Social Constructivism."
75. The International Organization for Standardization is a network of national standards institutes from 148 countries that collaborate with international organizations, governments, industry, business, and consumer representatives. The International Electrotechnical Commission focuses on standards for electronic and magnetic devices, and is now affiliated with the World Trade Organization.
76. Mitchell, Pennebaker, Fogg, and LeGall, MPEG *Video Compression Standard*, 1–4; Pohlmann, *Principles of Digital Audio*, 350. To my knowledge, this book is the first attempt at something like a full academic history of the MP3. The only thorough journalistic history currently available is Bruce Haring's *Beyond the Charts*. Taking his cues from the intellectual property debates, Haring presents the MP3 as a part of a longer story about digital audio, online distribution, and the music industry.
77. Sumner and Gooday, "Introduction," 3–4.
78. As Geoff Bowker and Susan Leigh Starr have noted, this appears to be an issue across fields, and Paul David has made the same point regarding the role of standards in economics; Bruno Latour has estimated that more money and energy is spent making standards than conducting "pure" research. See Bowker and Star, *Sorting Things Out*, 14; David and Rothwell, "Standardization, Diversity and Learning"; Latour, *Science in Action*. For the policy-as-government argument, see, e.g., McChesney, *The Problem of the Media*. McChesney is right that greater citizen participation in governmental policy-making around media is essential; but government alone is not enough and should not be mistaken as the only place where crucial policy decisions are made. Standards do come up in literature about communication infrastructure and policy, so it is a short step to targeting standards as themselves political. See, e.g., Lentz, "Media Infrastructure Policy and Media Activism."

79. Sumner and Gooday, "Introduction," 3.
80. Katz, *Capturing Sound*, 163; Gillespie, *Wired Shut*, 25–26.
81. There is no universally agreed-upon definition of an "open standard," for instance, whether a standard qualifies as "open" if someone owns a patent on it and charges for the rights. Joel West argues that most actually existing standards operate somewhere in between "open" and "closed." See West, "The Economic Realities of Open Standards"; Krechmer, "Open Standards Requirements." Thomson's revenue schemes and industrial arrangements are amply detailed at their website, http://www.MP3licensing.com. Revenue figures come from author interview with Karlheinz Brandenburg; UK Foreign and Commonwealth Office, "The Fraunhofer Society," Factsheet no. 5 (July 2007), http://ukingermany.fco.gov.uk.
82. There is an ever-growing body of work in this area, increasingly merging with scholarship on mobile telephony, intellectual property, or other larger concerns. See, e.g., Bettig, "The Enclosure of Cyberspace"; Vaidhyanathan, *Copyrights and Copywrongs*; Vaidhyanathan, *The Anarchist in the Library*; Jones, "Music and the Internet"; Lessig, *Code Version 2.0*; Ayers, *Cybersounds*; Bull, *Sound Moves*; Burkart, *Music and Cyberliberties*; Sinnreich, *Mashed Up*; McLeod and Kuenzli, *Cutting across Media*; McLeod and DiCola, *Creative License*; Karagangis, *Media Piracy in Emerging Economies*; Magaudda, "When Materiality Bites Back"; Shiga, "Translations"; Gunn and Hall, "Stick It in Your Ear"
83. Grossberg, *Cultural Studies in the Future Tense*, 20.

1. PERPETUAL TECHNICS

1. Boring, "Perspective," in Stevens and Davis, *Hearing*, vi–vii. In addition to Boring being known for his own psychological research, his historical work helped to consolidate the field. See Boring, *A History of Experimental Psychology*. (A few social psychologists of my acquaintance have confirmed that Boring's text has a history of being understood within the field as, well, boring.)
2. Peters, "Helmholtz, Edison and Sound History," 179.
3. Davis and Merzbach, *Early Auditory Studies*, 21.
4. Ibid., 32.
5. Ibid., 31–32.
6. Fletcher and Wegel, "The Frequency-Sensitivity of Normal Ears," 554–56.
7. Békésy and Rosenblith, "The Early History of Hearing—Observations and Theories," 728.
8. Davis, "Psychological and Physiological Acoustics," 264.
9. Danziger, *Constructing the Subject*, 45.

10. *Oxford English Dictionary*, s.v. "audiometer"; Bruce, *Bell*, 394; Mills, "Deafening," 12.
11. Jastrow, "An Apparatus for the Study of Sound Intensities," 544–45.
12. Ibid., 546.
13. Davis and Merzbach, *Early Auditory Studies*, 23–25; Mills, "Deafening" 14.
14. Boring, "Perspective," in Stevens and Davis, *Hearing*, vi–vii; Reich, *The Making of American Industrial Research*, 160–70. See also Boring, *A History of Experimental Psychology*.
15. Fagen, *A History of Engineering and Science in the Bell System*, 936; Reich, *The Making of American Industrial Research*, 180–81.
16. Davis and Merzbach, *Early Auditory*, 27; Reich, *The Making of American Industrial Research*, 180–81.
17. Thompson, *The Soundscape of Modernity*; Fagen, *A History of Engineering and Science in the Bell System*, 928–29.
18. Davis, "Psychological and Physiological Acoustics," 265.
19. Thompson, *The Soundscape of Modernity*, 97, 146.
20. Fletcher and Wegel, "The Frequency-Sensitivity of Normal Ears," 554.
21. Fletcher, *Speech and Hearing*, v.
22. Foucault, *The History of Sexuality*, 143. Despite my citation of Foucault here, I am not arguing that Bell's project is a form of biopower.
23. Fletcher, *Speech and Hearing*, xiv–xv.
24. Western Union's experiments with the harmonic telegraph, which funded both Alexander Graham Bell and Thomas Edison, were also based on extracting maximum use from limited infrastructure. The difference is that AT&T extended this project into a study of its users.
25. Baran and Sweezy, *Monopoly Capital*, 52–78; Chandler, *The Visible Hand*, 203–4. See also Adorno, "Culture and Administration."
26. The narrative I provide here has fallen out of fashion in business history, as many scholars have taken up a social constructivist approach, which allows them to be less focused on writing the history from the perspective of the winners and to move away from realist categories of political economy. While there are clear benefits to this approach, social constructivism is not as powerful a methodology for addressing systematic political questions, which are at the heart of my inquiry here. Wade Rowland has proposed conceiving of the corporation itself as a technology, as a way to bring systemic critique into social constructionist literature, though there are still many unresolved questions of how one would conceptualize interests, profits, markets, and economy itself in such a model. See Misa, "Toward an Historical Sociology of Business Culture"; Sterne and Leach, "The Point of Social Construction and the Purpose of Social Critique"; Rowland, "Recognizing the Role of the Modern Business Corporation in the Social Construction of Technology."

27. Brooks, *Telephone*, 102–7, 120–24.
28. Reich, *The Making of American Industrial Research*, 1–12, 151–52, quote at 3; Brooks, *Telephone*, 128–30; Fagen, *A History of Engineering and Science in the Bell System*, 32–58; Hounshell and Smith, *Science and Corporate Strategy*, 1–9. See also Millard, *Edison and the Business of Innovation*.
29. Fagen, *A History of Engineering and Science in the Bell System*, 35–36.
30. Brooks, *Telephone*, 132–40; Reich, *The Making of American Industrial Research*, 140, 179–80; Hounshell and Smith, *Science and Corporate Strategy*, 6.
31. Brooks, *Telephone*, 143.
32. Marx's classic discussions of the factory and the struggle between worker and machine outline this scenario quite well. See, e.g., Marx, *Capital*, 544–64.
33. Baran and Sweezy, *Monopoly Capital*, 337.
34. Nebeker, *Signal Processing*, 38–40; Fagen, *A History of Engineering and Science in the Bell System*, 279–80; O'Neill, *A History of Engineering and Science in the Bell System*, 158–61.
35. Twain, "A Telephonic Conversation"; Kafka, "The Neighbor"; Heffernan, "Funeral for a Friend," 22.
36. Here, I take the same liberty of terminological imprecision for rhetorical force taken by Baran and Sweezy. They use the term *monopoly* to include situations where oligopolies exist not only because the market conditions they describe are similar to monopoly conditions, but also because *monopoly capital* is a much catchier phrase than "oligopoly capital." Baran and Sweezy, *Monopoly Capital*, 6.
37. Marx, *Grundrisse*, 714–32.
38. This is also why perceptual capital does not inhere in individuals and is not a kind of symbolic capital in Pierre Bourdieu's sense of the term. For Bourdieu's definition of symbolic capital and his discussion of "species" of capital as they pertain to social fields, see Bourdieu, *Language and Symbolic Power*; Bourdieu and Wacquant, *An Invitation to Reflexive Sociology*. Foucault also talks about "human genetic capital," which is yet again a different use of the term. See Foucault, *The Birth of Biopolitics*, 229.
39. Marx, *Grundrisse*, 704–5. See also Negri, *Marx beyond Marx*, 140–50.
40. Mills, "Deafening," 130.
41. Marx, *Capital*, 320–39, 429–38; Marx, *Grundrisse*, 690–92.
42. Smythe, *Counterclockwise*, 268, quote from Livant at 74.
43. Jenkins, *Convergence Culture*, 249–60; Lazzarato, "Immaterial Labor," 133; Terranova, "Free Labor," 33, quotes at 36–37.
44. Deleuze and Guattari, *Anti-Oedipus*, 259. My reluctance to use labor to explain AT&T's generation of value opens out into a much larger debate regarding the status of value in contemporary economies. What if the labor theory of value only explains some formations or aspects of capitalism, such as industrial capitalism? Writers have advanced this line of thinking from a number

of perspectives. For instance, Moishe Postone claims that Marx already anticipated that capitalism would move beyond industrialism. Teresa Brennan proposes a theory of value where all natural resources and life processes can be used as energy in the processes of production, circulation, and reproduction. Lawrence Grossberg has argued for a radically contextualist theory of value that allows us to better grasp the current global crisis of labor. David Hesmondhalgh has argued that the free labor critique overextends the term and insufficiently engages with actual political pragmatics. Clearly, those of us interested in media and the history of technology have some thinking ahead of us; we need a more robust account of the range of ways that communication practices can produce and make use of value. Postone, *Time, Labor, and Social Domination*, 167; Brennan, "Why the Time Is Out of Joint," 266–68; Grossberg, *Cultural Studies in the Future Tense*, 157–58; Hesmondhalgh, "User-Generated Content, Free Labour and the Cultural Industries."

45. Deleuze and Guattari, *Anti-Oedipus*, 239. In rereading *Anti-Oedipus* for this point, I was struck by its authors' foresight in applying the concept of monetization to flows. While it appears there in a theoretical discussion of Marxist economics, today, that kind of thinking is central to and fashionable in the entrepreneurial economy of new media.

46. Münsterberg, *Psychology and Industrial Efficiency*, 17–20; Pickren and Rutherford, *A History of Modern Psychology in Context*, 195–98.

47. Kittler, *Gramophone-Film-Typewriter*, 160. For Kittler, the important connection is between psychology and media, whereas for me industry is the crucial third mediating term. See, e.g., Kittler, *Discourse Networks, 1800/1900*, 246.

48. Mumford, *Technics and Civilization*, 3. Jen E. Boyle has recently used the term "perceptual technics" as a descriptor for early modern media, but it appears mostly as a turn of phrase and is not used in the more precise sense I propose here. See Boyle, *Anamorphosis in Early Modern Literature*, 139–52. Also, my apologies for using the *techné* root for key neologisms in two consecutive books.

49. Stiegler, *Technics and Time 1*, 2; Meister, *History of Human Factors and Ergonomics*, 7, 19–20, 147–51, quotes at 147 and 150. In his classic work on design, Henry Dreyfuss does discuss sound, but in the form of auditory cues in the environment "as a substitute for visual displays to relieve tired eyes" (as opposed to designing media according to perception). Dreyfuss, *Human Factors in Design*, 11–12. Mills uses the term "ergonomopolitics of objects" (21) to describe the process I have named here as *perceptual technics*, and places it as a third type of biopower, against Foucault's "anatomo-politics," the disciplining of the individual, and "biopolitics," the regulation and transformation of populations. I've not used her formulation because I prefer to decouple perceptual technics from ergonomics and biopower. Although Michael Hardt and Tony Negri have, for instance, attempted to expand Michel Fou-

cault's conception of biopower to include new media, the term does not fit as well here. Biopower may be appropriate to some politics of contemporary communication—especially the contemporary interplay of biology and media in fields like genomics, biometrics, and motion capture—but the term does not describe corporations' or engineers' extraction of surplus value through gaming users' senses. For Foucault, biopower is clearly bound up with states. It is based on the management of life and death in individuals and populations, and it is bound up with his conceptions of governmentality and neoliberalism. As a theoretical construct, biopower offers no robust account of technology and has little to say about the modern business corporation. And inasmuch as perceptual technics is about the generation of surplus value through a combination of circulation and the general intellect, it is not about governmentality. See Foucault, *The History of Sexuality*, 140–41; Agamben, *Homo Sacer*, 7, 124; Chow, *The Protestant Ethnic and the Spirit of Capitalism*; Hardt and Negri, *Empire*, 22–24; Foucault, *The Birth of Biopolitics*, 21–22, 226–33; Gates, *Our Biometric Future*; Reardon, *Race to the Finish*; Rabinow and Rose, "Biopower Today."

50. See Mills, "The Dead Room," 330–63; Goggin and Newell, "Disabling Cell Phones," 158. I first learned about the "music problem" from Kate Crawford, who is currently conducting research on cochlear-implant wearers' sonic experiences.
51. Mills, "Deafening," 127. The argument about hardness-of-hearing is the central thread of Mills's unpublished manuscript.
52. Canguilhem, *The Normal and the Pathological*, 243.
53. Davis, *Enforcing Normalcy*, 61–71; Corker, "Sensing Disability." Foucault's interest in "normalization" in hospitals, prisons, and sexuality clearly expands his teacher Canguilhem's proposition about the life sciences into the human sciences.
54. Rose, *The Psychological Complex*, 39–61.
55. For a late example of philosophical psychology, see Watt, *The Psychology of Sound*.
56. Wundt, *Principles of Physiological Psychology*, 2–3; Danziger, *Constructing the Subject*, 27.
57. *Oxford English Dictionary*, s.v. "subject"; Danziger, *Constructing the Subject*, 54. While historical argument from semantics can be a dodgy enterprise, it is worth noting that the term *subject* itself contains the many ambiguities of the cultural, technological, and institutional situation I outline in this chapter. The OED notes a usage going back to 1541 to denote "an object with which a person's occupation or business is concerned or on which he exercises his craft; (one's) business; that which is operated upon manually or mechanically"—though clearly this is a thing and not a person. It also notes a usage as "the subject-matter of an art or science" from 1541 and as "a piece of prop-

erty" dating from 1754. Regarding human subjects, the earliest noted usage (1340) refers to "one who is under the dominion of a monarch or reigning prince; one who owes allegiance to a government or ruling power, is subject to its laws, and enjoys its protection" and referring back to English translations of Descartes, the OED notes a usage going back to 1682 as a "more fully *conscious or thinking subject*: The mind, as the 'subject' in which ideas inhere; that to which all mental representations or operations are attributed; the thinking or cognizing agent; the self or ego" (emphasis in the original). So while psychology journals follow the medical pattern from corpse to living body, the denoted subject of psychoacoustic experiments is at once a political being, a quality possessed, a self, an academic field, and the object or target of the psychoacoustician's business.

58. Danziger, *Constructing the Subject*, 87.
59. Ibid., 68–87; Quetelet, *A Treatise on Man and the Development of His Faculties*.
60. Fletcher, *Speech and Hearing*, 203.
61. Ibid., 214.
62. Ibid., 212–13.
63. Ibid., 246.
64. Ibid., 212–13; Mills, "Deafening."
65. Danziger, *Constructing the Subject*, 93.
66. Wegel and Lane, "The Auditory Masking of One Pure Tone by Another and Its Probable Relation to the Dynamics of the Inner Ear," 284–85; Fowler, "Historical Vignette."
67. Snyder, "Clarence John Blake and Alexander Graham Bell," 21; Fletcher, *Speech and Hearing*.
68. For a standard explanation of the telephone theory at the time, see Boring, "Auditory Theory, with Special Reference to Intensity, Volume and Localization."

2. NATURE BUILDS NO TELEPHONES

1. Wever and Bray, "Action Currents in the Auditory Nerve in Response to Acoustical Stimulation," 344.
2. Ibid., 345.
3. Ibid., 346.
4. Otis, *Networking*, 11.
5. Ibid., 121. See also Prescott, *History, Theory and Practice of the Electric Telegraph*, 56; Siemens, *Werner Von Siemens, Inventor and Entrepreneur*, 73.
6. Sterne, *The Audible Past*, 31–34, 84–85, 233–34.
7. Acousticians and biographers quoted in Thompson, *The Soundscape of Modernity*, 95.
8. Davis, "The Electrical Phenomena of the Cochlea and the Auditory Nerve,"

206. See also Saul and Davis, "Action Currents in the Central Nervous System"; Stevens and Davis, *Hearing*, 310–32.

9. Blume, "Cochlear Implantation," 99. See also Mathews, "The Ear and How It Works"; Mills, "The Dead Room," 330–63; Mills, "Do Signals Have Politics?"
10. Arthur Norberg, "An Interview with Robert M. Fano." Oral History transcript, Charles Babbage Institute, History of Information Processing, University of Minnesota, Minneapolis. Interview conducted 20–21 April 1989.
11. Avital Ronnell has also explored the significance of cats in telephone history. See Ronell, *The Telephone Book*, e.g., 239, 43, 49. Cats also appeared in my various work spaces as I wrote this chapter. Two cats spent a great deal of time on my home-office desk during the initial composition of this chapter, sometimes looking at me with disapproval. The day I made final my revisions at the Center for Advanced Study in the Behavioral Sciences was the only day of the year that the center's cat walked through my office. He seemed uninterested.
12. Eli W. Blake Jr. to Alexander Graham Bell, 10 July 1877, Alexander Graham Bell Collection, Library of Congress, Series Subject Files, Folder "The Telephone, The Providence Group, 1877–1912." Available online at American Memory from the Library of Congress: http://rs6.loc.gov/mss/magbell/297/29700210/0004.jpg.
13. Walsh, *Physiology of the Nervous System*, 317; Fletcher, *Speech and Hearing*. See also Bazett and Penfield, "A Study of the Sherrington Decerebrate Animal in the Chronic as Well as the Acute Condition"; Forbes and Sherrington, "Acoustic Reflexes in the Decerebrate Cat." Walsh does mention experiments done on monkeys and even a study of a human born without a cerebral cortex (318), but it appears the majority of the work was done on cats.
14. Forbes and Sherrington, "Acoustic Reflexes in the Decerebrate Cat," 374–75. See also Schantz, *Gossip, Letters, Phones*, 147.
15. Lederer, *Subjected to Science*, 31; Ryder, *Animal Revolution*, 77–119.
16. Bazett and Penfield, "A Study of the Sherrington Decerebrate Animal in the Chronic as Well as the Acute Condition," 186.
17. Cats remained popular research animals in psychoacoustics. Joseph Hall, who helped write the first well-known paper to propose the technique later called perceptual coding, studied auditory response in cats. When I asked him about the persistence of cats down to the present day, he acknowledged the ethical controversies around animal experimentation (coming down on the side that sometimes a living being is necessary, "but when I screwed up something and lost a cat, I felt terrible") and then said "the reason [cats] show up so often in this kind of research is that their auditory system is mammalian, it's reasonably similar to humans, and they are convenient to work with: a convenient size; and readily available. . . . Monkeys are also a similar size, and for some stuff they are certainly closer than humans, but they are much more expensive and I would really be concerned about doing something like that

on another primate." Hall, author interview. See Hall, "Binaural Interaction in the Accessory Superior-Olivary Nucleus of the Cat."

18. Békésy, *Experiments in Hearing*, 508; Stevens and Warshofsky, *Sound and Hearing*, 54–56.
19. Darnton, *The Great Cat Massacre and Other Episodes in French Cultural History*, 89–94, quote at 89; Leach, "Animal Categories and Verbal Abuse," 333; Berland, "Cat and Mouse," 435.
20. Darnton, *The Great Cat Massacre and Other Episodes in French Cultural History*, 74–78. The concept would be revisited—with rodents—in Monty Python's "mouse organ" sketch (1969).
21. Weber, *The Protestant Ethic and the Spirit of Capitalism*, 180–83; Darnton, *The Great Cat Massacre and Other Episodes in French Cultural History*, 89–94; Berland, "Cat and Mouse," 439–40. This kind of disenchantment is also visible in contemporary philosophy. When Deleuze and Guattari write that "*anyone who likes cats and dogs is a fool*" because these animals "invite us to regress, draw us into a narcissistic contemplation," their cats and dogs steal souls, turn adults into children, and in essence mesmerize their human companions. Once again, we have an iron cage for kitty: this time the magic is dressed up in modern psychoanalytic language, but is it any less rooted in superstition and mysticism than the common sense that told medieval builders to encase live cats in the walls of a newly built house? Centuries later, cats remain category violators. See Deleuze and Guattari, *A Thousand Plateaus*, 240.
22. Though this is largely mythology, as the famous phrase was spoken in the middle of a series of tests on the telephone. See Bruce, *Bell*.
23. Schmidgen, "Silence in the Laboratory," 61, emphasis in original.
24. Ibid., 50.
25. Mills, "The Dead Room," 332.
26. Fletcher and Wegel, "The Frequency-Sensitivity of Normal Ears," 556.
27. Wever, "The Nature of Acoustic Response," 375.
28. Boring, "Auditory Theory, with Special Reference to Intensity, Volume and Localization," 181. See also Wever, "The Upper Limit of Hearing in the Cat."
29. Lederer, *Subjected to Science*, 8–9, 43, 45–46, 120–21.
30. Peters, *Speaking into the Air*, 242.
31. Wever, "The Nature of Acoustic Response," 375.
32. For a discussion of this moment, see Peters, *Speaking into the Air*, 10–21. In fact, the "effects" paradigm had been in place by the mid-1920s. See, e.g., Schoen, *The Effects of Music*.
33. Haraway, *Simians, Cyborgs, and Women*, 149.
34. Rosenblueth, Wiener, and Bigelow, "Behavior, Purpose and Teleology," 23. Elsewhere in the article, cats share the spotlight with dogs, frogs, mice, rats, snakes, and people.
35. Borger, "Project." Thanks to Doug Kahn for pointing me to this story.

36. Marx, *Capital*, 544–64.
37. Moore, *An Introduction to the Psychology of Hearing*, 33–34, 51; Lindsay and Norman, *Human Information Processing*, 10–11.
38. De Landa, *War in the Age of Intelligent Machines*, 146; Bowker, "How to Be Universal," 119; Bohr, "Quantum Physics and Biology," 4; Galison, "The Ontology of the Enemy," 229.
39. Edwards, *The Closed World*, 225.
40. John Durham Peters writes that cybernetic models of information developed from "the 'information practice' of telecommunications, specifically from research on telephony at Bell Labs . . . and on cryptography" during the First World War. See Peters, *Speaking into the Air*, 23.
41. Haraway, *Simians, Cyborgs, and Women*, 59. That argument, originally published in 1979, was later expanded into the more sweeping thesis of Haraway's "Cyborg Manifesto," which again argues for the mutual implication of cybernetic modes of thought, biopower, and corporate capitalism: "Modern states, multinational corporations, military power, welfare state apparatuses, satellite systems, political processes, fabrication of our imaginations, labour-control systems, medical constructions of our bodies, commercial pornography, the international division of labor, and religious evangelism depend intimately upon electronics. . . . Microelectronics mediates the translations of labor into robotics and word processing, sex into genetic engineering and reproductive technology, and mind into artificial intelligence and decision procedures" (165).
42. Wiener, *The Human Use of Human Beings*, 16–17. Wiener had serious misgivings about the ethical and political implications of command and control as the guiding values for communication and for systems theory.
43. Ibid., 22.
44. Bowker, "How to Be Universal," 109.
45. Shannon, "A Symbolic Analysis of Relay and Switching Circuits," 743–51; Kahn, *The Codebreakers*; Sterne, "Claude Shannon."
46. Shannon and Weaver, *The Mathematical Theory of Communication*, 9–10, see also 32–33. Although the book is officially coauthored, it actually contains two entirely separate essays: Shannon's mathematical theory, and Weaver's popularization of it. I thus credit each author (or both) depending on which section of the book I cite.
47. Burks, "Logic, Biology and Automata—Some Historical Reflections," 303; Friedman, *Electric Dreams*, 26; Gardner, *Logic Machines and Diagrams*, 106–9, 116.
48. Shannon and Weaver, *The Mathematical Theory of Communication*, 31.
49. Fagen, *A History of Engineering and Science in the Bell System*, 279, 766; Hamilton, Nyquist, Long, and Phelps, "Voice-Frequency Carrier Telegraph System for Cables." It is worth noting that the first widely used bandpass filter, G. A.

Campbell's ladder filter, is an antecedent of the ladder filter that gave Bob Moog's synthesizer its distinctive sound. Thus, the two innovations that allowed for steady and long-distance telephone transmission — the ladder filter and the vacuum tube — are also the two technological innovations central to the innovation and commercialization of the most paradigm-changing musical instruments of the twentieth century, the synthesizer and the electric guitar. On the histories of these devices, see Pinch and Trocco, *Analog Days*; Waksman, *Instruments of Desire*.

50. Shannon and Weaver, *The Mathematical Theory of Communication*, 31.
51. Fagen, *A History of Engineering and Science in the Bell System*, 776–77.
52. Ibid., 776–79.
53. Ibid.
54. Nyquist, "Certain Topics in Telegraph Transmission Theory," 618.
55. The notion of telegraphic code as an on-off proposition probably predates Morse code, but the *Cyclopedia*'s rendition is striking for both its graphical representation and its proposition that "the dot" is the fundamental unit of time. The dot as a fundamental component of telegraphy probably descends from the point as a fundamental component of geometrical thought. See Schäffner, "The Point."
56. American School of Correspondence, *Cyclopedia of Applied Electricity*, 151; Nyquist, "Certain Topics in Telegraph Transmission Theory." See also Nyquist, "Certain Factors Affecting Telegraph Speed" and Pierce, "The Early Days of Information Theory." Pierce cites Morse code as a sort of first seed of information theory.
57. Hartley, "Transmission of Information," 535.
58. Ibid., 536, 537 (quotes), 540.
59. Shannon and Weaver, *The Mathematical Theory of Communication*, 33.
60. Hartley, "Transmission of Information," passim.
61. Shannon and Weaver, *The Mathematical Theory of Communication*, 34–35.
62. Buxton, "From Radio Research to Communications Intelligence," 295–99.
63. Hartley, "Transmission of Information," 535.
64. Shannon and Weaver, *The Mathematical Theory of Communication*, 7–8, emphasis in original.
65. Neal Verma points out that radio followed a homologous path during the 1940s. Verma, *Theater of the Mind*.
66. Galison and Hevly, *Big Science*.
67. Danziger, *Constructing the Subject*, 13; Bourdieu, *The Logic of Practice*, 107. Bourdieu's full quote here is instructive: "It is the curse of objectivism that, here as in all cases where it confronts collective belief, it can only establish, with great difficulty, truths that are not so much unknown as repressed; and that it cannot include, in the model it produces to account for practice, subjective illusion against which it has had to win its truth, in other words the

illusio, belief, and the conditions of production and functioning of this collective denial."
68. Napster.com, *Partner Brand Usage Guide*, 1. Available online at http://www.napster.com. Melissa Heckscher, "And about That Strange Logo . . ." Ironically, while it was subject to litigation from the RIAA for violating its intellectual property claims, Napster sued a California company for selling T-shirts with the Kittyhead logo on them. See "Napster Cat Attack."
69. This feline line of descent even enters back into recent theoretical discussions of sound, as in Aden Evens's claim that "the forces compressed in sound cry out as they crash headlong or rub up against each other like cats; and their cry answers to the problems of their collusion" (58).
70. Marx, *Grundrisse*, 706, emphasis in original.

3. PERCEPTUAL CODING

1. For instance, in describing his perceptual coder in a paper published in 1988, Johnston used the phrases "Transform Coding of Audio Signals" and "perceptual entropy." Johnston, "Transform Coding of Audio Signals Using Perceptual Noise Criteria," 314. On anachronism in historiography, see Rée, "The Vanity of Historicism." "There is no good reason why the thought of one period should not be judged by the standards of another; indeed we would be fools if we did not make use of the best concepts and criteria we can find, regardless of the period they come from" (979).
2. Bogert, "A Dynamical Theory of the Cochlea"; Millman, *A History of Engineering and Science in the Bell System*, 107; Nebeker, *Signal Processing*, 45.
3. Merton, *The Sociology of Science*, 371.
4. Author interview with JJ Johnston.
5. Merton, *The Sociology of Science*, 371.
6. Hall, "On Postmodernism and Articulation"; Williams, *Culture and Society, 1780–1950*.
7. Author interview with Marina Bosi; Born, *Rationalizing Culture*, 32, 344; Beranek, "Revised Criteria for Noise in Buildings."
8. Adorno, *Negative Dialectics*, 166. A host of recent writers from a variety of perspectives have argued for the need for alternatives to traditional realist historiographies of technology in order to better describe the mutual entanglements of heterogeneous, human, and nonhuman elements. I have avoided using Bruno Latour's "network" concept because I want to place a greater emphasis on power and difference than he does. While Jennifer Daryl Slack and J. Macgregor Wise make a compelling argument for using Deleuze's and Guattari's notion of "assemblage" in a case like this, I am adopting Adorno's more poetic "constellation" to foreground my own work of interpretation (and because my work in this chapter does not exactly follow their suggested

approach). For more on these perspectives, see Latour, *Aramis, or the Love of Technology*; Slack and Wise, *Culture + Technology*; Deleuze and Guattari, *A Thousand Plateaus*.

9. Zwislocki, "Masking," 284.
10. Mayer, "Researches in Acoustics," 502.
11. Inventors' self-experimentation is a common step in the development of new audio technologies. There are also notable cases from the history of psychoacoustics, as when Hallowell Davis and S. S. Stevens intentionally deafened themselves in one of Harvard's soundproof rooms.
12. Wegel and Lane, "The Auditory Masking of One Pure Tone by Another and Its Probable Relation to the Dynamics of the Inner Ear," 284. Wegel's and Lane's tone generator was an early subtractive synthesizer, using the basic synthesis technique that would be later popularized in the Moog synthesizer and similar analog devices. See Pinch and Trocco, *Analog Days*; Rodgers, "Synthesizing Sound," 19.
13. Wegel and Lane, "The Auditory Masking of One Pure Tone by Another and Its Probable Relation to the Dynamics of the Inner Ear," 283; Fletcher, *Speech and Hearing*, 167.
14. Wegel and Lane, "The Auditory Masking of One Pure Tone by Another and Its Probable Relation to the Dynamics of the Inner Ear," 268.
15. Fletcher, *Speech and Hearing*, 174, 187.
16. Fletcher, "Auditory Patterns," 47. See also Zwicker, "Subdivision of the Audible Frequency Range into Critical Bands."
17. Fletcher, *Speech and Hearing*, 62.
18. Ibid., 55.
19. Schafer, Gales, Shewmaker, and Thompson, "The Frequency Selectivity of the Ear as Determined by Masking Experiments"; Scharf, "Loudness of Complex Sounds as a Function of the Number of Components"; Zwicker, Flottorp, and Stevens, "Critical Band Width in Loudness Summation." It is worth noting that there was enough international ferment for the International Organization for Standardization to get involved in defining terms of acoustic measurement. See Rusakov, "Conference of the Group for Forumulating International Recommendations for Tables of Quantities and Units in the Division of Sound"; Zwicker, "Subdivision of the Audible Frequency Range into Critical Bands."
20. Erlmann, *Reason and Resonance*, 314.
21. Author interviews with Louis Thibault, Karlheinz Brandenburg, JJ Johnston, and Joseph Hall.
22. E-mail from Roland Wittje to author, 18 May 2006 (thank you, Roland). This was in response to my question as to whether Feldtkeller had any connection with Nazi acoustics.
23. Karin Bijsterveld notes that Zwicker was important for measurement of

loudness of sounds for noise-abatement purposes. See Bijsterveld, *Mechanical Sound*. See also Schultz, *Community Noise Ratings*, 18–19. Author interview with Joseph Hall. Hall told me, "Zwicker gets a lot of the credit for critical bands but, really, that's based on Fletcher's stuff."

24. Zwicker and Feldtkeller, *The Ear as a Communication Receiver*, xvii.
25. Herman A. O. Wilms, the European secretary of the Audio Engineering Society European Secretary, recalled, "In 1967 I bought the just released book *Das Ohr als Nachrichtenempfänger*, written by E. Zwicker and R. Feldtkeller, in which I learned the wonderful mechanism of our human hearing, especially the now famous concept of 'Frequenzgruppen,' the critical bandwidths in which our hearing system analyzes the frequency spectrum." Wilms, "In Memoriam," 199.
26. Boring, "The Beginning and Growth of Measurement in Psychology"; Shapiro, "What Is Psychophysics?"; Mills, "Deafening," 23.
27. Stevens, "A Scale for the Measurement of a Psychological Magnitude"; Fletcher and Munson, "Loudness of a Complex Tone, Its Definition, Measurement and Calculation"; Fletcher and Munson, "Relation between Loudness and Masking"; Fletcher and Wegel, "The Frequency-Sensitivity of Normal Ears."
28. Zwicker and Feldtkeller, *The Ear as a Communication Receiver*, 54–55.
29. Barthes, *Image-Music-Text*, 179.
30. Rheinberger, *Toward a History of Epistemic Things*; Rheinberger, "A Reply to David Bloor 'Toward a Sociology of Epistemic Things,'" 406–7. As in psychoacoustics, Barthes's cultural theory also presents us with the idea that listeners must be trained in particular techniques of audition, which will then yield access to characteristics of hearing itself that would otherwise be inaccessible to the listener. Barthes's notion of "grain," his solution to the problem of the adjective, also begins from the individual but seeks to transcend it by going below the subject to move beyond it. For Barthes, the grain is "the body in the voice as it sings, the hand as it writes, the limb as it performs," which is clearly meant as a quality that sits beyond the realm of subjective listening. But it can only be reached by going through the subject. Listening for the grain is individual and even erotic, he writes, but "in no way 'subjective' (it is not the psychological 'subject' in me who is listening; the climactic pleasure hoped for is not going to reinforce—to express—that subject but, on the contrary, to lose it). The evaluation will be made outside of any law, outplaying not only the law of culture but equally that of anticulture, developing beyond the subject all the value hidden behind 'I like' or 'I don't like.'" Barthes, *Image-Music-Text*, 188.
31. Zwicker and Feldtkeller, *The Ear as a Communication Receiver*, 66, 70.
32. Wilms, "In Memoriam."
33. Dudley, "Remaking Speech," 169–70; author interview with Bishnu Atal.

Schroeder, "Interview with Manfred Schroeder, Conducted by Frederik Nebeker," 55, 57. For much richer accounts of the vocoder than my truncated discussion, see Mills, "The Dead Room," 83–153; Tompkins, *How to Wreck a Nice Beach*.

34. Schroeder, "Interview with Manfred Schroeder, Conducted by Frederik Nebeker," 54, quotes at 57, 58. See also Atal and Schroeder, "Adaptive Predictive Coding of Speech Signals." Atal and Schroeder built on the work of Peter Elias, who first worked out prediction schemes for speech and coined the term "predictive coding" (which could be either linear or nonlinear); Elias's work was in turn based on Norbert Wiener's attempt to develop a general mathematical model of prediction. See Elias, "Predictive Coding I"; Elias, "Predictive Coding II"; Wiener, *Extrapolation, Interpolation, and Smoothing of Stationary Times Series*; Atal, "The History of Linear Predictive Coding," 154.

35. Atal, "50 Years of Progress in Speech Waveform Coming," 4. The beginnings of the mobile phone industry reignited interest in digital compression of speech. Atal continued: "It was clear, however, that with the coming digital age, the ability to code speech at low bit rates will be important. By 1980, the developing digital cellular telephone systems in Europe, Japan, and the United States became the major drivers for low-bit-rate coding of speech." Since mobile phones transmit their signals over the air, bandwidth considerations were preeminent from the beginning. The less bandwidth used by a mobile phone, the better. In other words, mobile phones had the same bandwidth problems as broadcasters, but with the added need for encoding and security.

36. Mills, "Deafening," 6, 28; Bijsterveld, *Mechanical Sound*, 37. See also Thompson, *The Soundscape of Modernity*; Picker, *Victorian Soundscapes*; Hegarty, *Noise/Music*; Radovac, "The 'War on Noise.'"

37. Shannon and Weaver, *The Mathematical Theory of Communication*; Wiener, *Cybernetics, or Control and Communication in the Animal and the Machine*, 10; Serres, *Hermes*, 67. See also Terranova, *Network Culture*, 10–27. Informational definitions of noise would also come to suffuse cultural theory. Writers like Michel Serres characterize noise as a kind of constitutive ambiance from which human action emerges. See Serres, *The Parasite*.

38. Author interview with Bishnu Atal. Schroeder, "Interview with Manfred Schroeder, Conducted by Frederik Nebeker," 58.

39. Author interview with Joseph Hall. Hall also made use of smaller computers to control experiments.

40. Author interview with Joseph Hall. Hellman, "Asymmetry of Masking between Noise and Tone." Although Hellman influenced Hall, she was not the first to propose the idea. Richard Ehmer's earlier work evidenced a similar approach and provided a theoretical basis for Oscar Bonello's coder in Argentina. See, e.g., Ehmer, "Masking by Tones vs. Noise Bands"; Ehmer, "Masking Patterns of Tones." Schroeder claimed on his website that "in 1972, work-

ing with Joseph Hall, I proposed perceptual coding of audio signals, taking into account the masking properties of the human ear" (http://www.physik3.gwdg.de/~mrs/). Schroeder may indeed have proposed the idea in 1972, but the earliest documentation of this thread of research at Bell Labs was Atal and Schroeder, "Predictive Coding of Speech Signals and Subjective Error Criteria," presented in 1978.

41. Schroeder, Atal, and Hall, "Optimizing Digital Speech Coders by Exploiting Masking Properties of the Human Ear," 1647. Author interviews with Bishnu Atal and Joseph Hall.
42. Krasner, "Digital Encoding of Speech and Audio Signals Based on the Perceptual Requirements of the Auditory System," 13, 16–17, 19.
43. Ibid., 70, 126–27.
44. Ibid., 128. See also Krasner, "The Critical Band Coder—Digital Encoding of Speech Signals Based on the Perceptual Requirements of the Auditory System."
45. Atal and Schroeder, "Predictive Coding of Speech Signals and Subjective Error Criteria"; Krasner, "Digital Encoding of Speech and Audio Signals Based on the Perceptual Requirements of the Auditory System," 14–15, 126.
46. Sterne, *The Audible Past*, 31–85, 185–96.
47. Pinch and Trocco, *Analog Days*, 14–17; Glinsky, *Theremin*; Kahn, *Noise, Water, Meat*; Weidenaar, *Magic Music from the Telharmonium*.
48. I am currently researching the use and status of models in audio signal processing as part of a larger project on the politics and aesthetics of signal processing. Several papers will be forthcoming.
49. Author interviews with Bishnu Atal, JJ Johnston, and Joseph Hall.
50. Author interviews with Karlheinz Brandenburg and Harald Popp; Sametband, "La historia de un pionero del audio digital"; Krasner, "Digital Encoding of Speech and Audio Signals Based on the Perceptual Requirements of the Auditory System," 109–10. Regarding Johnston's comment about CDs, it is interesting to note that Sony's first commercial CD player was released in 1982. The repertoire was still quite limited in the mid-1980s, and both CDs and players were expensive and remained marginal to the audio industry—marginal enough, apparently, that nobody at Bell Labs lent one to Johnston for his experiments.
51. Born, *Rationalizing Culture*, 197.
52. Ibid.
53. "It is possible to invent a *single machine* which can be used to compute *any* computable sequence." Turing, "On Computable Numbers, with an Application to the Entscheidungsproblem," 128, emphasis in original.
54. In conversation, some more technically minded people have suggested that Dolby noise reduction—released for professionals in 1965 and consumers in 1968—was a successful form of analog perceptual coding. Dolby's origi-

nal tape system boosted the high frequencies of tape during recording and reduced them during playback so that the apparent noise of the tape system would decrease. Certainly, Dolby is part of the story of the domestication of noise, and the connection is worth further investigation. But while Dolby's technique is clearly related to perceptual technics, it doesn't quite fit the mold for perceptual coding. The goal was increased definition for end-listeners; there was no particular savings of bandwidth. It didn't reduce the size of the signal for the benefits of perception; it manipulated the signal to put noise in its place (for the pleasure of listeners).

55. Bijsterveld, *Mechanical Sound*, 81–87, quote at 87.
56. For more on Licklider and psychoacoustics, see Edwards, *The Closed World*, 209–37.
57. Gardner and Licklider, "Auditory Analgesia in Dental Operations," 1145.
58. Ibid., 1146.
59. Maisel, "'Who's Afraid of the Dentist?'" 60–61.
60. Weisbrod, "Audio Analgesia Revisited," 13–14. See also Carlin, Ward, Gershon, and Ingraham, "Sound Stimulation and Its Effect on the Dental Sensation Threshold."
61. Thompson, *The Soundscape of Modernity*, 317–24.
62. Doelle, *Environmental Acoustics*, 158.
63. Ibid., 20, 159, 170–71.
64. Ibid., 185–88.
65. Beranek, "Criteria for Office Quieting Based on Questionnaire Rating Studies"; Beranek, Reynolds, and Wilson, "Apparatus and Procedures for Predicting Ventilation System Noise"; Beranek, "Revised Criteria for Noise in Buildings."
66. This practice would be derided by R. Murray Schafer in 1977, who called Doelle's ideas "memorial drool": "Architects and acoustical engineers have often conspired to make modern buildings noisier. It is a well-known practice to add Moozak or white noise (its proponents prefer to call it 'white sound' or 'acoustic perfume') to mask mechanical vibrations, footsteps and human speech. . . . There may indeed be times when masking techniques can be useful in soundscape design but they will never succeed in rescuing the botched architecture of the present. No amount of perfumery can cover up a stinking job." Schafer, *The Soundscape*, 223–24.
67. Attali, *Noise*, 6–7, 19.
68. Foucault, *The Archaeology of Knowledge and the Discourse on Language*.
69. Thompson, *The Soundscape of Modernity*, 317–24; Zak, *The Poetics of Rock*, 14–16; Doyle, *Echo and Reverb*, 226–27. Regarding the paradigm that opposes noise and power, noise and order, one can probably take it even further back, but the complaints about noise listed for London in the 1600s are essentially the same claims one finds about noise in New York in the 1970s, even though

the noises and their sources are different. See Cockayne, *Hubbub*, 106–30; Bijsterveld, *Mechanical Sound*; Radovac, "The 'War on Noise.'"

70. Of course, the history of noise in art music goes at least back to the Italian Futurists, and it could probably be found on the fringes of discussions of loudness from Bach to Wagner. But the application of noise to art music-making became more and more pervasive and mainstream during this period across a range of audio arts. Artists like John Cage ("Silence") and Alvin Lucier ("I Am Sitting in a Room") used room ambiance as a compositional resource; popular musicians sought out new methods of distorting sounds; and recordists moved from stereo to multitrack recordings to create impossible ambiances. See Kahn, *Noise, Water, Meat*; Waksman, *Instruments of Desire*; Zak, *The Poetics of Rock*; Gracyk, *Rhythm and Noise*; Hegarty, *Noise/Music*; Hainge, "Of Glitch and Men."

71. Attali, *Noise*, 133–48, quotes at 133, 147.

72. For talk of new musical technologies, practices, economies, and aesthetics as instantiations of composition, see e.g., Gunderson, "Danger Mouse's Grey Album, Mash-Ups, and the Age of Composition"; Johnson, "Machine Songs. 1. Music and the Electronic Media"; Johnson, "Technology, Commodity, Power"; Lysloff, "Musical Community on the Internet"; Ragland, "Mexican Deejays and the Transnational Space of Youth Dances in New York and New Jersey"; Swiss, Sloop, and Herman, *Mapping the Beat*; Taylor, *Strange Sounds*. Despite its brevity, Andrew Leyshon's critique of Attali's stages and proposed alternative "network of creativity" is one of the most conceptually robust analyses of the problems and promises of Attali's concepts of repetition and composition. See Leyshon, "Time-Space (and Digital) Compression," 58–60.

73. Mastering is the final stage of sound editing applied to commercial recordings before they are mass produced and released. A mastering engineer adjusts the recording's tonal balance, overall volume, dynamic range, and a number of other features to make it consistent with other, similar recordings, and the prevailing production aesthetics of the medium or genre in which it will be heard. The mastering engineer ensures a record sounds like "how a recording is supposed to sound." Mastering is still a largely specialized profession, and while cheap digital equipment has allowed for a proliferation of mastering services in recent years, only a handful of mastering engineers working at a few elite studios have much impact on popular expectations of what recorded audio should sound like.

74. There is no question that this post-noise regime was only emergent in the 1970s and was not anything approaching a total phenomenon, since there were plenty of examples of the "old way" of dealing with noise—as something to be eradicated and as interference—afoot in 1977. Most notable is Schafer, *The Soundscape*. Originally published in 1977, Schafer's book basically argued for an aesthetic of soundscapes where signal trumps noise, and where a "hi

fi" soundscape is one that is the least cluttered or complex. Although as of this writing Schafer still exerts a great deal of personal influence over soundscape art and the critical discourse in studies of soundscape, it is also fair to note that subsequent generations of authors have been able to make deft and interesting use of the concept of soundscape without the political and aesthetic baggage that Schafer attached to the term. Schafer's own opposition to noise, however, follows closely what Thompson calls "the modern sound." In his case, it was not buildings but whole environments.

75. Dudley, "Synthesizing Speech," 98. See also Mills, "The Dead Room," 86–87.
76. Attali's characterization of repetition has at least two problems: it is a simple dismissal of mass culture and mass production, and it fundamentally mischaracterizes the aesthetic function of repetition in music. In addition to Leyshon's critique cited above, see the following authors for more robust discussions of repetition in music: Small, *Music-Society-Education*; Keil and Feld, *Music Grooves*; Kivy, *The Fine Art of Repetition*.
77. Attali, *Noise*, 19.

4. MAKING A STANDARD

1. Hoeg and Lauterbach, *Digital Audio Broadcasting*, 4–5; O'Neill, "DAB Eureka-147," 7–8. See also Faller, Juang, Kroon, Lou, Ramprashad, and Sundberg, "Technical Advances in Digital Audio Radio Broadcasting."
2. Sametband, "La historia de un pionero del audio digital." Brandenburg, "High Quality Sound Coding at 2.5 Bit/Sample." Brandenburg and Seitzer, "OCF," 4; Brandenburg, Kapust, Eberlein, Gerhäuser, Brill, Krägeloh, Popp, and H. Schott, "Low Bit Rate Codecs for Audio Signals: Implementation in Real Time." Spurlin, "Rate-Reduced Digital Audio in the Broadcast Environment," 531, 33. Author interviews with Karlheinz Brandenburg, Harald Popp, and Bill Spurlin.
3. Kittler, "Observation on Public Reception," 75.
4. Faulkner, "FM."
5. See, e.g., Douglas, *Inventing American Broadcasting*; Gomery, "Theater Television"; Tang, "Sound Decisions"; Brock, *Telecommunication Policy for the Information Age*; Brooks, *Telephone*.
6. It is also notable that the MPEG participants sent engineers rather than other corporate representatives to participate in the standards-making process. They could have just as easily sent executives or lawyers. JTC rules shaped this in part (and the fact that MPEG was an initiative by two engineers), but it also underscores that no one company already had a standard that could dominate the market.
7. Jakobs, "Information Technology Standards, Standards Setting and Standards Research," np; Faulkner, "FM."

8. Bar, Borrus, and Steinberg, "Interoperability and the NII," 243.
9. Tsutsui, Suzuki, Shimoyoshi, Sonohara, Akagiri, and Heddle, "ATRAC"; Lokhoff, "Precision Adaptive Subband Coding (Pasc) for the Digital Compact Cassette (DCC)"; Hoogendoorn, "Digital Compact Cassette."
10. The logics behind these approaches were developed early in the twentieth century, and newer networks, like mobile phone networks, partake of some but not all of their characteristics: a mobile phone must work with a particular network but different providers will have their own sets of protocols that are only loosely regulated by a central government.
11. Bar, Borrus, and Steinberg, "Interoperability and the NII," 235–38. See also the very helpful typology of standards on page 241.
12. Gillespie, *Wired Shut*, 280.
13. In addition to the ISO and the ITU, companies involved in developing technologies associated with the internet might engage with the Internet Engineering Task Force (IETF), the Institute of Electrical and Electronics Engineers (IEEE), and several others in an effort to find a ruling that is most favorable to their interests—and that assumes companies don't just publish the standards themselves or keep them as a trade secret. Jakobs, "Information Technology Standards, Standards Setting and Standards Research," 5–9. Since the 1980s, the number of regulatory bodies for standards has continued to grow, with the founding of industry consortia like the World Wide Web Consortium (W3C) and the Internet Corporation for Assigned Names and Numbers (ICANN).
14. Ibid., 9; Wallace, "The JPEG Still Picture Compression Standard."
15. Cargill, *Information Technology Standardization*, 137–38; Piatier, "Transectorial Innovations and the Transformation of Firms," 212, 219.
16. Jakobs, "Information Technology Standards, Standards Setting and Standards Research," n.p; Musmann, "Genesis of the MP3 Audio Coding Standard."
17. Seckler and Yostpille, "Electronic Switching Control Techniques"; Noam, *Interconnecting the Network of Networks*; Abbate, *Inventing the Internet*; Schiller, *Digital Capitalism*.
18. Immink, "The Compact Disc Story."
19. Musmann, "Genesis of the MP3 Audio Coding Standard," 1043. See also Morris, "Understanding the Digital Music Commodity."
20. Musmann, "The ISO Audio Coding Standard," 1. Author interviews with Karlheinz Brandenburg and Bernhard Grill. The original VCD specification would eventually make use of MPEG 1-layer 2 audio (recall that MP3 stands for layer 3 of the MPEG-1 standard), and this is certainly one of the unheralded successes of the group, despite the greater press garnered by MP3.
21. Musmann, "The ISO Audio Coding Standard," 1. Author interviews with Karlheinz Brandenburg and Bernhard Grill. As mentioned in the introduction,

the 128 kbps specification comes in part from ISDN. ISDN caught on in some fields, like videoconferencing, though ADSL and other protocols have superseded it in many applications.
22. The report was composed roughly around the same time that the first code for a web browser was being written.
23. De Sonne, *Digital Audio Broadcasting*, 3, 7–8, 14–21; McLeod, "MP3s Are Killing Home Taping."
24. Musmann, "The ISO Audio Coding Standard," 1 (quoted); Musmann, "Genesis of the MP3 Audio Coding Standard," 1043. VLSI stands for Very Large Scale Integration, the process whereby thousands of transistor-based circuits were combined into a single chip, for example a microprocessor. The phrase *VLSI manufacturers* is no longer in common use, replaced by the much simpler *chip manufacturers*.
25. Fuller's analysis is based on a reading of Alfred North Whitehead's concepts of "simple location" and the "fallacy of misplaced concreteness." Fuller critiques Whitehead's holism as itself a "fallacy of misplaced concreteness," a broader point analogous to my argument in the introduction (and its notes) against the fallacy of assuming sound is "ontologically continuous" prior to its reproduction. Whitehead, *Science and the Modern World*, 58, 72; Fuller, *Media Ecologies*, 97, 210.
26. Author interviews with Karlheinz Brandenburg, JJ Johnston, and Marina Bosi.
27. ASPEC used a 512-band Modified Discrete Cosine Transform (MDCT) at this stage. Although discovered in a different context, the MDCT was probably the second most important enabling technical innovation for MP3 after perceptual coding. The added advantage of MDCT it that its "windows" overlap so that in filtering audio, fewer distortions are introduced by the boundaries between the bands in the filter. Debates within MPEG centered on whether it was a filterbank and if it required too much processing power. "It's not a transform, it's a filterbank," JJ Johnston told me. "Politically, [advocates of MUSICAM] had said subband coders are better," because they were simpler and used less processing power. Author interviews with JJ Johnston and Marina Bosi. Although it was not named until later, the concept was first worked out in Princen, Johnson, and Bradley, "Subband/Transform Coding Using Filter Bank Designs Based on Time Domain Aliasing Cancellation." See also Pohlmann, *Principles of Digital Audio*, 338–41.
28. On technical differences between layer 2 and layer 3, see Bosi and Goldberg, *Introduction to Digital Audio Coding and Standards*, 266–73; Pohlmann, *Principles of Digital Audio*, 370–78. Participants also were keenly aware of competition within MPEG at the time that the standards were set. Author interviews with JJ Johnston, Karlheinz Brandenburg, Harald Popp, and Bernhard Grill.

5. OF MPEG, MEASUREMENT, AND MEN

1. Bergman, Grewin, and Rydén, "MPEG/Audio Subjective Assessments Test Report"; Bergman, Grewin, and Rydén, "The SR Report on the MPEG/Audio Subjective Listening Test April/May 1991."
2. Pinch, "'Testing—One, Two, Three . . . Testing!'" 25, 26. On witnessing, see Rentschler, "Witnessing."
3. Pinch, "'Testing—One, Two, Three . . . Testing!'" 25, 28.
4. Author interview with Louis Thibault. Even with the first working perceptual coder, designed by Michael Krasner, there is careful attention to the testing situation as something that must be more rigorous than real-world use.
5. A full specification can be found in "Methods for the Subjective Assessment of Small Impairments in Audio Systems including Multichannel Audio Systems," ITU-R BS.1116.
6. See, e.g., Brandenburg, "High Quality Sound Coding at 2.5 Bit/Sample." A notable exception is Michael Krasner's dissertation. Krasner spends considerable effort designing testing scenarios, borrowing from both telecommunications and audiophile practice.
7. Kant, *Critique of Judgment*, 149. It would be possible to object that Kant's distinction between "beautiful objects" and "beautiful views of objects" (80–81) renders sound reproduction outside the realm of the beautiful. This is why I say that MPEG's listening tests are at best neo-Kantian in orientation. They depart from Kant in that they directly subject sound-reproduction systems (or more accurately, the artifacts thereof) to questions of beauty and revulsion. Even though he remains a quintessential thinker of mediation, electronic media lie just outside the conceptual and historical purview of Kant's philosophy.
8. Author interview with Joseph Hall; Krasner, "Digital Encoding of Speech and Audio Signals Based on the Perceptual Requirements of the Auditory System," 127. Krasner's experiment appears earlier in the thesis (62–63). Schubert, "History of Research on Hearing," 64; Zwislocki, "Masking."
9. Bonello's group's choice of Ehmer was particularly interesting because Ehmer's relatively simpler scheme provided a less intensive computational problem, one better suited to the digital signal-processing power available to them at the time through commercial personal computers and home-brew sound cards. Author interviews with Karlheinz Brandenburg, JJ Johnston, and Joseph Hall; Bonello to author 8 July 2008. Brandenburg, "Evaluation of Quality for Audio Encoding at Low Bit Rates"; Johnston, "Transform Coding of Audio Signals Using Perceptual Noise Criteria"; Bonello, "PC-Controlled Psychoacoustic Audio Processor." Brandenburg drew from Zwicker, *Psychoakustik*. Johnston consulted with Joe Hall, though neither could recall the exact publication—probably Fletcher, "Auditory Patterns"; Fletcher and

Galt, "The Perception of Speech and Its Relation to Telephony." Bonello used Ehmer, "Masking by Tones vs. Noise Bands."; Ehmer, "Masking Patterns of Tones."

10. Author interview with JJ Johnston.
11. Author interview with Karlheinz Brandenburg.
12. Bergman, Grewin, and Rydén, "MPEG/Audio Subjective Assessments Test Report," 4–5; Bergman, Grewin, and Rydén, "The SR Report on the MPEG/Audio Subjective Listening Test April/May 1991," 6.
13. The final test report has the same level of detail for Norddeutscher Rundfunk Studio 9. Bergman, Grewin, and Rydén, "The SR Report on the MPEG/Audio Subjective Listening Test April/May 1991," appendix 2, passim; Fuchs, "Report on the MPEG/Audio Subjective Listening Test in Hannover (Draft)," appendix 3.
14. "CD quality" clearly means two different things in the test. In the case of the DAT machine, it means audio that adheres to the 16 bit 44.1 kHz CD standard shared by digital audio tape and compact disc. In the case of listening tests "CD quality" refers to the sound of 16 bit 44.1kHz audio.
15. Bergman, Grewin, and Rydén, "MPEG/Audio Subjective Assessments Test Report," 4–5; Bergman, Grewin, and Rydén, "The SR Report on the MPEG/Audio Subjective Listening Test April/May 1991," 6.
16. "Anaesthetic" may describe the self-presentation of psychoacoustic research, but the term also conceals a part of its history. As the first two chapters discuss, from the moment AT&T took an active interest in the field in the 1910s, very particular notions of good sound came to dominate psychoacoustic measurements and testing. This notion of the good was rooted in the successful functioning of the phone system, but the functional language used to describe the aesthetic concealed a whole set of ongoing judgments regarding how telephones—and human speech—*ought* to sound.
17. E.g., Harvey Fletcher, *Speech and Hearing*, 141.
18. MPEG test documentation cites the earlier CCIR standard; later tests cite an IEEE Standard. Bergman, Grewin, and Rydén, "The SR Report on the MPEG/Audio Subjective Listening Test April/May 1991," appendix 1. "Subjective Assessment of Sound Quality," CCIR Recommendation 562-2, vol. 10, Part 1, 16th Plenary Assembly, Dubrovnik, 1986, 275–79; "Methods for the Subjective Assessment of Small Impairments in Audio Systems Including Multichannel Audio Systems," ITU-R BS.1116 (1994).
19. Sousa, "The Menace of Mechanical Music." For more on metaphoric connections between signal processing and food, see Sterne and Rodgers, "The Poetics of Signal Processing."
20. Peryam and Girardot, "Advanced Taste Test Method"; Bech and Zacharov, *Perceptual Audio Evaluation*, 74; Beebe-Center, *The Psychology of Pleasantness and Unpleasantness*; Fechner, *Elements of Psychophysics*; *Oxford English Dictionary*,

s.v. "hedonic." The entry from 1952 refers to a "hedonic scale method" that features a nine-point scale from "dislike" to "like" with a neutral term in the middle. It thus more closely resembles the CCIR comparison scale, which runs from "-3, much worse" to "+3, much better," with 0 as "the same." The absolute scale of "imperceptible" to "annoying" appears somewhat later, though it still conforms to the hedonic principle. CCIR 562-2 also refers to a number of studies done in the USSR, which I was unable to acquire, and which would likely provide an even richer intellectual genealogy for the approach.

21. Elias, *The Civilizing Process*, 98. The analogy is not exact: one could conceivably argue that modern life continues to present many physical threats to the ear, from the loudness of city life to the incredibly damaging sound-pressure levels available from most portable audio players. It is clear, however, that an ear for timbre and for "sound" has become part of the repertoire of the cultivated bourgeois listener, almost without regard for musical genre or sensibility (though in some aesthetics, like punk, this refinement is turned on its head). This development has followed the increasing profusion and proliferation of sound-reproduction technologies in everyday life.

22. Immink, "The Compact Disc Story"; Bergman, Grewin, and Rydén, "The SR Report on the MPEG/Audio Subjective Listening Test April/May 1991," 4–5. Although compact disc was officially the standard, the tests used digital audio tape (DAT) for their reference recordings; DAT had the same bit depth and sample rate as CD, but it stored its data differently.

23. AES Technical Council, "Perceptual Audio Coders: What to Listen For," CD-ROM, November 2001.

24. Grewin and Rydén, "Subjective Assessment of Low Bit-Rate Audio Codecs," 96–97. Author interview with Bernhard Grill. Oscar Bonello to author, 26 July 2008.

25. Author interviews with JJ Johnston and Bernhard Grill; Krasner, "Digital Encoding of Speech and Audio Signals Based on the Perceptual Requirements of the Auditory System," 128. See also Krasner, "The Critical Band Coder—Digital Encoding of Speech Signals Based on the Perceptual Requirements of the Auditory System."

26. Author interview with Louis Thibault.

27. Author interviews with Karlheinz Brandenburg and Bernhard Grill.

28. Grewin and Rydén, "Subjective Assessment of Low Bit-Rate Audio Codecs," 93.

29. Jakobs, "Information Technology Standards, Standards Setting and Standards Research," 13.

30. Author interviews with Louis Thibault and Karlheinz Brandenburg.

31. Professional musicians are clearly a categorical exception here, since one can find greater relative representation of women and nonwhite people, however not so much so that it outweighs the heavy biases in the other fields.

I had hoped to conduct a more thorough study of test subjects, but despite my repeated requests, sources at Fraunhofer, AT&T, and the Communication Research Centre in Ottawa did not provide me with any contact information for expert listeners used in their own tests. It may be the case that such information is private, proprietary, or not deemed important enough to keep for long periods of time. Sources did tell me that one of Fraunhofer's most frequently used expert listeners in the late 1980s and early 1990s was a woman musician, though nobody I spoke with was able to recall her name.

32. Of the eighty-eight unique participants in the two Swedish Radio tests (sixty people took each test), twenty-three were appointed by Swedish Radio, twenty-four were appointed by the four development groups whose codecs were being tested, and the rest were appointed by groups like the European Broadcasting Union and the Audio Engineering Society. Beyond names and locations on the list, no clear demographic data are available on this group, other than that many were employed in one or another audio business. Grewin and Rydén, "Subjective Assessment of Low Bit-Rate Audio Codecs," 93.
33. Born, *Rationalizing Culture*, 201–2.
34. Bech, "Selection and Training of Subjects for Listening Tests on Sound-Reproducing Equipment," 591.
35. Ibid., 593. Kirk, "Learning, a Major Factor Influencing Preferences for High-Fidelity Reproducing Systems."
36. For discussions of gender and musical cultures, see, e.g., Whitely, *Sexing the Groove*; Sandstrom, "Women Mix Engineers and the Power of Sound"; Meintjes, *Sound of Africa!*
37. Bech, "Selection and Training of Subjects for Listening Tests on Sound-Reproducing Equipment," 593–94.
38. This renders the listening panel susceptible to the critique of panels in studies of broadcast audiences. Even if a panel is selected according to careful social science criteria, if you change the panel or measurement tool, you get different results. See Meehan, "Why We Don't Count"; Meehan, "Heads of Households and Ladies of the House."
39. Brandenburg, "MP3 and AAC Explained," 9; Kirk, "Learning, a Major Factor Influencing Preferences for High-Fidelity Reproducing Systems"; Bech and Zacharov, *Perceptual Audio Evaluation*. I discuss the matter of preference *for* MP3 audio in the introduction.
40. Heidegger, *Being and Time*, 102; Ahmed, *Queer Phenomenology*, 49.
41. Author interview with JJ Johnston.
42. This last idea was first put to me in a conversation with the historian Susan Schmidt Horning, who heard it from the recording engineer Walter Sear. Sear told Horning, "When I mix with a woman client I mix differently. . . . You hear best in your own vocal range, women are almost an octave higher than

men, so of course they hear better there. And if I mix for me, my ears, I push 10k, you're gonna jump up and scream, it's gonna sound too high." Walter Sear, interview with Susan Schmidt Horning, New York City, 21 January 1999. Many thanks to Susan Schmidt Horning for sharing this material with me.

43. "Link to Lesbianism Found"; Corso, "Age and Sex Differences in Pure-Tone Thresholds," 498; Cassidy and Ditty, "Gender Differences among Newborns on a Transient Otoacoustic Emissions Test for Hearing," 28; Pearson et al., "Gender Differences in a Longitudinal Study of Age-Associated Hearing Loss," 1196; Osterhammel and Osterhammel, "High-Frequency Audiometry"; Sato, Sando, and Takahashi, "Sexual Dimorphism and Development of the Human Cochlea," 1037–39; McFadden and Pasanen, "Comparison of the Auditory Systems of Heterosexuals and Homosexuals," 2709, 2712; Deutsch, Henthorn, and Dolson, "Absolute Pitch, Speech, and Tone Language," 339; Levitin and Rogers, "Absolute Pitch"; Zatorre, "Absolute Pitch"; Tsai, "Tone Language, Absolute Pitch, and Cultural Differences [Online Discussion Thread]"; Steinberg, Montgomery, and Gardner, "Results of the World's Fair Hearing Tests," 291–97; Scott, "Hearing Research."
44. JJ Johnston, interview with author.
45. Bourdieu, *Distinction*, 41. I am not arguing that Johnston here represents "popular taste" as characterized in Bourdieu's study of France, only that his description is an interesting counterpoint to the rhetoric of disinterest in the construction of listening tests.
46. The SQAM is currently available online at: http://tech.ebu.ch/publications/sqamcd.
47. Bourdieu, *Distinction*, 16, 26–28.
48. Author interviews with Karlheinz Brandenburg, Bernhard Grill, and JJ Johnston.
49. Grossberg, *Dancing in Spite of Myself*, 65.
50. Katz, *Capturing Sound*, 1–2. See also Anderson, *Making Easy Listening*, 105–50.
51. Suzanne Vega, "How to Write a Song and Other Mysteries," *Measure for Measure* blog, *New York Times*, 23 September 2008. Available online at http://measureformeasure.blogs.nytimes.com.
52. Salomon, *Data Compression*, 846.
53. Author interview with Karlheinz Brandenburg.
54. Bergman, Grewin, and Rydén, "MPEG/Audio Subjective Assessments Test Report," appendix 5; Bergman, Grewin, and Rydén, "The SR Report on the MPEG/Audio Subjective Listening Test April/May 1991," appendix 6; Fuchs, "Report on the MPEG/Audio Subjective Listening Test in Hannover (Draft)," appendix 8. For the first test in July 1990, the Vega recording was only used in the "high-quality" tests. For the tests in May and November 1991, the Vega recording appears in lists for both "high-quality" and "intermediate-quality" tests.

55. Suzanne Vega, "Tom's Diner," *Solitude Standing*, A&M Records, 1987.
56. Perlman, "Golden Ears and Meter Readers," 798–800.
57. Ibid.
58. Author interview with Marina Bosi.
59. Vega, "How to Write a Song and Other Mysteries."
60. Harley, "From the Editor." Although this story has achieved wide circulation among audiophiles online, I have been unable to verify any aspect of the Locanthi story in Harley's account (except that Locanthi was very concerned with audio quality). It therefore reproduces audio's objectivist-relativist issue in historiography. With no way of confirming the story, do I take Harley at his word or do I discount it without verification? When I asked Harley in the comment thread to the piece for an exact citation, he could not give one, noting only that "if I recall correctly, Ron Streicher was the workshop chairman" and "I would say the convention was in Los Angeles between 1993 and 1996" (the ninety-first convention, cited in his article, occurred in New York) and directing me to the Audio Engineering Society. Unfortunately, the AES does not list programs that far back on its website, and convention recordings do not go that far back. In e-mail correspondence, Ron Streicher told me he has no recollection of being involved with the panel. It is therefore difficult to know exactly what Locanthi said or what transpired from his saying it. Streicher to author, 19 July 2009; the online comment thread for his editorial cited above can currently be found at AVguide.com: http://www.avguide.com/forums/blind-listening-tests-are-flawed-editorial?page=1.
61. Harley, "The Role of Critical Listening in Evaluating Audio Equipment Quality," 16.
62. Haraway, *Simians, Cyborgs, and Women*, 191, 194.
63. Bourdieu, *Distinction*, 12–13.

6. IS MUSIC A THING?

1. Haring, *Beyond the Charts*, 11; McLeod, "MP3s Are Killing Home Taping"; Burkart and McCourt, *Digital Music Wars*. Sandy Pearlman, personal conversation with author.
2. Haring, *Beyond the Charts*, 11; Frith, "Video Pop," 96–99, 114–19; McLeod, "MP3s Are Killing Home Taping."
3. Pater, *The Renaissance*, 111, but see 111–15.
4. Durant, *Conditions of Music*, 9. Durant contrasts this tradition in the philosophy of music with the theatrical tradition of concealing musicians, where the music seems to come from nowhere and provides a dramatic context for the action happening on the stage, a practice that was well established in Pater's time.
5. Brown, "Thing Theory," 18.

6. Loughlan, "Pirates, Parasites, Reapers, Sowers, Fruits, Foxes . . . the Metaphors of Intellectual Property," 218–19. Thanks to Nancy Baym for challenging me to take more care in my use of the term and suggesting some reading.
7. Johns, *Piracy*, 6–7.
8. Larkin, *Signal and Noise*, 240.
9. Chanan, *Musica Practica*, 54–55, 111–12; Taylor, *Strange Sounds*, 3–4.
10. Eisenberg, *The Recording Angel*, 17; Suisman, *Selling Sounds*.
11. Eisenberg, *The Recording Angel*, 24. Eisenberg's argument is thus a standard riff on reification. See Adorno, "On the Fetish Character of Music and the Regression of Listening"; Lukàcs, *History and Class Consciousness*.
12. The quote is from Keil and Feld, *Music Grooves*, 248. Keil's work embodies this disciplinary ambivalence quite profoundly. He is at once suspicious of recordings while acknowledging their centrality and importance for the vitality of the music cultures he studies (e.g., 21–22). For a discussion of the ambivalence at anthropology's first encounter with recording, see Brady, *A Spiral Way*; Shulemay, "Recording Technology, the Record Industry and Ethnomusicological Scholarship"; Sterne, *The Audible Past*, 311–33. Some of the best new ethnographic work on music is more likely to cast recording as a central part of musical process, rather than a disruption of liveness: Fox, *Real Country*; Greene and Thomas, *Wired for Sound*; Meintjes, *Sound of Africa!*
13. Small, *Musicking*. Both Keil and Small formed their positions in a period when capitalist culture industries still sought bigness and totality. Small's earlier *Music-Society-Education*, where he develops his normative program, explicitly draws upon Ivan Illich's critique of big technology and his argument for smaller, more "human-scale" or "convivial" forms of technology that would allow for greater democracy. Small and Keil also echoed midcentury mass-culture criticism, which emphasized the supposed passivity of audiences as the basis of the political problem behind mass culture. Leaving aside decades of cultural studies work that has criticized the notion of the passive audience, it is not entirely clear that activity, engagement, and participation are inherent goods in themselves. The increasingly coercive imperative to participate or interact—in social-networking culture online, in avant-garde art, and in political culture—begs the question of its political valence. As Darin Barney has noted in the context of politics, we need to rethink our basic assumptions: Is participation automatically good and passivity automatically bad? Is it meaningful and democratic participation, or might it lead to the same kinds of uncritical, accepting stances that the mass-culture critics worried after? Barney, "Politics and Emerging Media." For classic arguments in favor of participatory culture, see Small, *Music-Society-Education*, 182–83, 214; Illich, *Tools for Conviviality*; Illich, *Energy and Equity*; Keil, "Participatory Discrepancies and the Power of Music." For a discussion of midcentury mass-culture critique and the question of activity and passivity in audiences, see,

e.g., Hay, Grossberg, and Wartella, *The Audience and Its Landscape*; Morley, *Television, Audiences, and Cultural Studies*; Ross, *No Respect*. For an analysis of the ways in which the imperative of participation can be as coercive as passivity, see Bourdieu, "Program for a Sociology of Sport"; Peters, *Speaking into the Air*, 157–60. This is an argument I will fully develop elsewhere.

14. Frith, "Video Pop," 105. See also Burkart and McCourt, *Digital Music Wars*, 23.
15. Yar, "The Rhetorics of Anti-Piracy Campaigns," 611–12; Carey and Wall, "MP3"; Vaidhyanathan, *Copyrights and Copywrongs*, 18; Gillespie, *Wired Shut*, 26–27. As to the mattress tags, they are for the vendors, not the customers. Feel free to remove them as you see fit.
16. Gillespie, *Wired Shut*, 14, 50–64, quote at 105; Wallis and Malm, *Big Sounds from Small Peoples*, 316; Frith, "Video Pop," 117–18.
17. Adorno, *Towards a Theory of Musical Reproduction*, 3, 166.
18. The term *affordances* originally came out of psychology, design, and human-computer interaction to refer to the ways things can be used. Scholars in science and technology studies and communication have used the term as a middle ground between pure determinism and pure instrumentalism, which is how I intend to use it here. Music is not simply interchangeable with other modes of expression (and you'll have a hell of a time transmitting a long-playing record down a DSL line). At the same time, the possibilities afforded by MP3s, LPs, or music itself are nothing more than conditional at best. Just because it is possible to do something with a technology does not mean that it will be taken up that way. See Norman, *The Design of Everyday Things*; Gibson, *The Ecological Approach to Visual Perception*.
19. Heidegger, "The Thing." I realize I am taking some license in applying Heidegger's notion of the thing to digital files given his generally unfriendly relationship with the popular media of his day. But his conception of things does not require that we also accept his understanding of media, or anything else for that matter.
20. Plato, "Republic," 666 [424c]; Adorno, "On Popular Music," 439.
21. Sherburne, "Digital DJing App That Pulls You In," 46.
22. Heidegger, "The Thing," 169.
23. I am not using the term here as it is used in computer science parlance. *Container formats* are formats that can contain different types of data within them; for instance, a QuickTime movie is a container format because it contains audio, video, and synchronization data and each of these streams may itself be encoded in a different format.
24. Gitelman, *Scripts, Grooves, and Writing Machines*, 3.
25. Mumford, *The Myth of the Machine*, 4–5; see Mumford, "An Appraisal of Lewis Mumford's Technics and Civilization."
26. Sofia, "Container Technologies," 185.
27. Ibid., 188. These three categories indicate three different kinds of container

technology for Sofia. Parsing them out is not necessary for the current argument, but interested readers should refer to her article.

28. Ibid., 189.
29. "No matter how sharply we just *look* at the 'outward appearance' of Things in whatever form this takes, we cannot discover anything ready-to-hand. If we look at Things just 'theoretically,' we can get along without understanding readiness-to-hand. But when we deal with them by using them and manipulating them, this activity is not a blind one; it has its own kind of sight, by which our manipulation is guided and from which it acquires its specific Thingly character." Heidegger, *Being and Time*, 98.
30. Bull, *Sound Moves*, 148.
31. However, this is true of technologies in general, since they are always parts of systems. To channel Heidegger a little further, we encounter a room first as a room, "equipment for residing": "it is in this that any 'individual' item of equipment shows itself. *Before* it does so, a totality of equipment has already been discovered." Heidegger, *Being and Time*, 98.
32. Chun, "Software, or the Persistence of Visual Knowledge," 47.
33. Bosi and Goldberg, *Introduction to Digital Audio Coding and Standards*, 298; Hacker, MP3, 42–44; Pohlmann, *Principles of Digital Audio*, 350–54.
34. See chapters 4 and 5.
35. Karlheinz Brandenburg, author interview. Digital encoding of an ownership structure was standard procedure at the time. Other attempts to add some kind of digital rights management scheme to music have histories that stretch back into the 1980s and earlier. The RIAA's ill-fated Secure Digital Music Initiative (SDMI) also grew out of projects initiated in the early 1990s, in anticipation of some kind of online music business, but because they were in a period of relative success, they were slow to act. See Haring, *Beyond the Charts*, 42–43.
36. Author interviews with Karlheinz Brandenburg, Harald Popp.
37. My discussion is drawn from Jeffrey Tang's analysis of analog compatibility issues in component audio systems. See Tang, "Sound Decisions," 14–18. See also Cowan and Gunby, "Sprayed to Death"; David, "Clio and the Economics of QWERTY"; Liebowitz and Margolis, "Path Dependence, Lock-in and History."
38. Harald Popp, author interview. Today, MP3 benefits from the same path dependency issues that set it back against layer 2. When Apple and Nokia banded together to eliminate Ogg Vorbis compatibility from the .html specification, they were promoting proprietary audio compression schemes (including Fraunhofer's AAC, a successor to MP3) over Ogg, which is free and open source, offers comparable sound quality, and does not require developers to pay a fee for implementing it in new software. See the online discussions at Slashdot (http://yro.slashdot.org/article.pl?sid=07/12/11/1339251), Wikipe-

dia (http://en.wikipedia.org/wiki/Ogg_controversy), and WHATWG (http://lists.whatwg.org/pipermail/whatwg-whatwg.org/2007-December/013154.html). At the same time, MP3 has been able to edge out its competitors in part because of its ubiquity and compatibility. See Burkart and McCourt, *Digital Music Wars*, 47.

39. Author interviews with Karlheinz Brandenburg, Harald Popp, Bernhard Grill; Cook, "Telos Aims to Make Some Noise on Net"; Church, Grill, and Popp, "ISDN and ISO/MPEG Layer III Audio Coding"; Van Tassel, "Olympia, Telos Take on Realaudio"; Harald Popp, "Frequently Asked Questions about MPEG Audio Layer-3, Fraunhofer IIS, and All the Rest . . ." FAQ published December 1996, version 2.83, available online at ftp://ftp.comp.hkbu.edu.hk/pub/packages/layer3/l3faq.html.

40. Author interviews with Karlheinz Brandenburg and Harald Popp. Popp, "Frequently Asked Questions"; Adar, "Copyright Control for Audio-on-Demand." Apple Computer, "Apple's QuickTime 4 Goes Primetime," press release, 8 June 1999; Microsoft, "IIS Delivers Leading Multimedia Capabilities with NetShow," press release, 10 December 1996; Myrberg, "Windows Media in Consumer Electronics." The Cerberus story was pieced together from a Factiva search: Pope, "The Death of Discs"; "UK: Music Down the Telephone Live"; Pride, "U.K. Bands Attack Convention thru Internet"; "Dawn of the Electronic Our Price"; "Multimedia Business Analyst"; "EMI Signs Internet License with Cerberus"; Rawsthorn, "Internet Jukebox Company Signs US Deal"; "Cerberus Wins Industry Support." The version of WinPlay3 from 1997 is still available online at: http://www.sonicspot.com. Jeremy Morris argues that Cerberus ultimately failed because it courted industry favor by making kiosks in record stores the center of its business plan—a strategy that failed once CD burners became cheap enough that users could simply burn them at home. See Morris, "Understanding the Digital Music Commodity."

41. Popp, "Frequently Asked Questions"; author interviews with Karlheinz Brandenburg and Harald Popp. Denegri-Knott and Tadajewski, "The Emergence of MP3 Technology."

42. Haring, *Beyond the Charts*, 35–37.

43. E-mail translated by Janice Denegri-Knott in Denegri-Knott, "The Emergence of MP3s." Author interviews with Karlheinz Brandenburg and Harald Popp.

44. Knopper, *Appetite for Self-Destruction*; Haring, *Beyond the Charts*, 9, 24, 113–24; Burkart and McCourt, *Digital Music Wars*, 46. Author interview with Karlheinz Brandenburg. Indeed, all my Fraunhofer interviewees reiterated the almost militant lack of interest among labels in online distribution during the early and mid-1990s.

45. Cargill, *Information Technology Standardization*, 137–38; Piatier, "Transectorial Innovations and the Transformation of Firms," 212, 219; Théberge, *Any Sound*

You Can Imagine; Sterne, "What's Digital in Digital Music?"; Morris, "Understanding the Digital Music Commodity."

46. Although RealAudio was an early perceptual codec, I have largely left aside its story in this work. For a full discussion of early record industry attempts to sell music online and to develop other business models for digital music, see Morris, "Understanding the Digital Music Commodity."

47. "IBM Ultimedia Teams with Montage Group to Offer Low-Cost, Feature-Rich Non-Linear Video Editing Solution"; "Storage Life/Aging of Space Systems and Materials at MP3"; "Tender—Construction, Mining, Excavative and Highway Equipment"; Ward, "Scitech Ships Upgraded Multiplexer and Bandwidth Processor Cards." A *Factiva* database search of newspapers returns 13 hits for "MP3" related to music in 1997, 338 hits in 1998, and 3,447 hits in 1999.

48. Statistics for the United States indicate that as of 2000 approximately 5 million American households had broadband internet connections; by 2003, that number would surpass 50 million. Most users began with dial-up: whereas 15 percent of American homes were online in mid-1995, by 2000 over 50 percent of households were online. By mid-2006 that number would approach 80 percent. Studies done by the OECD (Organization for Economic Co-operation and Development) list 100 million internet subscribers in member nations in 2000, but only a small fraction using broadband. Another measure, by the ITU, has 400,000 computers connected directly to the internet worldwide in 1994, which rises to 29.7 million in 1998. OECD, "Information Technology Outlook 1997"; Horrigan, "Pew Internet Project Data Memo."

49. "Writable Drive Sales Set to Surge"; "CD Recordable Drives Enjoy Continued Popularity and Growth"; Mooradian, "Music."

50. Karlheinz Brandenburg, author interview.

51. Mooradian, "Music"; Haring, *Beyond the Charts*, 82–94; Burkart and McCourt, *Digital Music Wars*, 52, 89–91, 134; Leyshon, "Scary Monsters?" 546–47. Burkart and McCourt note the MP3.com's initial public offering in 1999 yielded $6.9 billion, which was half a billion dollars more than the valuation of EMI at the time. In 1997 the artists Thomas Dolby and Todd Rundgren also began to experiment with online delivery of their music. See also Morris, *Understanding the Digital Music Commodity*.

52. Burkart and McCourt, *Digital Music Wars*, 55–68; Leyshon, "Scary Monsters?" 548–53; Gillespie, *Wired Shut*, 43–50. Steve Knopper argues that Napster was the "last chance" for the RIAA labels to take control of the online distribution of music. Since all the users were in one place, if the RIAA had worked with Napster, it would have been able to reach a coherent audience. This is doubtful for a number of reasons. First, as Burkart and McCourt show, BMG did attempt to work with Napster, but its business plan ultimately didn't pan out. Further, Napster's success was in part predicated on the lack of a business plan: the lack of user fees, advertisements, digital rights management,

and data mining was the reason that users were attracted to the service. Finally, Napster was the most celebrated part of the file-sharing upsurge, but its dominance may have been overstated. Burkart and McCourt cite a study showing that Napster was only the fourth-largest online file-sharing service, even though it was the most famous. See also Knopper, *Appetite for Self-Destruction*.

53. Harald Popp, author interview.
54. Burkart and McCourt, *Digital Music Wars*, 74–75. Sony actually filed briefs on both sides of the case.
55. It has also been argued that Apple's iTunes music store was originally in part Apple's attempt to allay RIAA fears that the iPod was an inducement to piracy. Ibid., 74; Wakefield, "Millions Turn to Net for Pirate TV." See also Morris, "Understanding the Digital Music Commodity."
56. Boyd, "Pirate Radio in Britain"; Coase, *British Broadcasting*; Jones, "Making Waves"; Johns, "Piracy as a Business Force."
57. Johns, "Piracy as a Business Force." Quote at 53, emphasis in original. Boyd, "Pirate Radio in Britain"; Coase, *British Broadcasting*; Jones, "Making Waves."
58. There is an interesting parallel between the legitimacy of so-called legitimate music (to use Bourdieu's term) and the legitimacy of so-called legitimate media institutions, since both require official sanction of states (or in some cases now, supernational bodies). Once granted, legitimacy can be treated as natural or inherent, and legitimate institutions can themselves become tools of endorsement or marginalization, as when an industry seeks to define a practice as piracy, either through a PR campaign ("home taping is killing music"), a legal remedy (as in the RIAA's attempt to get a legal injunction to prevent sales of the Diamond Rio by labeling it an inducement to infringement), or through legislation (as in the United States Digital Millennium Copyright Act, which criminalized the circumvention of digital rights management schemes that otherwise would have been allowable under fair-use provisions of existing copyright law). We might therefore follow Bourdieu from the aesthetic to the political realm, since in his account politics operate according to a "logic of transubstantiation," where groups or individuals may come to "speak for" a constituency even if that constituency does not actually have any organic connection to its representative. "The mystery of the process of transubstantiation, whereby the spokesperson becomes the group he expresses, can only be explained by a historical analysis of the genesis and functioning of *representation*, through which the representative creates the group which creates him." Bourdieu, *Language and Symbolic Power*, 248, emphasis in original.
59. Wallis and Malm, *Big Sounds from Small Peoples*, 77–78, 288–89; Frith, "Video Pop," 115–19; Manuel, *Cassette Culture*, 30–31, 78–79, 81.
60. Manuel, *Cassette Culture*, 68–69, 86–87.

61. Barbrook, "The Hi-Tech Gift Economy"; Benkler, *The Wealth of Networks*. The classic argument is probably advanced by Erik Raymond in "Homesteading the Noosphere," though it echoes in important ways earlier work by Lewis Hyde. See Raymond, *The Cathedral and the Bazaar*; Hyde, *The Gift*.
62. Mauss, *The Gift*, 65–66, 128. This point is reiterated by Arjun Appadurai (via Bourdieu) in his analysis of commodity circulation. See Appadurai, "Introduction," 11–12. See also Bourdieu, *The Logic of Practice*, 98–111.
63. Leyshon, "Scary Monsters?" 542, 554.
64. Weber, *The Success of Open Source*, 149–53.
65. Kelty, *Two Bits*, 210–12, emphasis in the original. Kelty also argues that the case of open source does not conform very well with the anthropological literature on the gift (245).
66. Regnier, "Digital Music for Grown-Ups."
67. Benjamin, *Illuminations*, 67.
68. Appadurai, "Introduction," 13.
69. Gibson-Graham, *The End of Capitalism (as We Knew It)*, 261.
70. Ibid., 262.
71. Kopytoff, "The Cultural Biography of Things," 90.
72. For a sample (thank you, I'm here all week) of the existing Danger Mouse literature, see Best and Wilson, "In Celebration of the Gray Zone"; Ayers, "The Cyberactivism of a Dangermouse"; Collins, "Amen to That"; Gunkel, "Rethinking the Digital Remix"; McLeod, "Confessions of an Intellectual (Property)"; Remmer, "The Grey Album"; Serazio, "The Apolitical Irony of Generation Mash-Up." For a more general aesthetic and political account of digital sampling, see Rodgers, "On the Process and Aesthetics of Sampling in Electronic Music Production."
73. Benkler, *The Wealth of Networks*, 2.
74. Terranova, "Free Labor," 54. My argument here dovetails with at least two other perspectives on the politics of new media: Barbrook's and Cameron's claim that the "Californian Ideology" pairs hippie atomism with free market neoliberalism, which characterizes utopian discourse surrounding new media. Fred Turner takes it a step further by setting the problem in historical context. Given the 1960s left's identification of computers with the large capitalist and state institutions it resisted, it is a wonder that digital media are paired with utopianism at all. Thus, Terranova, Barbrook, and Cameron and Turner offer three angles on a crucial insight: just because digital media are experienced as radical and disruptive does not automatically render them socially progressive, anticapitalist, or more egalitarian. See Barbrook and Cameron, "The Californian Ideology"; Turner, *From Counterculture to Cyberculture*.
75. As John Durham Peters points out, the terms "communication" and "intercourse" traded meanings at some point in the nineteenth century, and the

OED entry for the term also includes references to "communion between man and that which is spiritual and unseen." Peters, *Speaking into the Air*, 8. *Oxford English Dictionary*, s.v. "Intercourse." For examples of "promiscuity" applied to file-sharing by otherwise sympathetic commentators, see, e.g., Bennett, Shank, and Toynbee, "Introduction," 4; Benkler, *The Wealth of Networks*, 423. The analogy is clearly meant to suggest that operating within the bounds of the legitimate music economy is like having relatively few sexual partners: more moral, more honorable, more faithful, more respectable, more legitimate and so on, even if used ironically. There is probably a fuller story to tell about the gender and sexual dynamics of the "promiscuous" epithet, since sexual traffic is social traffic, given feminist and queer critiques of the politics of monogamy (serial or otherwise). See Rubin, "The Traffic in Women"; Warner, *The Trouble with Normal*.
76. Burkart and McCourt, *Digital Music Wars*, 136–37.
77. Leyshon, "Scary Monsters?" 553.
78. My argument here follows C. B. MacPherson's definition of property in liberal societies as in the first instance, the right to manage (and, by extension, alienate) one's own "productive energies and life." MacPherson, *The Political Theory of Possessive Individualism*, 142–48, quote at 143.
79. Szendy, *Listen*, 8, 10, passim, emphasis in original.
80. Simmel, "The Metropolis and Mental Life," 325–32; Simmel, *The Philosophy of Money*, 255–58; Gitlin, *Media Unlimited*, 36–44.
81. The definition comes from Augé, *Non-Places*, 76. But the idea has considerably more history. Wolfgang Schivelbusch shows that the rail car was thought of as a non-place in the early nineteenth century, and several other authors have commented on this distinctive feature of modern life. Although my argument is about interaction inside a mediatic space, there is a tradition of talking about mediatic experience and perception as taking the perceiver out of place into a kind of non-place (or as a feature of non-places). Margaret Morse and Anna McCarthy both connect television with non-places. Joshua Meyerowitz offers a more Simmelian account to argue that mediatic experience actually destroys rootedness in space; Schivelbusch notes the connections between train travel and reading practices and Augé notes the same phenomenon in air travel; some of Michael Bull's interviewees also characterize their experience of walking in the city with iPods as moving through non-places. See Morse, "An Ontology of Everyday Distraction"; McCarthy, *Ambient Television*; Schivelbusch, *The Railway Journey*; Bull, *Sound Moves*; Meyerowitz, *No Sense of Place*.
82. Augé, *Non-Places*, 76.
83. "Against the backdrop of the metro our individual acrobatics thus seem to play a fortuitously calming effect in the destiny of everyone's daily lives, in the laws of human action summed up by a few commonplaces and symbol-

ized by a strange public place — an interlacing of routes whose several explicit prohibitions ('no smoking,' 'no entry') underscore its collective and ruled character. It is thus quite obvious that everyone has his or her 'life to live' in the metro, that life cannot be lived in a total freedom, not simply because no freedom could ever be totally lived in society at large, but more precisely because the coded and ordered character of subway traffic imposes on each and every person codes of conduct that cannot be transgressed without running the risk of sanction." Augé, *In the Metro*, 29.
84. Leyshon, "Scary Monsters?" 554.
85. Kassabian, "Ubisub," paragraph 15.
86. See Turner, "Burning Man at Google."

THE END OF MP3

1. AAC was the consensus choice over MP3 for everyone I interviewed who worked on the original MPEG standard.
2. Sumner and Gooday, "Introduction," 6. See also Edgerton, *The Shock of the Old*; Lipartito, "Picturephone and the Information Age."
3. Bowker and Star, *Sorting Things Out*, 14; Tang, "Sound Decisions."
4. Swalwell, "The Remembering and the Forgetting of Early Digital Games," 264.
5. This discussion is based on a personal conversation with Samuel Brylawski, former head of the Recorded Sound Division at the Library of Congress and Mark Katz, "The Second Digital Revolution in Music," Music Library Association Meeting (Pittsburgh, 2007).
6. I develop this line of argument in Sterne, "The Preservation Paradox."
7. Gopinath, "Ringtones, or the Auditory Logic of Globalization." For a discussion of audio compression technologies and mobile telephony, see Autti and Biström, "Mobile Audio." The claims about usage come from the NPD Group, "Convergence Drives Entertainment Opportunities of Cell Phones: Growing Usage of Multimedia Continues to Trail Voice," press release available at the NPD Group's website, http://www.npd.com.
8. That pattern describes historical arcs charted in this book as well as *The Audible Past*.
9. Kuhn, *The Structure of Scientific Revolutions*.
10. Jon Gertner, "Our Ratings, Ourselves," 37. On the history and politics of ratings, see Meehan, "Why We Don't Count"; Meehan, "Heads of Households and Ladies of the House."
11. Mowitt, "The Sound of Music in the Era of Its Electronic Reproducibility"; Rose, *Black Noise*; Toop, "Replicant"; Waksman, *Instruments of Desire*; Walser, *Running with the Devil*; Manuel, *Cassette Culture*, 82.
12. In 1975 the Aphex corporation introduced a device called the *aural exciter*, which added clarity and sharpness to a sound signal essentially by adding a

bit of distortion to a specific frequency range in the original signal, then mixing together the distorted and original signals. It was the closest thing in the recording business to a "better" button and aimed to overcome some of the problems introduced in tape recording. In the first few years of its existence, the unit was only available to rent, for $30 per minute of finished audio. Today, a wide range of such devices is now available under the generic name of *psychoacoustic enhancers*, which generally aim to shape the listener's experience of the frequency content of recorded audio. As with the audio compression and distortion techniques, their use helped to define the sound of modern mass-marketed music. Hugh Robjohns, "Enhancers: Frequently Asked Questions," *Sound on Sound* (January 2000), Sound on Sound (http://www.soundonsound.com). The word *psychoacoustics* itself has become a marketing buzzword in pro audio. For instance, ad copy for Omnisphere, a software synthesizer by Spectrasonics now in heavy use for film and television soundtracks, touts its "psychoacoustic sampling." Most of the examples given—like amplifying the filament of a light bulb to make an audible tone—are clever, but they have no clear connection with actual psychoacoustic principles in their conception and design. See the Omnisphere page of the Spectrasonics website, http://spectrasonics.net/products/omnisphere.php. On the avant-garde, see Born, *Rationalizing Culture*.

13. The "death of the author" is an old thesis, but recent work on remixing and reuse suggests its renewed urgency as a proposition. Barthes, *Image-Music-Text*; Lee and LiPuma, "Cultures of Circulation"; MacKenzie, "The Performativity of Code"; Sinnreich, *Mashed Up*; McLeod and Kuenzli, *Cutting across Media*.

14. Author interviews with Bosi, Johnston. Johnston quote from a PowerPoint presentation titled "Perceptual Coding—A Wide-Ranging Tutorial on Perceptual Audio Coding (3 Hours Minimum)," available at the Audio Engineering Society's website, http://www.aes.org.

15. Krauss, "Two Moments from the Post-Medium Condition," 55–56. Thanks to Christine Ross and Amelia Jones for the primer on "medium" in visual art and to tobias van Veen for demonstrating the different ways that records, .wavs, and .mp3s break down.

16. Schivelbusch, *The Railway Journey*, 64–60; Augé, *Non-Places*, 2–5, 76–77. Anna McCarthy makes a similar point about television in McCarthy, *Ambient Television*.

17. The uses of the iPod are a cross section from Bull, *Sound Moves*. The quote is from Grossberg, *Dancing in Spite of Myself*, 77.

18. Michael Bull to author, 5 February 2009. He identifies the interviewee as a thirty-four-year-old woman. Many thanks to Michael for sharing this outtake from his book with me.

19. Grevatt, "Pros and Cons Have Their Say on K. C. Speech," 14.

20. Keightley, "'Turn It Down!' She Shrieked," 150, 156.
21. Anderson, *Making Easy Listening*.
22. Bemis, "The Digital Devolution."
23. De Avila, "The Home Stereos That Refuse to Die." Speaker researchers have told me that their impression is that in the 1990s companies felt they could sell more speakers if there was only one bass speaker and the rest of the components of a surround system could be smaller, and more heavily designed to be integrated into home fashions. The claim is anecdotal at this point, but it would be interesting to know how much the "bass is nondirectional" proposition is based on ideology and marketing, rather than fact.
24. As Karin Bijsterveld has pointed out, the car has been promoted as a "mobile living room" since the 1940s. Bijsterveld, "Acoustic Cocooning," 198. See also, Cleophas and Bijsterveld, "Selling Sound"; Bull, "Automobility and the Power of Sound."
25. Mansell and Raboy, "Introduction"; Vaidhyanathan, *The Googlization of Everything (and Why We Should Worry)*; Gillespie, *The New Gatekeepers*.
26. Plomp, *The Intelligent Ear*, quote at 152–53, passim.
27. Shannon, "The Bandwagon," 3.
28. James Gleick's *The Information* is the latest popular text to do it. As Cory Doctorow puts in his review, Gleick treats information theory "as a toolkit for disassembling the world." Doctorow, "James Gleick's Tour-de-force: *The Information*, a Natural History of Information Theory," BoingBoing, http://www.boingboing.net. James Gleick, *The Information*.
29. Minsky, "Steps toward Artificial Intelligence," 27.
30. Bolter, *Turing's Man*, 221–22.
31. It is worth noting that there is still very little truly cross-cultural research in auditory psychology and the psychology of music (speech perception is somewhat more advanced because of the obvious connections between speech and language).
32. Cavarero, *For More Than One Voice*, 7, 9.
33. Radovac, "The 'War on Noise'"; Ostertag, "Why Computer Music Sucks"; Mills, "Do Signals Have Politics?"

INTERVIEWS

In addition to consulting documentary sources, I conducted a handful of interviews for this book. My goal was to flesh out some details left vague in the written record, and also to get at some of the more tacit forms of knowledge that would not have been written down. Although I was able to talk with a number of important people in the history of perceptual coding, another handful of interview requests were declined or in some cases simply did not receive a response. Thus, these interviews are more of a supplement to the written material I consulted than any kind of complete oral history in themselves. Below is a list of my interviewees and brief descriptions of who they are in my story. My thanks go out to each of them.

BISHNU ATAL, interviewed 21 February 2007. Atal worked at AT&T Bell Labs from 1962 to 2002, where he researched speech analysis, synthesis, and coding, as well as speech recognition. Atal is one of the inventors of adaptive predictive coding, and is coauthor of one of the first articles to propose perceptual coding. His work was especially important for the development of cellular telephony.

OSCAR BONELLO, interviewed over a series of e-mail exchanges, June–July 2008. Bonello founded Solidyne, an Argentinian company dedicated to research on professional audio, and was a professor of engineering at the University of Buenos Aires. With a team in the 1980s, Bonello developed a PC-based automation system for radio called Audicom, which used perceptual coding to save bandwidth. They did not submit to MPEG's competition.

MARINA BOSI, interviewed 9 February 2009. Bosi was an early employee of Digidesign—which built equipment to allow computers to be used in recording studios. She was also the first woman engineer at Dolby, where she worked on their AC-3 audio standard (which includes perceptual coding). She worked at MPEG during the 1990s in a number of capacities, and served as editor for the MPEG-2 AAC (Advanced Audio Coding) standard, a successor to MP3.

KARLHEINZ BRANDENBURG, interviewed 23 August 2006 and 3 October 2007. While at school in Erlangen, Brandenburg met Dieter Seitzer, a professor who had organized a research group on what is now called perceptual coding. That became the subject of Brandenburg's graduate and postdoctoral work (including a stint with JJ Johnston at AT&T in 1989–90). He played a central role in developing OCF (Optical Coding in the Frequency Domain—a predecessor of MP3) and ASPEC (one of the competing standards considered by MPEG, which became the core of MP3), as well as AAC (Advanced Audio Coding). After the establishment of the MPEG standard, he was also involved in Fraunhofer's attempts at marketing MP3.

JOEL FLEISCHER, interviewed 23 June 2005 and 4 March 2011. Fleischer is vice president of sales and marketing at BigChampagne, an audience measurement company. BigChampagne developed techniques for tracking peer-to-peer file-sharing traffic online and was discreetly employed by major labels seeking to better understand their audiences.

BERNHARD GRILL, interviewed 27 May 2008. After meeting Dieter Seitzer and Karlheinz Brandenburg in school, Grill joined Fraunhofer in 1988 and became heavily involved with the development of audio coding algorithms and perceptual testing, co-writing the code for their first real-time perceptual coder. Grill later participated in MPEG's listening tests at Swedish Radio and went on to write Fraunhofer's proposal for what would become AAC.

JOSEPH HALL, interviewed 24 October 2008. Hall was a psychoacoustician at Bell Labs from 1966, where he studied auditory perception. Hall developed a working model of the cochlea and he is coauthor on one of the first papers to propose perceptual coding. In the 1980s, he consulted with JJ Johnston on psychoacoustics as Johnston built Bell's first working perceptual coder.

JJ JOHNSTON, interviewed 22 February 2007. Johnston was an engineer at Bell Labs, where he developed their first working perceptual coder as a way to test out a new Unix supercomputer in the early 1980s. He also contributed to ASPEC, as well as AAC. He has also worked on perceptual coding of images, as well as a wide range of other areas related to digital audio and signal processing.

HARALD POPP, interviewed 18 April 2007. Popp has worked for Fraunhofer, where he developed a range of different real-time processors, and helped build the ASPEC coder for the MPEG competition. Following the establishment of the MPEG standard, Popp moved to marketing, where he worked to commercialize Fraunhofer's technology (in a hardware box called ASPEC '91, then in software called L3Enc and L3Dec and finally as MP3).

BILL SPURLIN, interviewed 10 July 2007. While working as operations manager at Christian Science radio in the late 1980s, Spurlin oversaw the installation of a Fraunhofer OCF system. When the network ran into financial trouble and scaled back its radio operations in 1993, Spurlin took a proposal to commercialize perceptual coding in the United States to the weekly meeting of a group of venture capitalists at MIT. None of them saw any future in the technology.

LOUIS THIBAULT, interviewed 9 August 2005. As an engineer at the Communication Research Centre in Ottawa, Thibault became the Canadian delegate to major international radio standards organizations (the CCIR, which became the ITU-R). He participated in the development of standards for integrated digital audio broadcasting systems, and organized listening tests on perceptual audio codecs, including MUSICAM (which became layer 2 of the MPEG-1 standard) as well as AAC.

BIBLIOGRAPHY

Abbate, Janet. *Inventing the Internet*. Cambridge: MIT Press, 2000.

Adar, Ricky. "Copyright Control for Audio-on-Demand." In *Audio Engineering Society UK 11th Conference: Audio for New Media*. London, March 1996.

Adorno, Theodor. "Culture and Administration." *Telos* 37 (1978): 93–111.

———. *Negative Dialectics*. Translated by E. B. Ashton. New York: Continuum, 1973.

———. "On Popular Music." In *Adorno: Essays on Music*, edited by Richard Leppert, 437–69. Berkeley: University of California Press, 2002.

———. "On the Fetish Character of Music and the Regression of Listening." In *Adorno: Essays on Music*, edited by Richard Leppert, 288–317. Berkeley: University of California Press, 2002.

———. "Opera and the Long Playing Record." *October* 55 (1990): 62–66.

———. "A Social Critique of Radio Music." In *Radiotext(E)*, edited by Neil Strauss, 272–79. New York: Semiotext(e), 1993.

———. *Towards a Theory of Musical Reproduction*. Translated by Wieland Hoban. Malden, Mass.: Polity Press, 2006.

Agamben, Giorgio. *Homo Sacer: Sovereign Power and Bare Life*. Translated by Daniel Heller-Roazen. Stanford: Stanford University Press, 1998.

Ahmed, Murad. "Young Music Fans Deaf to iPod's Limitations." *London Times Online*, 5 March 2009.

Ahmed, Sara. *Queer Phenomenology: Orientations, Objects, Others*. Durham: Duke University Press, 2006.

Akrich, Madeleine. "The De-Scription of Technical Objects." In *Shaping Technology, Building Society: Studies in Sociotechnical Change*, edited by Wiebe Bijker and J. Law, 205–24. Cambridge: MIT Press, 1992.

American School of Correspondence. *Cyclopedia of Applied Electricity*. Vol. 6. Chicago: American School of Correspondence at Armour Institute of Technology, 1911.

Anderson, Tim. *Making Easy Listening: Material Culture and Postwar American Recording*. Minneapolis: University of Minnesota Press, 2006.

Appadurai, Arjun. "Introduction: Commodities and the Politics of Value." In *The Social Life of Things: Commodities in Cultural Perspective*, edited by Arjun Appadurai, 3–63. Cambridge: Cambridge University Press, 1988.

Arnheim, Rudolf. *Radio*. Translated by Margaret Ludwig and Herbert Read. London: Faber and Faber, 1936.

Arrango, Tim. "Judge Tells LimeWire, the File-Trading Service, to Disable Its Software." *New York Times*, 27 October 2010.

Atal, Bishnu S. "50 Years of Progress in Speech Waveform Coming." *Echoes: Newsletter of the Acoustical Society of America* 15, no. 1 (2005): 1, 4.

———. "The History of Linear Predictive Coding." *IEEE Signal Processing Magazine* 49, no. 10 (2006): 154–61.

Atal, Bishnu S., and Manfred R. Schroeder. "Adaptive Predictive Coding of Speech Signals." *Bell Systems Technical Journal* 49, no. 8 (1970): 1973–86.

———. "Predictive Coding of Speech Signals and Subjective Error Criteria." Paper presented at the International Conference on Acoustics, Speech and Signal Processing, Tulsa, Oklahoma, 1978.

Atkinson, John. "MP3 and the Marginalization of High End Audio." *Stereophile*, February 1999.

Attali, Jacques. *Noise: The Political Economy of Music*. Translated by Brian Massumi. Minneapolis: University of Minnesota Press, 1985.

Augé, Marc. *In the Metro*. Translated by Tom Conley. Minneapolis: University of Minnesota Press, 2002.

———. *Non-Places: An Introduction to Supermodernity*. 2nd ed. London: Verso, 2008.

Auslander, Philip. *Liveness: Performance in a Mediatized Culture*. New York: Routledge, 1999.

Autti, Henri, and Johnny Biström. "Mobile Audio: From MP3 to AAC and Further." Telecommunications Software and Multimedia Laboratory, Helsinki University of Technology, http://people.arcada.fi/~johnny/MobileAudio7.pdf.

Ayers, Michael. "The Cyberactivism of a Dangermouse." In *Cybersounds: Essays on Virtual Music Culture*, edited by Michael Ayers, 127–36. New York: Peter Lang, 2006.

———, ed. *Cybersounds: Essays on Virtual Music Cultures.* New York: Peter Lang, 2006.

Bar, François, Michael Borrus, and Richard Steinberg. "Interoperability and the NII: Mapping the Debate." *Information Infrastructure and Policy* 4, no. 4 (1995): 235–54.

Baran, Paul, and Paul Sweezy. *Monopoly Capital: An Essay on the American Social and Economic Order.* New York: Monthly Review Press, 1966.

Barbrook, Richard. "The Hi-Tech Gift Economy." *First Monday* 3, no. 12 (1998), http://firstmonday.org/.

Barbrook, Richard, and Andy Cameron. "The Californian Ideology." *The Hypermedia Research Centre* (1995), http://www.hrc.wmin.ac.uk.

Barnett, Kyle. "Furniture Music: The Phonograph as Furniture, 1900–1930." *Journal of Popular Music Studies* 18, no. 3 (2006): 301–24.

Barney, Darin. "Politics and Emerging Media: The Revenge of Publicity." *Global Media Journal—Canadian Edition* 1, no. 1 (2008): 89–106.

Barry, Andrew. *Political Machines: Governing a Technological Society.* New York: Athelone, 2001.

Barthes, Roland. *Image-Music-Text.* Translated by Stephen Heath. New York: Hill and Wang, 1977.

Baudrillard, Jean. *For a Critique of the Political Economy of the Sign.* Translated by Charles Levin. St. Louis: Telos, 1981.

Bazett, H. C., and W. G. Penfield. "A Study of the Sherrington Decerebrate Animal in the Chronic as Well as the Acute Condition." *Brain* 45, no. 2 (1922): 185–265.

Bech, Søren. "Selection and Training of Subjects for Listening Tests on Sound-Reproducing Equipment." *Journal of the Audio Engineering Society* 40, no. 7–8 (1992): 590–610.

Bech, Søren, and Nick Zacharov. *Perceptual Audio Evaluation: Theory, Method and Application.* Hoboken: John Wiley and Sons, 2006.

Beebe-Center, John G. *The Psychology of Pleasantness and Unpleasantness.* New York: D. Van Nostrand, 1932.

Békésy, Georg von. *Experiments in Hearing.* Translated by Ernest Glen Wever. Toronto: McGraw-Hill, 1960.

Békésy, Georg von, and Walter Rosenblith. "The Early History of Hearing—Observations and Theories." *Journal of the Acoustical Society of America* 20, no. 6 (1948): 727–48.

Bemis, Alec Hanley. "The Digital Devolution." *Wired*, July 2005, 56.

Benjamin, Walter. *Illuminations.* Translated by Harry Zohn. New York: Schocken, 1968.

Benkler, Yochai. *The Wealth of Networks: How Social Production Transforms Markets and Freedom.* New Haven: Yale University Press 2006.

Bennett, Andy, Barry Shank, and Jason Toynbee. "Introduction." In *The Popular*

Music Studies Reader, edited by Andy Bennett, Barry Shank, and Jason Toynbee, 1–9. New York: Routledge, 2005.

Beranek, Leo. "Criteria for Office Quieting Based on Questionnaire Rating Studies." *Journal of the Acoustical Society of America* 28, no. 5 (1956): 833–52.

———. "Revised Criteria for Noise in Buildings." *Noise Control* 3, no. 1 (1957): 19–27.

Beranek, Leo, J. J. Reynolds, and K. E. Wilson. "Apparatus and Procedures for Predicting Ventilation System Noise." *Journal of the Acoustical Society of America* 25, no. 2 (1953): 313–21.

Bergman, Sten, Christer Grewin, and Thomas Rydén. "MPEG/Audio Subjective Assessments Test Report." Stockholm, Sweden: International Organization for Standardization ISO/IEC JTC1/SC2/WG8, 1990.

———. "The SR Report on the MPEG/Audio Subjective Listening Test April/May 1991." Stockholm: International Organization for Standardization ISO/IEC JTC1/SC2/WG11, 1991.

Berland, Jody. "Cat and Mouse: Iconographies of Nature and Desire." *Cultural Studies* 22, no. 3–4 (2008): 431–54.

Best, Michael L., and Ernest J. Wilson. "In Celebration of the Gray Zone." *Information Technologies and International Development* 3, no. 2 (2006): iii–iv.

Bettig, Ronald. "The Enclosure of Cyberspace." *Critical Studies in Mass Communication* 14, no. 2 (1997): 138–58.

Bijker, Wiebe. *Of Bicycles, Bakelites, and Bulbs: Toward a Theory of Sociotechnical Change.* Cambridge: MIT Press, 1995.

Bijsterveld, Karin. "Acoustic Cocooning: How the Car Became a Place to Unwind." *Senses and Society* 5, no. 2 (2010): 189–211.

———. *Mechanical Sound: Technology, Culture, and Public Problems of Noise in the Twentieth Century.* Cambridge: MIT Press, 2008.

Blume, Stuart. "Cochlear Implantation: Establishing Clinical Feasibility, 1957–1982." In *Sources of Medical Technology: Universities and Industry*, edited by Nathan Rosenberg, Annetine C. Gelijns, and Holly Dawkins, 97–124. Washington: National Academy Press, 1995.

Boddy, William. *Fifties Television: The Industry and Its Critics.* Illinois Studies in Communications. Urbana: University of Illinois Press, 1990.

Bogost, Ian. *Unit Operations: An Approach to Video Game Criticism.* Cambridge: MIT Press, 2006.

Bohr, Niels. "Quantum Physics and Biology." *Symposia of the Society for Experimental Biology* 14, no. 1 (1960): 1–5.

Bolter, J. David. *Turing's Man: Western Culture in the Computer Age.* Chapel Hill: University of North Carolina Press, 1984.

Bolter, Jay, and Richard Grusin. *Remediation: Understanding New Media.* Cambridge: MIT Press, 2000.

Bonello, Oscar. "PC-Controlled Psychoacoustic Audio Processor." Paper presented at the 94th Annual Convention of the Audio Engineering Society, Berlin, 1993.

Borger, Julian. "Project: Acoustic Kitty." *Guardian Unlimited*, 11 September 2001.

Boring, Edwin. "Auditory Theory, with Special Reference to Intensity, Volume and Localization." *American Journal of Psychology* 37, no. 7 (1926): 157–88.

———. "The Beginning and Growth of Measurement in Psychology." *Isis* 52, no. 2 (1961): 238–57.

———. *A History of Experimental Psychology*. 2nd ed. New York: Century, 1950.

Born, Georgina. *Rationalizing Culture: IRCAM, Boulez, and the Institutionalization of the Musical Avant-Garde*. Berkeley: University of California Press, 1995.

Bosi, Marina, and Richard E. Goldberg. *Introduction to Digital Audio Coding and Standards*. New York: Springer, 2002.

Boudreau, John. "Illegal File Sharing Showing No Letup." *Seattle Times*, 3 July 2006, 1.

Bourdieu, Pierre. *Distinction: A Social Critique of the Judgement of Taste*. Translated by Richard Nice. Cambridge: Harvard University Press, 1984.

———. *Language and Symbolic Power*. Translated by Gino Raymond and Matthew Adamson. Cambridge: Harvard University Press, 1991.

———. *The Logic of Practice*. Translated by Richard Nice. Stanford: Stanford University Press, 1990.

———. "Men and Machines." In *Advances in Social Theory and Methodology: Toward an Integration of Micro- and Macro-Sociologies*, edited by K. Knorr-Cetina and A. V. Cicourel, 304–17. Boston: Routledge and Kegan Paul, 1981.

———. "Program for a Sociology of Sport." *Sociology of Sport Journal* 5, no. 2 (1988): 153–61.

Bourdieu, Pierre, and Loic J. D. Wacquant. *An Invitation to Reflexive Sociology*. Chicago: University of Chicago Press, 1992.

Bowker, Geoffrey C. "How to Be Universal: Some Cybernetic Strategies, 1943–1970." *Social Studies of Science* 23, no. 1 (1993): 107–27.

Bowker, Geoffrey C., and Susan Leigh Star. *Sorting Things Out: Classification and Its Consequences*. Cambridge: MIT Press, 1999.

Boyd, Douglas. "Pirate Radio in Britain: A Programming Alternative." *Journal of Communication* 36, no. 2 (1986): 83–94.

Boyle, Jen E. *Anamorphosis in Early Modern Literature: Mediation and Affect*. London: Ashgate, 2010.

Brady, Erika. *A Spiral Way: How the Phonograph Changed Ethnography*. Jackson: University Press of Mississippi, 1999.

Braman, Sandra. *Change of State: Information, Policy, and Power*. Cambridge: MIT Press, 2007.

Braman, Sandra, and Stephanie Lynch. "Advantage ISP: Terms of Service as Media Law." *New Media and Society* 5, no. 3 (2003): 422–48.

Brandenburg, Karlheinz. "Evaluation of Quality for Audio Encoding at Low Bit

Rates." Paper presented at the 82nd Annual Convention of the Audio Engineering Society, London, 1987.

———. "High Quality Sound Coding at 2.5 Bit/Sample." Paper presented at the Audio Engineering Society, Los Angeles, 3–6 November 1988.

———. "MP3 and AAC Explained." Paper presented at the AES 17th International Conference on High Quality Audio Encoding, Florence, Italy, 1999.

Brandenburg, K., R. Kapust, D. Seitzer, E. Eberlein, H. Gerhäuser, B. Brill, S. Krägeloh, H. Popp, and H. Schott. "Low Bit Rate Codecs for Audio Signals: Implementation in Real Time." Paper presented at the Audio Engineering Society, Los Angeles, 3–6 November 1988.

Brandenburg, Karlheinz, and Dieter Seitzer. "OCF: Coding High Quality Audio with Data Rates of 64 Kbit/Sec." Paper presented at the Audio Engineering Society, Los Angeles, 3–6 November 1988.

Braudel, Fernand. "History and Social Science." In *Economy and Society in Early Modern Europe*, edited by Peter Burke, 11–42. London: Routledge and Kegan Paul, 1972.

Brennan, Teresa. "Why the Time Is out of Joint: Marx's Political Economy without the Subject." *South Atlantic Quarterly* 97, no. 2 (1998): 263–80.

Brock, Gerald. *Telecommunication Policy for the Information Age: From Monopoly to Competition*. Cambridge: Harvard University Press, 1994.

Brooks, John. *Telephone: The First Hundred Years*. New York: Harper and Row, 1976.

Brown, Bill. "Thing Theory." *Critical Inquiry* 28, no. 1 (2001): 1–22.

Bruce, Robert V. *Bell: Alexander Graham Bell and the Conquest of Solitude*. Boston: Little, Brown, 1973.

Bull, Michael. "Automobility and the Power of Sound." *Theory, Culture and Society* 21, no. 4–5 (2004): 243–59.

———. *Sounding Out the City: Personal Stereos and Everyday Life*. New York: NYU Press, 2000.

———. *Sound Moves: iPod Culture and Urban Experience*. New York: Routledge, 2008.

Burkart, Patrick. *Music and Cyberliberties*. Middletown, Conn.: Wesleyan University Press, 2010.

Burkart, Patrick, and Tom McCourt. *Digital Music Wars*. New York: Rowman and Littlefield, 2006.

Burks, A. W. "Logic, Biology and Automata—Some Historical Reflections." *International Journal of Man-Machine Studies* 7, no. 3 (1975): 297–312.

Buxton, William. "From Radio Research to Communications Intelligence: Rockefeller Philanthropy, Communications Specialists, and the American Policy Community." In *Commmunication Researchers and Policy-Making*, edited by Sandra Braman, 295–346. Cambridge: MIT Press, 2003.

Canguilhem, Georges. *The Normal and the Pathological*. Translated by Carolyn R. Fawcett and Robert S. Cohen. New York: Zone Books, 1991.

Cantril, Hadley, and Gordon Allport. *The Psychology of Radio*. New York: Harper and Row, 1935.

Carey, Mark, and David Wall. "MP3: The Beat Bytes Back." *International Review of Law, Computers and Technology* 15, no. 1 (2001): 35–58.

Cargill, Carl F. *Information Technology Standardization: Theory, Process, and Organizations*. Bedford, England: Digital Press, 1989.

Carlin, Sidney, W. Dixon Ward, Arthur Gershon, and Rex Ingraham. "Sound Stimulation and Its Effect on the Dental Sensation Threshold." *Science* 138 (1962): 1258–59.

Carterette, Edward C., and Morton P. Friedman, eds. *Handbook of Perception*. Vol. 8, *Perceptual Coding*. New York: Academic, 1978.

Cassidy, Jane W., and Karen M. Ditty. "Gender Differences among Newborns on a Transient Otoacoustic Emissions Test for Hearing." *Journal of Music Therapy* 38, no. 1 (2001): 28–35.

Cavarero, Adriana. *For More Than One Voice: Toward a Philosophy of Vocal Expression*. Stanford: Stanford University Press, 2005.

"CD Recordable Drives Enjoy Continued Popularity and Growth." *Data Storage Report*, 1 September 1995.

Chanan, Michael. *Musica Practica: The Social Practice of Western Music from Gregorian Chant to Postmodernism*. London: Verso, 1994.

Chandler, Alfred. *The Visible Hand: The Managerial Revolution in American Business*. Cambridge: Harvard University Press, 1977.

Chion, Michel. *Audio-Vision*. Translated by Claudia Gorbman. New York: Columbia University Press, 1994.

Chow, Rey. *The Protestant Ethnic and the Spirit of Capitalism*. New York: Columbia University Press, 2002.

Chun, Wendy Hui Kyong. "Software, or the Persistence of Visual Knowledge." *Grey Room* 18 (2004): 26–51.

Church, Steve, Bernhard Grill, and Harald Popp. "ISDN and ISO/MPEG Layer III Audio Coding: Powerful New Tools for Broadcast Audio Distribution." Paper presented at the Audio Engineering Society conference, New York City, 7–10 October 1993.

Cleophas, Eefje, and Karin Bijsterveld. "Selling Sound: Testing, Designing, and Marketing Sound in the European Car Industry." In *The Oxford Handbook of Sound Studies*, edited by Karin Bijsterveld and Trevor Pinch. New York: Oxford University Press, 2011.

Coase, Ronald. *British Broadcasting: A Study in Monopoly*. Cambridge: Harvard University Press, 1950.

Cockayne, Emily. *Hubbub: Filth, Noise, and Stench in England, 1600–1770*. New Haven: Yale University Press, 2007.

Collins, Steve. "Amen to That: Sampling and Adapting the Past." *M/C* 10, no. 2 (2007), http://journal.media-culture.org.

Cook, Bob. "Telos Aims to Make Some Noise on Net." *Crains Cleveland Business* 20 (1996), http://www.factiva.com.

Corker, Mairian. "Sensing Disability." *Hypatia* 16, no. 4 (2001): 34–52.

Corso, John F. "Age and Sex Differences in Pure-Tone Thresholds." *Journal of the Acoustical Society of America* 31, no. 4 (1959): 498–507.

Cowan, Robin, and Philip Gunby. "Sprayed to Death: Path-Dependence, Lock-in and Pest Control Strategies." *Economic Journal* 106, no. 436 (1996): 521–42.

Crafts, Susan D., Daniel Cavacci, Charles Keil, and the Music in Daily Life Project, eds. *My Music: Explorations of Music in Everyday Life.* Hanover, N.H.: Wesleyan University Press and the University Press of New England, 1993.

Danziger, Kurt. *Constructing the Subject: Historical Origins of Psychological Research.* New York: Cambridge University Press, 1990.

Darnton, Robert. *The Great Cat Massacre and Other Episodes in French Cultural History.* New York: Vintage Books, 1985.

David, Paul A. "Clio and the Economics of QWERTY." *American Economic Review* 75, no. 2 (1985): 332–37.

David, Paul, and Geoffrey S. Rothwell. "Standardization, Diversity and Learning: Strategies for the Coevolution of Technology and Industrial Capacity." Stanford: Center for Economic Policy Research, Stanford University, 1994.

Davis, Audrey B., and Uta C. Merzbach. *Early Auditory Studies: Activities in the Psychology Laboratories of American Universities.* Washington: Smithsonian Institution Press, 1975.

Davis, Hallowell. "The Electrical Phenomena of the Cochlea and the Auditory Nerve." *Journal of the Acoustical Society of America* 6, no. 4 (1935): 205–15.

———. "Psychological and Physiological Acoustics: 1920–1942." *Journal of the Acoustical Society of America* 61, no. 2 (1977): 264–66.

Davis, Lennard. *Enforcing Normalcy.* New York: Verso, 1995.

"Dawn of the Electronic Our Price." *Independent—London,* 9 June 1995, http://www.factiva.com.

De Avila, Joseph. "The Home Stereos That Refuse to Die." *Wall Street Journal,* 4 February 2010.

De Landa, Manuel. *War in the Age of Intelligent Machines.* New York: Zone Books, 1991.

Deleuze, Gilles. *Cinema 1: The Movement Image.* Translated by Hugh Tomlinson and Barbara Habberjam. New York: Continuum, 2001.

Deleuze, Gilles, and Felix Guattari. *Anti-Oedipus: Capitalism and Schizophrenia.* New York: Viking Books, 1977.

———. *A Thousand Plateaus: Capitalism and Schizophrenia.* Vol. 2. Translated by Brian Massumi. Minneapolis: University of Minnesota Press, 1987.

Denegri-Knott, Janice. "The Emergence of MP3s." Unpublished manuscript, Bournemouth University, 2008.

Denegri-Knott, Janice, and Mark Tadajewski. "The Emergence of MP3 Technology." *Journal of Historical Research in Marketing* 2, no. 4 (2010): 397–425.

Denora, Tia. *Music in Everyday Life*. New York: Cambridge University Press, 2000.

De Sonne, Marcia L. *Digital Audio Broadcasting: Status Report and Outlook*. Washington: National Association of Broadcasters, 1990.

Deutsch, Diana, Trevor Henthorn, and Mark Dolson. "Absolute Pitch, Speech, and Tone Language: Some Experiments and a Proposed Framework." *Music Perception* 21, no. 3 (2004): 339–56.

Doelle, Leslie. *Environmental Acoustics*. Toronto: McGraw-Hill, 1972.

Douglas, Susan. *Inventing American Broadcasting: 1899–1922*. Baltimore: Johns Hopkins University Press, 1987.

———. *Listening In: Radio and the American Imagination from Amos 'N Andy and Edward R. Murrow to Wolfman Jack and Howard Stern*. New York: Times Books/Random House, 1999.

Doyle, Peter. *Echo and Reverb: Fabricating Space in Popular Music Recording, 1900–1960*. Middletown, Conn.: Wesleyan University Press, 2005.

Dreyfuss, Henry. *Human Factors in Design*. New York: Whitney Library of Design, 1967.

Dudley, Homer. "Remaking Speech." *Journal of the Acoustical Society of America* 11, no. 2 (1939): 169–77.

———. "Synthesizing Speech." *Bell Laboratories Record* 15, no. 12 (1936): 98–102.

du Gay, Paul, Stuart Hall, Keith Negus, Hugh Mackay, and Linda Janes. *Doing Cultural Studies: The Story of the Sony Walkman*. Thousand Oaks, Calif.: Sage, 1997.

Durant, Alan. *Conditions of Music*. London: Macmillan, 1984.

Dyson, Frances. *Sounding New Media: Immersion and Embodiment in the Arts and Culture*. Berkeley: University of California Press, 2009.

Edgerton, David. *The Shock of the Old: Technology and Global History since 1900*. New York: Oxford University Press, 2007.

Edwards, Paul. *The Closed World: Computers and the Politics of Discourse in Cold War America*. Cambridge: MIT Press, 1996.

Ehmer, Richard H. "Masking by Tones vs. Noise Bands." *Journal of the Acoustical Society of America* 31, no. 9 (1959): 1253–56.

———. "Masking Patterns of Tones." *Journal of the Acoustical Society of America* 31, no. 8 (1959): 1115–20.

Eisenberg, Evan. *The Recording Angel: The Experience of Music from Aristotle to Zappa*. New York: Penguin, 1987.

Elias, Norbert. *The Civilizing Process: Sociogenetic and Psychogenetic Investigations*. Translated by Edmund Jephcott. Maltham, England: Basil Blackwell, 2000.

Elias, Peter. "Predictive Coding I." *IRE Transactions on Information Theory* 1, no. 1 (1955): 16–24.

———. "Predictive Coding II." *IRE Transactions on Information Theory* 1, no. 1 (1955): 24–33.

Elmer, Greg. "The Vertical (Layered) Net: Interrogating the Conditions of Network Connectivity." In *Critical Cyberculture Studies: Current Terrains, Future Directions*, edited by David Silver and Adrienne Massanari, 159–67. New York: NYU Press, 2006.

"EMI Signs Internet License with Cerberus." *ProSound News Europe*, 1 May 1997, http://www.factiva.com.

Erlmann, Veit. *Reason and Resonance: A History of Modern Aurality*. New York: Zone Books, 2010.

Evens, Aden. *Sound Ideas: Music, Machines, and Experience*. Minneapolis: University of Minnesota Press, 2005.

Fagen, M. D., ed. *A History of Engineering and Science in the Bell System: The Early Years (1875–1925)*. Indianapolis: Bell Telephone Laboratories, 1975.

Faller, Christof, Biing-Hwang Juang, Peter Kroon, Hui-Ling Lou, Sean Ramprashad, and Carl-Erik W. Sundberg. "Technical Advances in Digital Audio Radio Broadcasting." *Proceedings of the IEEE* 90, no. 8 (2002): 1303–33.

Faulkner, Tony. "FM: Frequency Modulation or Fallen Man?" In *Radiotext(E)*, edited by Neil Strauss, 61–65. New York: Semiotext(e), 1993.

Fechner, Gustav Theodor. *Elements of Psychophysics*. New York: Holt, Rinehart and Winston, 1966.

Fletcher, H. "Auditory Patterns." *Review of Modern Physics* 12, no. 1 (1940): 47–66.

———. *Speech and Hearing, with an Introduction by H. D. Arnold*. New York: Van Nostrand, 1929.

Fletcher, H., and R. H. Galt. "The Perception of Speech and Its Relation to Telephony." *Journal of the Acoustical Society of America* 22, no. 2 (1950): 89–151.

Fletcher, H., and W. A. Munson. "Loudness of a Complex Tone, Its Definition, Measurement and Calculation." *Journal of the Acoustical Society of America* 5, no. 1 (1933): 65.

———. "Relation between Loudness and Masking." *Journal of the Acoustical Society of America* 9, no. 1 (1937): 1–10.

Fletcher, H., and R. L. Wegel. "The Frequency-Sensitivity of Normal Ears." *Physical Review* 19, no. 6 (1922): 553–65.

Forbes, A., and C. S. Sherrington. "Acoustic Reflexes in the Decerebrate Cat." *American Journal of Physiology* 35, no. 4 (1914): 367–76.

Foucault, Michel. *The Archaeology of Knowledge and the Discourse on Language*. Translated by A. M. Sheridan Smith. New York: Pantheon Books, 1972.

———. *The Birth of Biopolitics: Lectures at the Collège De France, 1978–1979*. Translated by Graham Burchell. New York: Palgrave Macmillan, 2008.

———. *The History of Sexuality*. Vol. 1, *An Introduction*. Translated by Robert Hurley. New York: Vintage Books, 1978.

Fowler, Edmund Prince. "Historical Vignette: Harvey Fletcher." *Archives of Otolaryngology* 85, no. 1 (1967): 107–9.

Fox, Aaron. *Real Country: Music and Language in Working-Class Culture*. Durham: Duke University Press, 2004.

Frasca, Gonzalo. "Simulation versus Narrative: Introduction to Ludology." In *The Video Game Theory Reader*, edited by Mark J. P Wolf and Bernard Perron, 221–35. New York: Routledge, 2003.

Friedberg, Anne. *The Virtual Window: From Alberti to Microsoft*. Cambridge: MIT Press, 2006.

Friedman, Ted. *Electric Dreams: Computers in American Culture*. New York: NYU Press, 2005.

Frith, Simon. "Video Pop: Picking Up the Pieces." In *Facing the Music*, edited by Simon Frith, 88–130. New York: Pantheon, 1988.

Fuchs, Hendrick. "Report on the MPEG/Audio Subjective Listening Test in Hannover (Draft)." Hanover, Germany: International Organization for Standardization ISO/IEC JTC1/SC2/WG11, 1991.

Fukuyama, Francis. "All Hail . . . Analog? When It Comes to the Quality of Photos and Music, the Digital Revolution May Be Failing Us." *Wall Street Journal*, 26 February 2011.

Fuller, Matthew. *Media Ecologies: Materialist Energies in Art and Technoculture*. Cambridge: MIT Press, 2005.

Galison, Peter. "The Ontology of the Enemy: Norbert Wiener and the Cybernetic Vision." *Critical Inquiry* 21, no. 1 (1994): 228–66.

Galison, Peter, and Bruce Hevly, eds. *Big Science: The Growth of Large-Scale Research*. Stanford: Stanford University Press, 1992.

Galloway, Alexander R. *Gaming: Essays on Algorithmic Culture*. Minneapolis: University of Minnesota Press, 2006.

Gardner, Martin. *Logic Machines and Diagrams*. 2nd ed. Chicago: University of Chicago Press, 1982.

Gardner, Wallace, and J. C. R Licklider. "Auditory Analgesia in Dental Operations." *Journal of the American Dental Association* 59 (1959): 1144–49.

Gates, Kelly A. *Our Biometric Future: Facial Recognition Technology and the Culture of Surveillance*. New York: NYU Press, 2011.

Gertner, John. "Our Ratings, Ourselves." *New York Times Magazine*, 10 April 2005, 34–41.

Gibson, James J. *The Ecological Approach to Visual Perception*. Boston: Houghton-Mifflin, 1979.

Gibson-Graham, J. K. *The End of Capitalism (as We Knew It)*. Cambridge: Blackwell, 1996.

Gillespie, Tarleton. *The Politics of Platforms*. New Haven: Yale University Press (forthcoming).

———. *Wired Shut: Copyright and the Shape of Digital Culture*. Cambridge: MIT Press, 2007.

Gitelman, Lisa. *Always Already New: Media, History, and the Data of Culture*. Cambridge: MIT Press, 2006.

———. *Scripts, Grooves, and Writing Machines: Representing Technology in the Edison Era*. Stanford: Stanford University Press, 1999.

Gitlin, Todd. *Media Unlimited: How the Torrent of Images and Sounds Overwhelms Our Lives*. New York: Metropolitan Books, 2001.

Gleick, James. *The Information: A History, a Theory, a Flood*. New York: Pantheon, 2011.

Glinsky, Albert. *Theremin: Electronic Music and Espionage*. Urbana: University of Illinois Press, 2000.

Goffman, Erving. *Frame Analysis: An Essay on the Organization of Experience*. New York: Harper and Row, 1974.

Goggin, Gerard, and Christopher Newell. "Disabling Cell Phones." In *The Cell Phone Reader*, edited by Anandam Kavoori and Noah Arceneaux, 155–72. New York: Peter Lang, 2006.

Gomery, Douglas. *The Coming of Sound: A History*. New York: Routledge, 2005.

———. *Shared Pleasures: A History of Movie Presentation in the United States*. London: BFI, 1992.

———. "Theater Television: A History." *Society of Motion Picture and Television Engineers (SMPTE) Journal* 92, no. 2 (1989): 120–23.

Goodman, David. "Distracted Listening: On Not Making Sound Choices in the 1930s." In *Sound in the Age of Mechanical Reproduction*, edited by David Suisman and Susan Strasser, 15–46. Philadelphia: University of Pennsylvania Press, 2009.

Gopinath, Sumanth. "Ringtones, or the Auditory Logic of Globalization." *First Monday* 10, no. 12 (2005), http://firstmonday.org.

Gould, Glenn. *The Glenn Gould Reader*. New York: Vintage Books, 1990.

Gracyk, Theodore. *Rhythm and Noise: An Aesthetics of Rock*. Durham: Duke University Press, 1996.

Grau, Oliver. *Virtual Art: From Illusion to Immersion*. Cambridge: MIT Press, 2004.

Greene, Paul D., and Thomas Porcello, eds. *Wired for Sound: Engineering and Technologies in Sonic Cultures*. Middletown, Conn.: Wesleyan University Press, 2005.

Grevatt, Ren. "Pros and Cons Have Their Say on K. C. Speech." *Billboard*, 24 March 1958, 14.

Grewin, Christer, and Thomas Rydén. "Subjective Assessment of Low Bit-Rate Audio Codecs." Paper presented at the 10th International Conference of the Audio Engineering Society: Images of Audio, London, September 1991.

Grossberg, Lawrence. *Cultural Studies in the Future Tense*. Durham: Duke University Press, 2010.

———. *Dancing in Spite of Myself: Essays on Popular Culture*. Durham: Duke University Press, 1997.

Gunderson, Philip A. "Danger Mouse's Grey Album, Mash-Ups, and the Age of Composition." *Postmodern Culture* 15, no. 1 (2004), http://muse.jhu.edu/journals/postmodern_culture.

Gunkel, David J. "Rethinking the Digital Remix: Mash-Ups and the Metaphysics of Sound Recording." *Popular Music and Society* 31, no. 4 (2008): 489–510.

Gunn, Joshua, and Mirko M. Hall. "Stick It in Your Ear: The Psychodynamics of iPod Enjoyment." *Communication and Critical/Cultural Studies* 5, no. 2 (2008): 135–57.

Hacker, Scot. *MP3: The Definitive Guide*. Cambridge: O'Reilly, 2000.

Hainge, Greg. "Of Glitch and Men: The Place of the Human in the Successful Integration of Failure and Noise in the Digital Realm." *Communication Theory* 17, no. 1 (2007): 26–42.

Hall, Edward T. *Beyond Culture*. Garden City, N.Y.: Anchor Press, 1976.

Hall, J. L. "Binaural Interaction in the Accessory Superior-Olivary Nucleus of the Cat." *Journal of the Acoustical Society of America* 37, no. 5 (1965): 814–23.

Hall, Stuart. "On Postmodernism and Articulation: An Interview with Stuart Hall." *Journal of Communication Inquiry* 10, no. 2 (1986): 45–60.

Hamilton, B. P., H. Nyquist, M. B. Long, and W. A. Phelps. "Voice-Frequency Carrier Telegraph System for Cables." *Transactions of the AIEE* 44, no. 1 (1925): 327–32.

Hankins, Thomas L., and Robert J. Silverman. *Instruments and the Imagination*. Princeton: Princeton University Press, 1995.

Hansen, Mark B. N. *New Philosophy for a New Media*. Cambridge: MIT Press, 2004.

Haraway, Donna. *Simians, Cyborgs, and Women*. New York: Routledge, 1991.

Hardt, Michael, and Antonio Negri. *Empire*. Cambridge: Harvard University Press, 2000.

Hargreaves, David J., and Adrian North, eds. *The Social Psychology of Music*. New York: Oxford University Press, 1997.

Haring, Bruce. *Beyond the Charts: MP3 and the Digital Music Revolution*. Los Angeles: OTC Books, 2000.

Harley, Robert. "From the Editor: The Blind (Mis-) Leading the Blind." *Absolute Sound*, August 2008, 12.

———. "The Role of Critical Listening in Evaluating Audio Equipment Quality." Paper presented at the 91st Annual Convention of the Audio Engineering Society, Santa Fe, 1991.

Hartley, John. *Tele-Ology: Studies in Television*. New York: Routledge, 1992.

Hartley, R. V. L. "Transmission of Information." *Bell System Technical Journal* 7, no. 3 (1928): 535–63.

Hay, James, Lawrence Grossberg, and Ellen Wartella, eds. *The Audience and Its Landscape*. Boulder: Westview Press, 1996.

Hecker, Tim. "Glenn Gould, the Vanishing Performer and the Ambivalence of the Studio." *Leonardo Music Journal* 118 (2009): 77–83.

Heckscher, Melissa. "And about That Strange Logo . . ." *Entertainment Weekly*, 9 February 2001.

Heffernan, Virginia. "Funeral for a Friend." *New York Times*, 31 October 2010.

Hegarty, Paul. *Noise/Music: A History*. New York: Continuum, 2007.

Heidegger, Martin. *Being and Time*. New York: Harper and Row, 1962.

———. "The Thing." In *Poetry, Language, Thought*, 165–82. New York: Harper and Row, 1971.

Hellman, Rhona. "Asymmetry of Masking between Noise and Tone." *Perception and Psychophysics* 11, no. 3 (1972): 2241–246.

Hesmondhalgh, David. "User-Generated Content, Free Labour and the Cultural Industries." *Ephemera* 10, no. 3–4 (2010): 267–84.

Hillis, Ken. *Digital Sensations: Space, Identity, and Embodiment in Virtual Reality*. Minneapolis: University of Minnesota Press, 1999.

———. *Online a Lot of the Time: Ritual, Fetish, Sign*. Durham: Duke University Press, 2009.

Hilmes, Michelle. *Radio Voices: America Broadcasting 1922–1952*. Minneapolis: University of Minnesota Press, 1997.

Hilmes, Michelle, and Jason Loviglio, eds. *The Radio Reader: Essays in the Cultural History of Radio*. New York: Routledge, 2001.

Hoeg, Wolfgang, and Thomas Lauterbach. *Digital Audio Broadcasting: Principles and Applications*. Chichester, England; New York: Wiley, 2001.

Hoogendoorn, A. "Digital Compact Cassette." *Proceedings of the IEEE* 82, no. 10 (1994): 1479–89.

Horrigan, John. "Pew Internet Project Data Memo: 55% of Adult Internet Users Have Broadband at Home or Work." Washington: Pew Internet and American Life Project, 2004.

Hosokawa, Shuhei. "The Walkman Effect." *Popular Music* 4 (1984): 165–80.

Hounshell, David A., and John K. Smith. *Science and Corporate Strategy: Du Pont R&D, 1902–1980*. New York: Cambridge University Press, 1988.

Hyde, Lewis. *The Gift: Imagination and the Erotic Life of Property*. New York: Vintage Books, 1983.

"IBM Ultimedia Teams with Montage Group to Offer Low-Cost, Feature-Rich Non-Linear Video Editing Solution." *Business Wire*, 20 April 1993.

Illich, Ivan. *Energy and Equity*. London: Calder and Boyars, 1974.

———. *Tools for Conviviality*. New York: Harper and Row, 1973.

Immink, Kees A. Schouhamer. "The Compact Disc Story." *Journal of the Audio Engineering Society* 46, no. 5 (1998): 458–65.

Jackson, Steven J., Paul N. Edwards, Geoffrey C. Bowker, and Cory P. Knobel. "Understanding Infrastructure: History, Heuristics, and Cyberinfrastructure Policy." *First Monday* 12, no. 6 (2007), http://firstmonday.org/.

Jakobs, Kai. "Information Technology Standards, Standards Setting and Stan-

dards Research." Paper presented at the Cotwolds Conference on Technology Standards and the Public Interest, Cotswolds, England, 2003.
Jastrow, Joseph. "An Apparatus for the Study of Sound Intensities." *Science* N.S. 3, no. 67 (1896): 544–46.
Jenkins, Henry. *Convergence Culture: Where Old and New Media Collide*. New York: NYU Press, 2006.
Johns, Adrian. *Piracy: The Intellectual Property Wars from Gutenberg to Gates*. Chicago: University of Chicago Press, 2009.
———. "Piracy as a Business Force." *Culture Machine* 10 (2009): 44–63.
Johnson, R. "Machine Songs. 1. Music and the Electronic Media." *Computer Music Journal* 15, no. 2 (1991): 12–20.
———. "Technology, Commodity, Power." *Computer Music Journal* 18, no. 3 (1994): 25–32.
Johnston, J. D. "Transform Coding of Audio Signals Using Perceptual Noise Criteria." *IEEE Journal on Selected Areas in Communications* 6, no. 2 (1988): 314–23.
Johnston, John. *Information Multiplicity: American Fiction in the Age of Media Saturation*. Baltimore: Johns Hopkins University Press, 1998.
Jones, Steve. "Making Waves: Pirate Radio and Popular Music." Paper presented at the Association for Education in Journalism and Mass Communication, Portland, Oreg., 2–5 July 1988.
———. "Music and the Internet." *Popular Music* 19, no. 2 (2000): 217–30.
———. "Music That Moves: Popular Music, Distribution and Network Technologies." *Cultural Studies* 16, no. 2 (2002): 213–32.
Kafka, Franz. "The Neighbor." The Kafka Project, http://www.kafka.org.
Kahn, David. *The Codebreakers*. New York: Simon and Schuster, 1996.
Kahn, Douglas. *Noise, Water, Meat: A History of Sound in the Arts*. Cambridge: MIT Press, 1999.
Kant, Immanuel. *Critique of Judgment*. Translated by J. H. Bernard. New York: Free Press, 1951.
Karagangis, Joe, ed. *Media Piracy in Emerging Economies*. New York: Social Sciences Research Council, 2011.
Kassabian, Anahid. "Ubisub: Ubiquitous Listening and Networked Subjectivity." *Echo: A Music-Centered Journal* 3, no. 2 (2001), http://www.echo.ucla.edu.
Katz, Bob. *Mastering Audio: The Art and the Science*. New York: Focal Press, 2002.
Katz, Mark. *Capturing Sound: How Technology Has Changed Music*. Berkeley: University of California Press, 2004.
Keightley, Keir. "'Turn It Down!' She Shrieked: Gender, Domestic Space, and High Fidelity, 1948–59." *Popular Music* 15, no. 2 (1996): 149–77.
Keil, Charles. "Participatory Discrepancies and the Power of Music." In *Music Grooves*, 96–108. Chicago: University of Chicago Press, 1994.
Keil, Charles, and Steven Feld. *Music Grooves*. Chicago: University of Chicago Press, 1996.

Kelty, Chris. *Two Bits*. Durham: Duke University Press, 2008.

Kirk, Roger. "Learning, a Major Factor Influencing Preferences for High-Fidelity Reproducing Systems." *Journal of the Acoustical Society of America* 28, no. 6 (1956): 1113–16.

Kirschenbaum, Matthew G. "Extreme Inscription: Towards a Grammatology of the Hard Drive." TEXT *Technology* 13, no. 2 (2004): 91–125.

———. *Mechanisms: New Media and the Forensic Imagination*. Cambridge: MIT Press, 2008.

Kittler, Friedrich. *Discourse Networks, 1800/1900*. Translated by Michael Metteer and Chris Cullens. Stanford: Stanford University Press, 1985.

———. *Gramophone-Film-Typewriter*. Translated by Geoffrey Winthrop-Young and Michael Wutz. Stanford: Stanford University Press, 1999.

———. "Observation on Public Reception." In *Radio Rethink: Art, Sound and Transmission*, edited by Daina Augaitis and Dan Lander, 74–85. Banff, Canada: Walter Phillips Gallery, 1994.

Kivy, Peter. *The Fine Art of Repetition: Essays in the Philosophy of Music*. New York: Cambridge University Press, 1993.

Knopper, Steve. *Appetite for Self-Destruction: The Spectacular Crash of the Record Industry in the Digital Age*. New York: Free Press, 2009.

Konecni, Vladimir. "Social Interaction and Musical Preference." In *The Psychology of Music*, edited by Diana Deutsch, 479–516. New York: Academic, 1982.

Kopytoff, Igor. "The Cultural Biography of Things." In *The Social Life of Things: Commodities in Cultural Perspective*, edited by Arjun Appadurai, 64–91. Cambridge: Cambridge University Press, 1988.

Koschorke, Albrecht. *Körperströme und Schriftverkehr: Mediologie des 18. Jahrhunderts*. Munich: Fink, 1999.

Krasner, Michael A. "The Critical Band Coder—Digital Encoding of Speech Signals Based on the Perceptual Requirements of the Auditory System." Paper presented at the Acoustics, Speech and Signal Processing: IEEE International Conference on ICASSP '80, April 1980.

———. "Digital Encoding of Speech and Audio Signals Based on the Perceptual Requirements of the Auditory System." PhD dissertation, Massachusetts Institute of Technology, 1979.

Krauss, Rosalind. "Two Moments from the Post-Medium Condition." *October*, no. 116 (2006): 55–62.

Krechmer, Ken. "Open Standards Requirements." *International Journal of IT Standards and Standardization Research* 4, no. 1 (2006): 43–61.

Kuhn, Thomas S. *The Structure of Scientific Revolutions*. Chicago: University of Chicago Press, 1962.

Labelle, Brandon. *Background Noise: Perspectives on Sound Art*. New York: Continuum, 2006.

Langer, Susanne Katherina Knauth. *Philosophy in a New Key: A Study in the Symbolism of Reason, Rite, and Art*. 3rd ed. Cambridge: Harvard University Press, 1957.

Larkin, Brian. *Signal and Noise: Media, Infrastructure, and Urban Culture in Nigeria*. Durham: Duke University Press, 2008.

Lastra, James. *Sound Technology and American Cinema: Perception, Representation, Modernity*. New York: Columbia University Press, 2000.

Latour, Bruno. *Aramis, or the Love of Technology*. Cambridge: Harvard University Press, 1996.

———. *Science in Action: How to Follow Scientists and Engineers through Society*. Milton Keynes: Open University Press, 1987.

Lazzarato, Maurizio. "Immaterial Labor." In *Radical Thought in Italy: A Potential Politics*, edited by Paulo Virno and Michael Hardt, 133–47. Minneapolis: University of Minnesota Press, 1996.

Leach, Edmund. "Animal Categories and Verbal Abuse." In *The Essential Edmund Leach*, edited by Stephen Hugh-Jones and James Laidlaw, 322–43. New Haven: Yale University Press, 2000.

Lederer, Susan. *Subjected to Science: Human Experimentation before the Second World War*. Baltimore: Johns Hopkins University Press, 1995.

Lee, Benjamin, and Edward LiPuma. "Cultures of Circulation: The Imaginations of Modernity." *Public Culture* 14, no. 1 (2002): 191–213.

Lentz, Becky. "Media Infrastructure Policy and Media Activism." In *Sage Encyclopedia of Social Movement Media*, edited by John D. Downing, 323–26. Thousand Oaks, Calif.: Sage, 2010.

Lessig, Lawrence. *Code Version 2.0*. New York: Basic Books, 2006.

Levitin, Daniel J., and Susan E. Rogers. "Absolute Pitch: Perception, Coding, and Controversies." *Trends in Cognitive Science* 9 (2005): 26–33.

Leyshon, A. "Scary Monsters? Software Formats, Peer-to-Peer Networks, and the Spectre of the Gift." *Environment and Planning D: Society and Space* 21 (2003): 533–58.

———. "Time-Space (and Digital) Compression: Software Formats, Musical Networks, and the Reorganisation of the Music Industry." *Environment and Planning A* 33, no. 1 (2001): 49–77.

Liebowitz, S. J., and Stephan E. Margolis. "Path Dependence, Lock-in, and History." *Journal of Law, Economics, and Organization* 11 (1995): 205–26.

Lindsay, Peter, and Donald Norman. *Human Information Processing: An Introduction to Psychology*. New York: Harcourt, Brace Janovich, 1977.

"Link to Lesbianism Found." *New York Times*, 3 March 1998.

Lipartito, Kenneth. "Picturephone and the Information Age: The Social Meaning of Failure." *Technology and Culture* 44, no. 1 (2003): 50–81.

Lokhoff, Gerard. "Precision Adaptive Subband Coding (Pasc) for the Digital Compact Cassette (DCC)." *IEEE Transactions on Consumer Electronics* 38, no. 4 (1992): 784–89.

Loughlan, Patricia Louise. "Pirates, Parasites, Reapers, Sowers, Fruits, Foxes . . . the Metaphors of Intellectual Property." *Sydney Law Review* 28, no. 2 (2006): 211–26.

Loviglio, Jason. *Radio's Intimate Public: Network Broadcasting and Mass-Mediated Democracy*. Minneapolis: University of Minnesota Press, 2005.

Lukàcs, György. *History and Class Consciousness: Studies in Marxist Dialectics*. Cambridge: MIT Press, 1971.

Lysloff, R. T. A. "Musical Community on the Internet: An On-Line Ethnography." *Cultural Anthropology* 18, no. 2 (2003): 233–63.

MacKenzie, Adrian. "The Performativity of Code: Software and Cultures of Circulation." *Theory, Culture and Society* 22, no. 1 (2005): 71–92.

MacKenzie, Donald. "Introduction." In *Knowing Machines: Essays on Technical Change*, edited by Donald MacKenzie, 1–22. Cambridge: MIT Press, 1996.

MacPherson, C. B. *The Political Theory of Possessive Individualism: Hobbes to Locke*. New York: Oxford University Press, 1962.

Magaudda, Paolo. "When Materiality 'Bites Back': Digital Music Consumption Practices in the Age of Dematerialization." *Journal of Consumer Culture* 11, no. 1 (2011): 15–36.

Maisel, Albert Q. "'Who's Afraid of the Dentist?'" *Popular Science*, August 1960, 59–62.

Manovich, Lev. *The Language of New Media*. Cambridge: MIT Press, 2001.

Mansell, Robin, and Marc Raboy. "Introduction: Foundations of the Theory and Practice of Global Media and Communication Policy." In *The Handbook of Global Media and Communication Policy*, edited by Robin Mansell and Marc Raboy, 1–20. Malden, Mass.: Blackwell, 2011.

Manuel, Peter. *Cassette Culture: Music and Technology in North India*. Chicago: University of Chicago Press, 1993.

Marvin, Carolyn. *When Old Technologies Were New: Thinking about Electrical Communication in the Nineteenth Century*. New York: Oxford University Press, 1988.

Marx, Karl. *Capital*. Vol. 1, *A Critique of Political Economy*. Translated by Ben Fowkes. New York: Penguin Classics, 1992.

———. *Grundrisse: Foundations of the Critique of Political Economy*. Translated by Martin Nicolaus. New York: Penguin, 1973.

Mathews, Max. "The Ear and How It Works." In *Music, Cognition, and Computerized Sound*, edited by Perry R. Cook, 1–10. Cambridge: MIT Press, 2001.

Mattelart, Armand. *The Invention of Communication*. Translated by Susan Emmanuel. Minneapolis: University of Minnesota Press, 1996.

———. *Networking the World: 1794–2000*. Translated by Liz Carey-Libbrecht and James A. Cohen. Minneapolis: University of Minnesota Press, 2000.

Mauss, Marcel. *The Gift: Forms and Functions of Exchange in Archaic Societies*. Translated by Ian Cunnison. New York: W. W. Norton, 1967.

Mayer, Alfred. "Researches in Acoustics." *Philosophical Magazine* 2 (1876): 500–507.

McCarthy, Anna. *Ambient Television: Visual Culture and Public Space*. Durham: Duke University Press, 2001.

McChesney, Robert. *The Problem of the Media: U.S. Communication Politics in the 21st Century*. New York: Monthly Review Press, 2004.

McFadden, Dennis, and Edward G. Pasanen. "Comparison of the Auditory Systems of Heterosexuals and Homosexuals: Click-Evoked Otoacoustic Emissions." *Proceedings of the National Academy of Sciences of the United States of America* 95, no. 5 (1998): 2709–13.

McLeod, Kembrew. "Confessions of an Intellectual (Property): Danger Mouse, Mickey Mouse, Sonny Bono, and My Long and Winding Path as a Copyright Activist-Academic." *Popular Music and Society* 28, no. 1 (2005): 79–93.

———. *Freedom of Expression: Tales from the Dark Side of Intellectual Property Law*. New York: Doubleday, 2005.

———. "MP3s Are Killing Home Taping: The Rise of Internet Distribution and Its Challenge to the Major Label Music Monopoly." *Popular Music and Society* 28, no. 4 (2005): 521–31.

McLeod, Kembrew, and Peter DiCola. *Creative License: The Law and Culture of Digital Sampling*. Durham: Duke University Press, 2011.

McLeod, Kembrew, and Rudolf Kuenzli, eds. *Cutting across Media: Appropriate Art, Interventionist Collage, and Copyright Law*. Durham: Duke University Press, 2011.

McLuhan, Marshall. *Understanding Media: The Extensions of Man*. New York: McGraw-Hill, 1964.

Meehan, Eileen. "Heads of Households and Ladies of the House: Gender, Genre and Broadcast Ratings 1929–1990." In *Ruthless Criticism: New Perspectives in U.S. Communication History*, edited by William Samuel Solomon and Robert Waterman McChesney, 204–21. Minneapolis: University of Minnesota Press, 1993.

———. "Why We Don't Count: The Commodity Audience." In *Logics of Television*, edited by Patricia Mellencamp, 117–37. Bloomington: Indiana University Press, 1990.

Meintjes, Louise. *Sound of Africa! Making Music Zulu in a South African Studio*. Durham: Duke University Press, 2003.

Meister, David. *History of Human Factors and Ergonomics*. London: Lawrence Erlbaum, 1999.

Merton, Robert K. *The Sociology of Science: Theoretical and Empirical Investigations*. Chicago: University of Chicago Press, 1973.

Meyer, Leonard B. *Emotion and Meaning in Music*. Chicago: University of Chicago Press, 1956.

Meyerowitz, Joshua. *No Sense of Place: The Impact of Electronic Media on Social Behavior*. New York: Oxford University Press, 1985.

Millard, Andre. *Edison and the Business of Innovation*. Baltimore: Johns Hopkins University Press, 1990.

Millman, S., ed. *A History of Engineering and Science in the Bell System: Communication Sciences (1925–1980)*. Indianapolis: AT&T Bell Laboratories, 1984.

Mills, Mara. "The Dead Room: Deafness and Communication Engineering." PhD dissertation, Harvard University, 2008.

———. "Deafening: Noise and the Engineering of Communication in the Telephone System." *Grey Room* 43 (2011): 118–43.

———. "Do Signals Have Politics? Inscribing Abilities in Cochlear Implants." In *Oxford Handbook of Sound Studies*, edited by Karin Bijsterveld and Trevor Pinch, 320–46. New York: Oxford University Press, 2012.

———. "On Disability and Cybernetics: Helen Keller, Norbert Wiener, and the Hearing Glove." *Differences* 22, no. 2–3 (2011): 74–111.

Milner, Greg. *Perfecting Sound Forever: An Aural History of Recorded Music*. New York: Faber and Faber, 2009.

Mindell, David. *Between Human and Machine: Feedback, Control, and Computing before Cybernetics*. Baltimore: Johns Hopkins University Press, 2002.

Minsky, Marvin. "Steps toward Artificial Intelligence." *Proceedings of the IRE* 49 (1961): 8–30.

Mirzoeff, Nicholas. "Introduction to Part Three." In *The Visual Culture Reader*, edited by Nicholas Mirzoeff, 181–88. New York: Routledge, 1998.

Misa, Thomas J. "Toward an Historical Sociology of Business Culture." *Business and Economic History* 25, no. 1 (1996): 55–64.

Mitchell, Joan L., William B. Pennebaker, Chad E. Fogg, and Didier J. LeGall, eds. *MPEG Video Compression Standard*. New York: Chapman and Hall, 1997.

Mooradian, Mark. "Music: Record Labels and the Imperative for Digital Distribution." New York: Jupiter Strategic Planning Services, 1998.

Moore, Brian C. J. *An Introduction to the Psychology of Hearing*. New York: Academic, 2003.

Morley, David. *Television, Audiences, and Cultural Studies*. London: Routledge, 1992.

Morris, Jeremy. "Understanding the Digital Music Commodity." PhD dissertation, McGill University, 2010.

Morse, Margaret. "An Ontology of Everyday Distraction: The Freeway, the Mall, and Television." In *Logics of Television*, edited by Patricia Mellencamp, 117–37. Bloomington: Indiana University Press, 1990.

Mowitt, John. "The Sound of Music in the Era of Its Electronic Reproducibility." In *Music and Society: The Politics of Composition, Performance and Reception*, edited by Richard Leppert and Susan McClary, 173–97. New York: Cambridge University Press, 1987.

"Multimedia Business Analyst." *Business Conference and Management Reports*, 13 September 1995, http://www.factiva.com.

Mumford, Lewis. "An Appraisal of Lewis Mumford's Technics and Civilization." *Daedalus* 88 (1959): 527–36.

———. *The Myth of the Machine: Technics and Human Development.* Vol. 1. New York: Harcourt, Brace and World, 1966.

———. *Technics and Civilization.* New York: Harcourt, Brace and World, 1934.

Münsterberg, Hugo. *Psychology and Industrial Efficiency.* Boston: Houghton Mifflin, 1913.

Musmann, Hans-Georg. "Genesis of the MP3 Audio Coding Standard." *IEEE Transactions On Consumer Electronics* 52, no. 3 (2006): 1043–49.

———. "The ISO Audio Coding Standard." Paper presented at the IEEE Global Telecommunications Conference (GLOBECOM '90): "Communications: Connecting the Future," San Diego, 2–5 December 1990.

Myrberg, Mats. "Windows Media in Consumer Electronics: Past, Present and Future." Paper presented at the AES UK 17th Conference, Audio Delivery, April 2002.

"Napster Cat Attack." *Interactive Week,* 8 January 2001.

Nebeker, Frederik. *Signal Processing: The Emergence of a Discipline.* New York: IEEE Press, 1998.

Negri, Antonio. *Marx beyond Marx: Lessons on the Grundrisse.* New York: Autonomedia, 1991.

Newman, Kathy M. *Radio Active: Advertising and Consumer Activism, 1935–1947.* Berkeley: University of California Press, 2004.

Noam, Eli M. *Interconnecting the Network of Networks.* Cambridge: MIT Press, 2001.

Norman, Donald. *The Design of Everyday Things.* New York: Doubleday, 1990.

Nyquist, Harry. "Certain Factors Affecting Telegraph Speed." *Bell System Technical Journal* 3, no. 2 (1924): 324–46.

———. "Certain Topics in Telegraph Transmission Theory." *Trans. AIEE* 47, no. 1 (1928): 617–44.

OECD. "Information Technology Outlook 1997." Paris: Organization for Economic Co-operation and Development, 1997.

O'Neill, Brian. "DAB Eureka-147: The European Vision for Digital Radio." Paper presented at The Long History of New Media: Contemporary and Future Developments Contextualized, preconference of the International Communication Association's annual convention, Montreal, 20–25 May 2008.

O'Neill, Eugene F. *A History of Engineering and Science in the Bell System: Transmission Technology (1925–1975).* Indianapolis: AT&T Bell Laboratories, 1985.

Osterhammel, Dorrit, and P. Osterhammel. "High-Frequency Audiometry: Age and Sex Variations." *Scandinavian Audiology* 8, no. 2 (1979): 73–80.

Ostertag, Bob. "Why Computer Music Sucks." *Resonance Magazine* 5, no. 1 (2001), 2.

Otis, Laura. *Networking: Communication with Bodies and Machines in the Nineteenth Century.* Ann Arbor: University of Michigan Press, 2001.

Pan, Davis. "A Tutorial on MPEG/Audio Compression." *IEEE Multimedia* 2, no. 2 (1995): 60–74.

Parks, Lisa. "Air Raids: Television and the War on Terror." In *America and the Misshaping of the New World Order*, edited by Giles Gunn and Carl Gutierrez-Jones, 152–68. Berkeley: University of California Press, 2010.

———. *Cultures in Orbit: Satellites and the Televisual*. Durham: Duke University Press, 2005.

Pater, Walter. *The Renaissance*. New York: Modern Library, 1873.

Pearson, Jay D., et al. "Gender Differences in a Longitudinal Study of Age-Associated Hearing Loss." *Journal of the Acoustical Society of America* 97, no. 2 (1995): 1196–205.

Perlman, Marc. "Golden Ears and Meter Readers: The Contest for Epistemic Authority in Audiophilia." *Social Studies of Science* 34, no. 5 (2004): 783–807.

Peryam, D. R., and N. F. Girardot. "Advanced Taste Test Method." *Food Engineering* 24, no. 194 (1952): 58–61.

Peters, John Durham. "Helmholtz, Edison, and Sound History." In *Memory Bytes: History, Technology, and Digital Culture*, edited by Lauren Rabinovitz and Abraham Geil, 177–98. Durham: Duke University Press, 2004.

———. *Speaking into the Air: A History of the Idea of Communication*. Chicago: University of Chicago Press, 1999.

Peterson, L. C., and B. P. Bogert. "A Dynamical Theory of the Cochlea." *Journal of the Acoustical Society of America* 22, no. 3 (1950): 369–81.

Piatier, André. "Transectorial Innovations and the Transformation of Firms." *Information Society* 5, no. 4 (1987/1988): 205–31.

Picker, John. *Victorian Soundscapes*. New York: Oxford University Press, 2003.

Pickren, Wade, and Alexandra Rutherford. *A History of Modern Psychology in Context*. Hoboken: Wiley and Sons, 2010.

Pierce, John R. "The Early Days of Information Theory." *IEEE Transactions in Information Theory* 19, no. 1 (1973): 3–8.

Pinch, Trevor. "'Testing—One, Two, Three . . . Testing!': Toward a Sociology of Testing." *Science, Technology and Human Values* 18, no. 1 (1993): 25–41.

Pinch, Trevor, and Wiebe Bijker. "The Social Construction of Facts and Artefacts: Or How the Sociology of Science and the Sociology of Technology Might Benefit Each Other." *Social Studies of Science* 14, no. 3 (1984): 399–441.

Pinch, Trevor, and Frank Trocco. *Analog Days: The Invention and Impact of the Moog Synthesizer*. Cambridge: Harvard University Press, 2002.

Pingree, Geoffrey B., and Lisa Gitelman. "Introduction: What's New about New Media?" In *New Media, 1740–1915*, edited by Lisa Gitelman and Geoffrey B. Pingree, xi–xxii. Cambridge: MIT Press, 2003.

Plato. "Republic." In *The Collected Dialogues of Plato*, edited by Edith Hamilton and Huntington Cairns, 575–844. Princeton: Princeton University Press, 1961.

Plomp, Reinier. *The Intelligent Ear: On the Nature of Sound Perception*. Mahwah, N.J.: Lawrence Erlbaum Associates, 2002.

Pohlmann, Ken C. *Principles of Digital Audio*. 5th ed. New York: McGraw-Hill, 2005.

Pope, Ivan. "The Death of Discs." *net* 4 (1995), http://minidisc.org/Death.html.

Poss, Robert. "Distortion Is Truth." *Leonardo Music Journal* 8, no. 1 (1998): 45–48.

Postone, Moishe. *Time, Labor, and Social Domination*. Cambridge: Cambridge University Press, 1993.

Prescott, George B. *History, Theory and Practice of the Electric Telegraph*. Boston: Ticknor and Fields, 1860.

Pride, Dominic. "U.K. Bands Attack Convention thru Internet." *Billboard* 32 (6 August 1994): 3.

Princen, J. P., A. W. Johnson, and A. B. Bradley. "Subband/Transform Coding Using Filter Bank Designs Based on Time Domain Aliasing Cancellation." *IEEE Transactions on Acoustics Speech and Signal Processing* ASSP-34, no. 5 (1987): 1153–61.

Quetelet, Lambert Adolphe. *A Treatise on Man and the Development of His Faculties*. Edinburgh: W. and R. Chambers, 1842.

Rabinow, Paul, and Nikolas Rose. "Biopower Today." *BioSocieties* 1 (2006): 195–217.

Radovac, Lilian. "The 'War on Noise': Sound and Space in La Guardia's New York." *American Quarterly* 63, no. 3 (2011): 733–60.

Ragland, C. "Mexican Deejays and the Transnational Space of Youth Dances in New York and New Jersey." *Ethnomusicology* 47, no. 3 (2003): 338–54.

Rawsthorn, Arlie. "Internet Jukebox Company Signs US Deal." *Financial Times*, 31 May 1997, http://www.factiva.com.

Raymond, Eric. *The Cathedral and the Bazaar: Musings of Linux and Open Source by an Accidental Revolutionary*. Sabastopol, Calif.: O'Reilly Press, 2001.

Reardon, Jenny. *Race to the Finish: Identity and Governance in an Age of Genomics*. Princeton: Princeton University Press, 2005.

Rée, Jonathan. "The Vanity of Historicism." *New Literary History* 22, no. 4 (1991): 961–83.

Regnier, Pat. "Digital Music for Grown-Ups." *Money*, July 2003, 113.

Reich, Leonard S. *The Making of American Industrial Research: Science and Business at GE and Bell, 1876–1926*. New York: Cambridge University Press, 1985.

Remmer, Matthew. "The Grey Album: Copyright Law and Digital Sampling." *Media International Australia Incorporating Culture and Policy*, no. 114 (2005): 40–53.

Rentschler, Carrie. "Witnessing: US Citizenship and the Vicarious Experience of Suffering." *Media, Culture and Society* 26, no. 2 (2004): 296–304.

The Representation and Rendering Project. "Survey and Assessment of Sources of Information on File Formats and Software Documentation: Final Report." University of Leeds, 2004.

Rheinberger, Hans-Jorg. "A Reply to David Bloor 'Toward a Sociology of Epistemic Things.'" *Perspectives on Science* 13, no. 3 (2005): 406–10.

---. *Toward a History of Epistemic Things: Synthesizing Proteins in the Test Tube.* Stanford: Stanford University Press, 1997.
Rodgers, Tara. "On the Process and Aesthetics of Sampling in Electronic Music Production." *Organised Sound* 8, no. 3 (2003): 313-20.
---. "Synthesizing Sound: Metaphor in Audio-Technical Discourse and Synthesis History." PhD dissertation, McGill University, 2011.
Rodowick, David. *The Virtual Life of Film.* Cambridge: Harvard University Press, 2007.
Ronell, Avital. *The Telephone Book: Technology-Schizophrenia-Electric Speech.* Lincoln: University of Nebraska Press, 1989.
Rose, Nikolas. *The Psychological Complex: Psychology, Politics and Society in England, 1869–1939.* London: Routledge and Kegan Paul, 1985.
Rose, Tricia. *Black Noise: Rap Music and Contemporary Culture in Black America.* Hanover, N.H.: Wesleyan University Press and University Press of New England, 1994.
Rosenblueth, Arturo, Norbert Wiener, and Julian Bigelow. "Behavior, Purpose and Teleology." *Philosophy of Science* 10, no. 1 (1943): 18-24.
Ross, Andrew. *No Respect: Intellectuals and Popular Culture.* New York: Routledge, 1989.
Rothenbuhler, Erik, and Tom McCourt. "Radio Redefines Itself, 1947-1962." In *The Radio Reader: Essays in the Cultural History of Radio*, edited by Michelle Hilmes and Jason Loviglio, 367-87. New York: Routledge, 2002.
Rothenbuhler, Erik, and John Durham Peters. "Defining Phonography: An Experiment in Theory." *Musical Quarterly* 81, no. 2 (1997): 242-64.
Rowland, Wade. "Recognizing the Role of the Modern Business Corporation in the Social Construction of Technology." *Social Epistemology* 19, no. 2-3 (2005): 287-313.
Rubin, Gayle. "The Traffic in Women: Notes on the 'Political Economy' of Sex." In *Feminist Anthropology*, edited by Ellen Lewin, 87-106. Malden, Mass.: Blackwell, 2006.
Rusakov, I. G. "Conference of the Group for Forumulating Internaional Recommendations for Tables of Quantities and Units in the Division of Sound." *Measurement Techniques* 1, no. 2 (1958): 239-40.
Russo, Alexander. *Points on the Dial: Golden Age Radio beyond the Networks.* Durham: Duke University Press, 2010.
Ryder, Richard. *Animal Revolution: Changing Attitudes toward Speciesism.* New York: Berg, 2000.
Salomon, David. *Data Compression: The Complete Reference.* Berlin: Springer, 2007.
Sametband, Ricardo. "La historia de un pionero del audio digital." *La nacion*, 5 February 2001.
Sandstrom, Boren. "Women Mix Engineers and the Power of Sound." In *Music*

and Gender, edited by Pirkko Moisala and Beverly Diamond, 289–305. Urbana: University of Illinois Press, 2000.

Sandvig, Christian. "Shaping Infrastructure and Innovation on the Internet: The End-to-End Network That Isn't." In *Shaping Science and Technology Policy: The Next Generation of Research*, edited by David H. Guston and Daniel Sarewitz, 234–55. Madison: University of Wisconsin Press, 2006.

———. "The Structural Problems of the Internet for Cultural Policy." In *Critical Cyberculture Studies*, edited by David Silver and Adrienne Massanari, 107–18. New York: NYU PRESS, 2006.

Sato, Hiroaki, Isamu Sando, and Haruo Takahashi. "Sexual Dimorphism and Development of the Human Cochlea." *Acta Otolaryngologica* 111 (1991): 1037–40.

Saul, L. J., and Hallowell Davis. "Action Currents in the Central Nervous System: I. Action Currents in the Auditory Nerve Tracts." *Archives of Neurology and Psychiatry* 28 (1932): 1104–16.

Schafer, R. Murray. *The Soundscape: Our Sonic Environment and the Tuning of the World*. Rochester, Vt.: Destiny Books, 1994.

Schafer, T. H., R. S. Gales, C. A. Shewmaker, and P. O. Thompson. "The Frequency Selectivity of the Ear as Determined by Masking Experiments." *Journal of the Acoustical Society Of America* 22 (1950): 490–96.

Schäffner, Wolfgang. "The Point: The Smallest Venue of Knowledge in the 17th Century." In *Collection-Laboratory-Theater: Scenes of Knowledge in the 17th Century*, edited by Helmar Schramm, Ludger Schwarte, and Jan Lazardzig, 57–74. Berlin: Walter de Gruyter, 2005.

Schantz, Ned. *Gossip, Letters, Phones: The Scandal of Female Networks in Film and Literature*. New York: Oxford University Press, 2008.

Scharf, B. "Loudness of Complex Sounds as a Function of the Number of Components." *Journal of the Acoustical Society of America* 31 (1959): 783–85.

Schiesel, Seth. "Ideas Unlimited, Built to Order." *New York Times*, 30 October 2003.

Schiffer, Michael B. *The Portable Radio in American Life*. Tucson: University of Arizona Press, 1991.

Schiller, Dan. *Digital Capitalism: Networking the Global Market System*. Cambridge: MIT Press, 1999.

Schivelbusch, Wolfgang. *The Railway Journey: The Industrialization of Time and Space in the 19th Century*. Berkeley: University of California Press, 1986.

Schmidgen, Henning. "Silence in the Laboratory: The History of Sound-Proof Rooms." Paper presented at the Sounds of Science: Schall im Labor (1800–1930), Max Planck Institute for the History of Science, Berlin, 6 October 2006.

Schoen, Max, ed. *The Effects of Music*. New York: Harcourt, Brace and Company, 1927.

Schroeder, Manfred. "Interview with Manfred Schroeder, Conducted by Frederik Nebeker." New Brunswick: IEEE History Center, 1994.

Schroeder, M. R., B. S. Atal, and J. L. Hall. "Optimizing Digital Speech Coders by Exploiting Masking Properties of the Human Ear." *Journal of the Acoustical Society of America* 66, no. 6 (1979): 1647–52.

Schubert, Earl D. "History of Research on Hearing." In *Handbook of Perception*. Vol. 4, *Hearing*, edited by Edward C. Carterette and Morton P. Friedman, 41–80. New York: Academic, 1978.

Schultz, Theodore John. *Community Noise Ratings*. London: Applied Science Publishers, 1972.

Schulze, Hendrik, and Klaus Mochalski. "Ipoque Internet Study 2008/2009." Leipzig: Ipoque GmbH, 2009.

Schwarz, David. *Listening Subjects: Music, Psychoanalysis, Culture*. Durham: Duke University Press, 1997.

Schwartz Shapiro, Danielle. "Modernism for the Masses: The Industrial Design of John Vassos." *Archives of American Art Journal* 46, no. 1–2 (2008): 4–23.

Scott, Diane M. "Hearing Research: Children from Culturally and Linguistically Diverse Populations." *ASHA Leader (American Speech-Language-Hearing Association)* (2005), http://www.asha.org.

Seashore, Carl E. "An Audiometer." *University of Iowa Studies in Psychology* 2 (1899): 158.

Seckler, H. N., and J. J. Yostpille. "Electronic Switching Control Techniques." *Bell Laboratories Record* 40, no. 4 (1962): 138–43.

Serazio, Michael. "The Apolitical Irony of Generation Mash-Up: A Cultural Case Study in Popular Music." *Popular Music and Society* 31, no. 1 (2008): 79–94.

Serres, Michel. *Hermes: Literature, Science, Philosophy*. Baltimore: Johns Hopkins University Press, 1983.

———. *The Parasite*. Translated by Lawrence R. Schehr. Baltimore: Johns Hopkins University Press, 1982.

Shannon, Claude. "The Bandwagon." *IRE Transactions on Information Theory* 2, no. 1 (1956): 3.

———. "A Symbolic Analysis of Relay and Switching Circuits." Unpublished Master's thesis, Massachusetts Institute of Technology, 1937.

Shannon, Claude, and Warren Weaver. *The Mathematical Theory of Communication*. Urbana: University of Illinois Press, 1949.

Shapiro, Lawrence. "What Is Psychophysics?" Paper presented at the Proceedings of the Biennial Meeting of the Philosophy of Science Association, 1994.

Shelemay, Kay Kaufman. "Recording Technology, the Record Industry, and Ethnomusicological Scholarship." In *Comparative Musicology and the Anthropology of Music*, edited by Bruno Nettl and Philip V. Bohlmann, 277–92. Chicago: University of Chicago Press, 1991.

Sherburne, Philip. "Digital DJing App That Pulls You In." *Grooves*, 2003, 46–47.

Shiga, John. "Translations: Artifacts from an Actor-Network Perspective." *Artifact* 1, no. 1 (2006): 19–34.

Siemens, Werner von. *Werner Von Siemens, Inventor and Entrepreneur: Recollections of Werner Von Siemens*. London: Lund, 1966.

Silverman, Kaja. *The Acoustic Mirror: The Female Voice in Psychoanalysis and Cinema*. Bloomington: Indiana University Press, 1988.

Simmel, Georg. "The Metropolis and Mental Life." In *Georg Simmel: On Individuality and Social Forms*, edited by Donald Levine. Chicago: University of Chicago Press, 1971.

———. *The Philosophy of Money*. Translated by Tom Bottomore and David Frisby. New York: Routledge, 1990.

Sinnreich, Aram. *Mashed Up: Music, Technology, and the Rise of Configurable Culture*. Amherst: University of Massachusetts Press, 2010.

Slack, Jennifer Daryl, and J. Macgregor Wise. *Culture + Technology: A Primer*. New York: Peter Lang, 2006.

Sloboda, John A., Susan A. O'Neill, and Antonia Ivaldi. "Functions of Music in Everyday Life: An Exploratory Study Using the Experience Sampling Method." *Musicae Scientiae* 1, no. 1 (2001): 9–32.

Small, Christopher. *Musicking: The Meanings of Performing and Listening*. Hanover, N.H.: Wesleyan University Press and University Press of New England, 1998.

———. *Music-Society-Education*. London: John Calder, 1977.

Smythe, Dallas. *Counterclockwise: Perspectives on Communication*. Boulder: Westview Press, 1994.

Snyder, Charles. "Clarence John Blake and Alexander Graham Bell: Otology and the Telephone." *Annals of Otology, Rhinology, and Laryngology* 83, no. 4, pt. 2 suppl. 13 (1974): 3–31.

Sobchack, Vivian. "Nostalgia for a Digital Object: Regrets on the Quickening of Quicktime." In *Memory Bytes: History, Technology, and Digital Culture*, edited by Lauren Rabinovitz and Abraham Geil, 305–29. Durham: Duke University Press, 2004.

Sofia, Zoe. "Container Technologies." *Hypatia* 15, no. 2 (2000): 181–219.

Sousa, John Philip. "The Menace of Mechanical Music." *Appleton's Magazine*, September 1906, 278–84.

Spurlin, William J. "Rate-Reduced Digital Audio in the Broadcast Environment." Paper presented at the NAB Engineering Conference, 1990.

Stabile, Carol. "Introduction." In *Turning the Century: Essays in Media and Cultural Studies*, edited by Carol Stabile, 1–8. Denver: Westview Press, 2000.

Star, Susan Leigh. "Infrastructure and Ethnographic Practice." *Scandanavian Journal of Information Systems* 14, no. 2 (2002): 107–22.

Steinberg, J. C., H. C. Montgomery, and M. B. Gardner. "Results of the World's Fair Hearing Tests." *Journal of the Acoustical Society of America* 12, no. 2 (1940): 291–301.

Sterne, Jonathan. *The Audible Past: Cultural Origins of Sound Reproduction*. Durham: Duke University Press, 2003.

———. "Claude Shannon." In *Encyclopedia of New Media*, edited by Steve Jones, 406–7. Thousand Oaks, Calif.: Sage, 2003.

———. "The Death and Life of Digital Audio." *Interdisciplinary Science Review* 31, no. 4 (2006): 338–48.

———. "The Preservation Paradox." In *Sound Souvenirs: Audio Technologies, Memory and Cultural Practices*, edited by Karin Bijsterveld and José Van Dijck, 55–65. Amsterdam: Amsterdam University Press, 2009.

———. "What's Digital in Digital Music?" In *Digital Media: Transformations in Human Communication*, edited by Paul Messaris and Lee Humphreys, 95–109. New York: Peter Lang, 2006.

Sterne, Jonathan, and Joan Leach. "The Point of Social Construction and the Purpose of Social Critique." *Social Epistemology* 19, no. 2–3 (2005): 189–98.

Sterne, Jonathan, and Tara Rodgers. "The Poetics of Signal Processing." *Differences* 22, no. 2–3 (2011): 31–53.

Stevens, S. S. "A Scale for the Measurement of a Psychological Magnitude: Loudness." *Psychological Review* 43, no. 5 (1936): 405–16.

Stevens, S. S., and Hallowell Davis. *Hearing: Its Psychology and Physiology*. New York: John Wiley and Sons, 1938.

Stevens, S. S., and Fred Warshofsky. *Sound and Hearing*. New York: Time, 1965.

Stiegler, Bernard. *Technics and Time 1: The Fault of Epimethus*. Translated by Richard Beadsworth and Geroge Collins. Stanford: Stanford University Press, 1998.

"Storage Life/Aging of Space Systems and Materials at MP3." *Composites and Adhesives Newsletter*, 1 October 1996.

Suisman, David. *Selling Sounds: The Commercial Revolution in American Music*. Cambridge: Harvard University Press, 2009.

Sumner, James, and Graeme J. N. Gooday. "Introduction: Does Standardization Make Things Standard?" In *By Whose Standards? Standardization, Stability and Uniformity in the History of Information and Electrical Technologies*, edited by Ian Insker, James Sumner, and Graeme J. N. Gooday, 1–15. London: Continuum, 2008.

Swallwell, M. "The Remembering and the Forgetting of Early Digital Games: From Novelty to Detritus and Back Again." *Journal of Visual Culture* 6, no. 2 (2007): 255–73.

Swiss, Thomas, John M. Sloop, and Andrew Herman. *Mapping the Beat: Popular Music and Contemporary Theory*. Malden, Mass.: Blackwell Publishers, 1998.

Symes, Colin. *Setting the Record Straight: A Material History of Classical Music Recording*. Middletown, Conn.: Wesleyan University Press, 2004.

Szendy, Peter. *Listen: A History of Our Ears*. Translated by Charlotte Mandell. New York: Fordham University Press, 2008.

Tang, Jeffrey Donald. "Sound Decisions: Systems, Standards and Consumers in American Audio Technology, 1945–1975." PhD dissertation, University of Pennsylvania, 2004.

Taylor, Timothy. *Strange Sounds: Music, Technology and Culture.* New York: Routledge, 2001.

"Tender—Construction, Mining, Excavative and Highway Equipment." *Tenders Electronic Daily*, 1996, http://www.factiva.com.

Terranova, Tiziana. "Free Labor: Producing Culture for the Digital Economy." *Social Text* 18, no. 2 (2000): 33–58.

———. *Network Culture: Politics for the Information Age.* London: Pluto Press, 2004.

Théberge, Paul. *Any Sound You Can Imagine: Making Music/Consuming Technology.* Hanover, N.H.: Wesleyan University Press and University Press of New England, 1997.

Thompson, Emily. *The Soundscape of Modernity: Architectural Acoustics and the Culture of Listening in America 1900–1930.* Cambridge: MIT Press, 2002.

Tompkins, David. *How to Wreck a Nice Beach: The Vocoder from World War II to Hip Hop, the Machine Speaks.* Chicago: Stop Smiling Books, 2010.

Toop, David. "Replicant: On Dub." In *Audio Culture: Readings in Modern Music*, edited by Christoph Cox and Daniel Warner, 355–57. New York: Continuum, 2004.

Tsai, Chen-Gia. "Tone Language, Absolute Pitch, and Cultural Differences [Online Discussion Thread]." Auditory.org (2004), http://www.auditory.org.

Tsutsui, Kyoya, Hiroshi Suzuki, Osamu Shimoyoshi, Mito Sonohara, Kenzo Akagiri, and Robert M. Heddle. "ATRAC: Adaptive Transform Acoustic Coding for Minidisc." Paper presented at the 93rd Annual Audio Engineering Society, San Francisco, 1–4 October 1992.

Turing, Alan. "On Computable Numbers, with an Application to the Entscheidungsproblem." In *The Undecidable: Basic Papers on Undecidable Propositions, Unsolvable Problems and Computable Functions*, edited by Martin Davis, 116–54. New York: Raven Press, 1965.

Turino, Thomas. "Signs of Imagination, Identity and Experience: A Peircean Semiotics Theory for Music." *Ethnomusicology* 43, no. 2 (1999): 221–55.

Turner, Fred. "Burning Man at Google: A Cultural Infrastructure for New Media Production." *New Media and Society* 11, no. 1–2 (2009): 73–94.

———. *From Counterculture to Cyberculture: Stewart Brand, the Whole Earth Network, and the Rise of Digital Utopianism.* Chicago: University of Chicago Press, 2006.

Twain, Mark. "A Telephonic Conversation." http://www.mtwain.com.

"UK: Music Down the Telephone Live." *Media Monitor*, 8 July 1994, http://www.factiva.com.

Vaidhyanathan, Siva. *The Anarchist in the Library: How the Clash between Freedom and Control Is Hacking the Real World and Crashing the System.* New York: Basic Books, 2004.

———. *Copyrights and Copywrongs: The Rise of Intellectual Property and How It Threatens Creativity*. New York: NYU Press, 2001.

———. *The Googlization of Everything (and Why We Should Worry)*. Berkeley: University of California Press, 2011.

Van Tassel, Joan. "Olympia, Telos Take on RealAudio." *Broadcasting and Cable*, no. 4 (1997), http://www.factiva.com.

Verma, Neil. *Theater of the Mind: Imagination, Aesthetics, and American Radio Drama*. Chicago: University of Chicago Press, 2012.

Wakefield, Jane. "Millions Turn to Net for Pirate TV." *BBC News*, 27 November 2006.

Waksman, Steve. *Instruments of Desire: The Electric Guitar and the Shaping of Musical Experience*. Cambridge: Harvard University Press, 1999.

Wallace, Gregory K. "The JPEG Still Picture Compression Standard." *IEEE Transactions On Consumer Electronics* 38, no. 1 (1992): xviii–xxxxiv.

Wallis, Roger, and Krister Malm. *Big Sounds from Small Peoples: The Record Industry in Small Countries*. New York: Pendragon, 1984.

Walser, Robert. *Running with the Devil: Power, Gender, and Madness in Heavy Metal*. Hanover, N.H.: Wesleyan University Press and University Press of New England, 1993.

Walsh, E. Geoffrey. *Physiology of the Nervous System*. 2nd ed. London: Longmans, 1964.

Ward, Mark. "Scitech Ships Upgraded Multiplexer and Bandwidth Processor Cards." *Computer Weekly*, 27 May 1993, 20.

Wark, McKenzie. *Gamer Theory*. Cambridge: Harvard University Press, 2007.

Warner, Michael. *The Trouble with Normal: Sex Politics and the Ethics of Queer Life*. Cambridge: Harvard University Press, 1999.

Watt, Henry J. *The Psychology of Sound*. Cambridge: Cambridge University Press, 1917.

Weber, Max. *The Protestant Ethic and the Spirit of Capitalism*. New York: Charles Scribner's Sons, 1958.

Weber, Steven. *The Success of Open Source*. Cambridge: Harvard University Press, 2004.

Wegel, R. L., and C. E. Lane. "The Auditory Masking of One Pure Tone by Another and Its Probable Relation to the Dynamics of the Inner Ear." *Physical Review* 23, no. 2 (1924): 266–85.

Weidenaar, Reynold. *Magic Music from the Telharmonium*. Metuchen, N.J.: Scarecrow Press, 1995.

Weisbrod, Richard. "Audio Analgesia Revisited." *Anesthesia Progress* 16, no. 1 (1969): 8–14.

West, Joel. "The Economic Realities of Open Standards: Black, White and Grey." In *Standards and Public Policy*, edited by Shane Greenstein and Victor Stango, 87–122. New York: Cambridge University Press, 2007.

Wever, Ernest Glen. "The Nature of Acoustic Response: The Relation between Sound Frequency and Frequency of Impulses in the Auditory Nerve." *Journal of Experimental Psychology* 13, no. 5 (1930): 373–87.

———. "The Upper Limit of Hearing in the Cat." *Journal of Comparative Psychology* 10, no. 2 (1930): 221–33.

Wever, Ernest Glen, and Charles W. Bray. "Action Currents in the Auditory Nerve in Response to Acoustical Stimulation." *Proceedings of the National Academy of Science* 16 (1930): 344–50.

Whitehead, Alfred North. *Science and the Modern World*. Reissue ed. New York: Free Press, 1997.

Whitely, Sheila, ed. *Sexing the Groove: Popular Music and Gender*. New York: Routledge, 1997.

Wiener, Norbert. *Cybernetics, or Control and Communication in the Animal and the Machine*. Cambridge: MIT Press, 1948.

———. *Extrapolation, Interpolation, and Smoothing of Stationary Times Series*. Cambridge: MIT Press, 1949.

———. *The Human Use of Human Beings: Cybernetics and Society*. New York: Doubleday, 1954.

Williams, Raymond. *Culture and Society, 1780–1950*. New York: Columbia University Press, 1958.

———. *Marxism and Literature*, Marxist Introductions. Oxford: Oxford University Press, 1977.

———. *Television: Technology and Cultural Form*. Hanover, N.H.: Wesleyan University Press and University Press of New England, 1992.

Wilms, Herman. "In Memoriam: Eberhard Zwicker." *Journal of the Audio Engineering Society* 39, no. 3 (1991): 199–200.

Winner, Langdon. "Social Constructivism: Opening the Black Box and Finding It Empty." *Science as Culture* 3, no. 3 (1993): 427–52.

Wolpert, Rita S. "Attention to Key in a Nondirected Music Listening Task: Musicians vs. Nonmusicians." *Music Perception* 18, no. 2 (2000): 225–30.

"Writable Drive Sales Set to Surge." *Multimedia Week* 5, no. 33 (1996): 7.

Wundt, Wilhelm. *Principles of Physiological Psychology*. Translated by Edward Titchener. New York: Macmillan, 1910.

Yar, Majid. "The Rhetorics of Anti-Piracy Campaigns: Criminalization, Moral Pedagogy and Capitalist Property Relations in the Classroom." *New Media and Society* 10, no. 4 (2008): 605–23.

Zak, Albin J. *The Poetics of Rock: Cutting Tracks, Making Records*. Berkeley: University of California Press, 2001.

Zatorre, Robert J. "Absolute Pitch: A Model for Understanding the Influence of Genes and Development on Neural and Cognitive Function." *Nature Neuroscience* 6 (2003): 692–95.

Zwicker, E. *Psychoakustik*. Berlin: Springer, 1982.

———. "Subdivision of the Audible Frequency Range into Critical Bands." *Journal of the Acoustical Society of America* 33, no. 2 (1961): 248.

Zwicker, E., and Richard Feldtkeller. *The Ear as a Communication Receiver*. Translated by Hans Müsch, Søren Buus, and Mary Florentine. Woodbury, N.Y.: Acoustical Society of America, 1967/1999.

Zwicker, E., G. Flottorp, and S. S. Stevens. "Critical Band Width in Loudness Summation." *Journal of the Acoustical Society of America* 29 (1957): 548–57.

Zwislocki, J. J. "Masking: Experimental and Theoretical Aspects of Simultaneous, Forward, Backward, and Central Masking." In *Handbook of Perception*. Vol. 4, *Hearing*, edited by Edward C. Carterette and Morton P. Friedman, 283–336. New York: Academic, 1978.

INDEX

Page references in italics indicate illustrations.

.aac, 8. *See also* Advanced Audio Coding
acoustics, 68, 102, 113; architectural, 12–22; electro-, 30; origin point of, 63; physiological, 19, 33, 34. *See also* psychoacoustics
adaptive predictive coding, 107–9, 271n34
Adorno, Theodor, 9, 12, 96, 192–93, 268n8
Advanced Audio Coding, 30, 134, 227. *See also* .aac
affordances, 136, 189, 193, 198, 214, 223
aesthetics, 4–6, 24, 25, 33, 94, 167, 169, 235; acclimatization, 166; annoying, 157–62, 173; commodity, 239; of compression technology, 236; debt to telephony, 3; of format, 15; "good sound," 26, 143, 183–84; in listening tests, 148–49, 152–53, 157, 171–73, 180–81; of MP3, 160, 244; musical, 59, 165, 173, 176, 185, 237; of recorded sound, 174, 176; of television, 8. *See also* anaesthetic
algorithm: ASPEC, 146; avant-garde composition, 244; compression, 174, 175; hearing research for, 154; of MP3, 7, 199; MUSICAM, 146; prediction, 107
algorithmic culture, 3
American Bell Telephone. *See* AT&T
American Federation of Organizations for the Hard of Hearing, 55, 59
anaesthetic, 2, 149, 157, 279n16
Armstrong, Howard, 132–33
Arnold, Harold, 38, 41–42, 45
artificial intelligence, 242

ASPEC (Adaptive Spectral Perceptual Entropy Coding), 142, 143, 145, 146, 150, 175, 199, 277n27
Atal, Bishnu, 107–12, 114, 120, 124
AT&T, 142; hearing research and, 3, 19, 41–42, 44, 45, 50, 52, 53, 56, 92, 117; hearing tests and, 54–55, 58–60, 162; infrastructure of, 41, 43–44, 48, 62, 80, 85; monopoly of, 33, 43–44; oligopoly of, 42; surplus frequencies of, 50–51; telegraph system and, 81–82; *See also* Bell Labs; Bell System
Attali, Jacques, 122–27
Audio Engineering Society (AES), 161
audiometer, 35, 37–49, 60; use of, in hearing tests, 36, 55, 56, 58–59, 71
audition: concept of, 101; direct, 5; effect of expectation on, 179; faculty of, 56, 98; measurement of, 38; ontologies of, 240–45; in psychology, 34; pure, 171
Augé, Marc, 222–23

bandwidth, 4, 15, 16, 93, 99, 139; AT&T's maximization of, 33, 40–47, 76, 80, 106; management of, 17, 19, 129; MP3 use of, 1, 7, 224, 231, 234; reduced, 112–13, 117, 129, 153; signal definition vs., 88; surplus, 19; teletype and, 82; use of, by voice, 107, 113–14. *See also* critical bands
Barthes, Roland, 94, 103–4, 108, 270n30
BASF, 28, 211
BBC-Radio One, 209–10
Bech, Søren, 165

Békésy, George von, 35, 40, 67, 84, 101
Bell, Alexander Graham, 34, 36, 55, 60, 64, 70, 92
Bell Labs, 3, 7, 71; hearing research and, 34, 40–41, 54; hearing tests and, 54–55, 58–60, 169; information theory research of, 20, 33, 79–80, 87; psychoacoustic research of, 18, 29, 30, 49, 76, 80, 84, 93, 97–101, 109–14, 154, 162; synthesized speech research of, 106–10, 113. *See also* AT&T; Bell System
Bell System, 33, 38, 42; monopoly of, 41, 43; telegraph system of, 81–82. *See also* AT&T; Bell Labs
Beranek, Leo L., 121
Berland, Jody, 68, 69
Bijesterveld, Karin, 118
birds, 66
BitTorrent, 1, 207, 212, 213, 214, 223, 247
Bolter, J. David, 242–43
Bonello, Oscar, 115, 154, 161
Boring, Edwin G., 32, 34, 72–73, 258n1
Born, Georgina, 30, 96, 115–16, 164
Bosi, Marina, 95, 167–68, 178
Boulez, Pierre, 95, 115
Bourdieu, Pierre, 89, 152, 171–73, 181, 267n67, 289n58
brain, 38, 60, 72, 73, 75; ability of, to distinguish between tones, 98; of cat, 61, 65, 66, 67; media and, 77; in psychoacoustic research, 241; sexuality and, 168; signals of, 62; translation work of, 64
Brandenburg, Karlheinz, 112, 115; on marketing MP3, 198, 199, 205; on MPEG, 140; on MPEG listening tests, 154, 162, 168, 173, 174–76, 197; on perceptual coding, 153. *See also* Fraunhofer IIS
Bray, Charles W., 61–75, 78, 90

British Broadcasting Corporation, 209–10
Bull, Michael, 195, 236–37

camera silencia, 72. *See also* soundproof rooms
capital: fixed, 50; perceptual, 48–49, 51, 161, 181, 183
capitalism, 30, 188, 193, 212, 215–17; anti-, 210; corporate, 42–45, 51, 80; nineteenth-century, 91
cassette, 8, 12–13, 114, 195, 204; piracy, 28, 192, 209–12, 217
Castel, Louis-Bertrand, 68
cat, 61–64, 73, 74, 75, 77, 233, 26n21; model brains of, 72. *See also* kitten
cat microphone, 63
cat piano, 68–69
cat telephone, 61–62, 63, 70–73, 74, 78
Cavarero, Andrea, 243–44
CD, 4, 6, 52, 129, 140, 219, 228; aesthetics of, 12; consumer demand for, 1, 13–14, 141–42, 184–85, 201; durability of, 230; MPEG and, 48, 134, 160–61; as property, 220; -quality sound, 151, 154, 156, 161, 176, 181; standard for digital music, 139; unauthorized, 27, 187, 205, 212, 223, 225
CD burners, 27, 205, 207
CD players, 10, 141, 207, 208, 239
celestial jukebox, 219–20, 223
Chiariglione, Leonardo, 132
Chion, Michael, 5, 249n13, 250n16
Chun, Wendy, 196
circulation, 261; of audiometers, 36; of media, 50, 161, 188, 191–92, 204; of MP3s, 1, 28, 137, 182, 195, 196, 216, 221, 223–25, 228, 235; of sound and music, 161, 176, 185, 208
cochlea, 63, 67, 98, 100, 109, 110, 116, 168

cochlear implant, 54, 64, 244
code, 8, 11, 15, 78–79, 169–70; copyright, 211; gender and, 194; of MP3, 2, 3, 17, 25, 26, 64, 89, 162, 194, 196, 208, 235; of music, 127, 175–76, 197, 208; of Portable People Meter, 233; of speech, 107; sonic, 124, 125; telegraphic, 6, 81–84; user, 192. *See also* adaptive predictive coding; codec; Morse code; OCF; perceptual coding
codec, 24–25, 134, 183; of ASPEC, 142, 145, 150, 200; of ATAC, 143; of mobile phones, 232; of MP3, 27, 112, 145, 150; of MPEG-1, 142; of MUSICAM, 142, 145; OCF, 175; SB/ADPCM, 143; testing, 142–64, 169, 173, 176, 177–78, 179–82
collection: archival, 229–30; music, 29, 190, 206, 214, 225, 228–29, 231
commodity, 41; aesthetics, 239; economies, 213, 215, 224; MP3 as, 218–21; music as, 27, 189, 190–93, 197–98, 220–21
communication: cats and, 66, 70, 73–75, 78; disability and, 54–55, 244; economics of, 30, 48, 50, 131, 132–35, 139–43, 145–47; engineering of, 115, 117, 124, 182; history of, 4–7, 19, 29, 30, 89, 93–94, 240, 245, 250n20; industries, 125, 131–32; policy and politics of, 24, 130–43, 145–47, 201, 209, 220–21, 257n78; systems of, 88, 92, 124, 182, 242; technology of, 7, 9–10, 26, 33, 36, 48, 72, 89, 147, 194, 240, 243, 244; theory of, 20, 64, 74, 76, 77, 78, 82–87, 90, 92, 102–6, 245. See also *The Ear as a Communication Receiver*; *The Mathematical Theory of Communication*
compact disc. *See* CD
compression, 28, 112, 208, 230–32, 251n22; aesthetics of, 6, 236–37;
audio, 15, 117, 130, 131–32, 174–76, 196–97, 286n38; history of, 5–6, 15, 21, 93, 250n20; lossless/lossy, 2; of speech, 107, 113, 271n35. *See also* perceptual coding
computer, 16, 26, 27, 88, 205, 206, 207, 208; as acoustic research tool, 109, 111, 114–18; audio and, 21, 22; as audio media, 95, 107, 182, 194, 195, 231; Boolean algebra and, 78; industry, 132, 198, 200, 203, 204; model of cochlea, 109, 110; MPEG standard for, 135, 136, 137, 142, 145; as multimedia device, 139; music, 95, 113, 116; perceptual capital and, 48
consumer electronics, 28, 208; industry, 49, 138, 142, 143, 198; market, 140, 141; MPEG standard for, 135, 137, 142–43; online music and, 204, 218, 238, 239; piracy and, 188, 211. *See also* CD players; CD burners; computer; home stereo; iPod; mobile phone; MP3 player; phonograph; smartphone; Walkman
container technology, 193–96
copy protection, 27, 187, 197, 200
copyright, 187, 191, 196–98, 209–11, 216–17, 221, 224, 225. *See also* intellectual property
corporations, 11, 24, 40–45, 50, 53, 55, 88, 89, 138, 212
critical bands, 153, 240; research into, 99–101, 110, 111, 114, 117, 232; theory of, 21, 22, 94–96, 102, 106, 111, 112, 114, 118. *See also* masking
cybernetics, 19, 64, 74–78, 83, 85; language, 122, 124; noise and, 88, 108

Danger Mouse, DJ, 216–17
Danziger, Kurt, 36, 57
DAT (digital audio tape), 139, 156, 160, 197

data, 4, 6, 114, 196, 206, 233; aggregate of, 35, 57; compression of, 21, 112, 128, 130, 131–32, 140, 145, 175, 237; management of, 52; materiality of, 194, 251n23; networks of, 16; psychoacoustic, 149, 154; redundant, 2, 20; storage of, 13, 15, 139, 280n22; system of, 109; test of, 149–51; transmission of, 14, 15
database, 29, 206–7
Davis, Audrey B., 34
Davis, Hallowell, 35, 40, 63
dead bodies, experiments on, 56, 67, 68
Deaf: hearing tests for, 36, 55; telephony and, 54, 244. See also deafness; disability; hard-of-hearing; hearing loss
deafness: testing for, 36, 55, 58; treatment for, 64; See also Deaf; disability; hard-of-hearing; hearing loss
death, 62, 63, 68
decerebration, 65–69, 73
decomposition, 128, 143
decompositionism, 122, 126–27
definition, 4–5, 46, 47, 48, 53, 113, 148, 154, 160, 230, 231, 237; high, 151, 185, 228, 238, 239; signal, 88; surplus, 46–51. See also perceptual capital
Digital Compact Cassette, 134
digital rights management (DRM), 192, 220, 229
disability, 54, 244. See also Deaf; deafness; hard-of-hearing
distortion, 6
Doelle, Leslie, 120–23
dog, 64, 66, 233, 255n64, 265n21
Dolby, 272n54, 273n54
Dudley, Homer, 107, 126
DVD (Digital Video Disk) 3, 6, 129, 145, 230, 233
DVD-A, 185

The Ear as a Communication Receiver, 102–6. See also Feldtkeller, Richard; Zwicker, Eberhard
Edison, Thomas, 34, 43, 92, 113
Edwards, Paul, 30, 76
Eisenberg, Evan, 190, 204, 214
elephant, 67
Elias, Norbert, 160
Ellis, Alexander, 97–98
encoder, MP3, 1–2, 17, 20, 21, 64, 183, 195
ergonomics, 19, 53, 55
Erlmann, Veit, 101
Eureka, 130–31, 140
everyday life, 99, 221, 233, 237, 241
experience: aesthetic, 6, 8, 15; of format, 7, 15, 162, 163, 194; of hearing, 58–59, 102, 104; interior, 35–36, 222; listener, 4, 177; previous, 165–66, 168; psychological, 57; sonic, 5, 153, 181; subjective, 104, 150, 157

Fano, Robert, 21, 64
Fast Fourier Transform, 7
FCC (Federal Communications Commission) 133, 140
Fechner, Gustav Theodor, 102–3, 160
Federal Communications Commission (FCC) 133, 140
Feldtkeller, Richard, 102–6, 108, 110, 117, 121
fidelity. See bandwidth; definition; verisimilitude
file-sharing, 1, 27–29, 125, 185, 187–88, 191, 198, 202–24, 229, 247n1, 289n52. See also Gnutella; music distribution: unauthorized file-sharing/copying; Napster; piracy
film sound, 10, 14, 35, 191
Fletcher, Harvey, 29, 34, 35, 38, 40–41, 55, 56, 58–60, 71, 81, 84, 98, 154; masking research of, 99–100, 102,

108, 110, 111, 114, 117. See also *Speech and Hearing*
Fourier Transform, 7
Fowler, Edmund, 38, 54
Fraunhofer IIS, 23, 26, 116, 204, 207; AAC, 134; ASPEC, 143, 145, 150; marketing of MP3 by, 27, 186, 198–202, 205–6, 208; MPEG and, 140; OCF (Optimum Coding in the Frequency Domain) and, 153; patents of, 208; perceptual coding and, 93, 112, 115, 131, 153; "thank you Fraunhofer," 202, 208; Suzanne Vega's visit to, 178–79. *See also* Brandenburg, Karlheinz; Grill, Bernhard; Popp, Harald
frequencies, 42, 105; surplus, 45, 46, 50, 51
Fukuyama, Francis, 4
Fuller, Matthew, 22–23, 144, 277n25

Gardner, Wallace, 118–20
gender, 165, 180, 194, 215, 224, 238
Gertner, Jon, 233
Gibson-Graham, J. K., 215–16
gift economy, 212–13, 215, 218, 220, 223–24
Gitelman, Lisa, 11, 194, 253n43
Gnutella, 1, 185, 207, 212, 227
Grill, Bernhard, 140, 146, 161, 162, 167, 171, 173

Hall, Joseph, 108–12, 114, 115, 116, 120, 124
Haraway, Donna, 77, 181, 266n41
hard-of-hearing, 30, 54–56, 59, 64, 102, 244. *See also* Deaf; deafness; disability; hearing loss
hardware, 10, 137, 145, 192, 196, 197, 221, 228; industry, 219; market for, 215; perceptual coding as problem of, 115; software vs., 139; technological shift in, 136, 240

Harley, Robert, 179–81
Hartley, R. V. L., 80–85, 87
header, 196–98
hearing: gaps in, 2, 19, 194; mediality of, 9, 99; models of, 21, 33, 101, 109, 243; normal/abnormal, 54–56, 58–59; threshold of human, 34, 45, 51–52, 97, 105, 157
hearing, telephone theory of, 60, 72–73
hearing aids, 54
hearing as such, 58, 59, 60, 72
hearing-in-itself, 19, 54, 90
hearing in its natural state, 19
hearing loss, 58, 59, 167, 168. *See also* Deaf; deafness; hard-of-hearing
hearing research, 32, 34–35, 61–91, 94, 241–45; telephony and, 2, 3, 7, 17, 19–22, 30, 33, 44–45, 58–59, 64, 70–72, 73, 109, 113, 249n11. *See also* hearing tests; masking: research into
hearing tests, 54–55, 58–60, 162; age and, 168; gender and, 168; race/class and, 169
Heidegger, Martin, 193–94, 197
Hellman, Rhona, 109, 110, 153
Helmholtz, Hermann von, 7, 32, 34, 56, 61, 101
home stereo, 231, 238, 239, 240
Huffman, David, 21, 64
Huffman coding, 20–21, 64

IEA (Institute for Economic Affairs), 209–10
IEC (International Electrochemical Commission) 136, 137
IEEE (Institute of Electrical and Electronics Engineers), 93, 139
Immink, Kees, 12–13
infrastructure, 2–3, 11, 15–16, 126, 161; of AT&T, 41–44, 50, 56; of Bell Sys-

infrastructure (*continued*)
    tem, 33; of music reproduction, 188, 204, 216, 223
intellectual property, 125, 187–88, 191–92, 225, 229
intelligibility, 42, 110, 121; threshold of, 84, 99
International Electrochemical Commission (IEC), 136, 137
International Organization for Standardization (ISO), 16, 23, 136–37, 143–45, 146, 147, 207, 257n75
internet: as end-to-end network, 16; file-sharing on, 27, 28, 135, 203, 206, 209; free labor on, 51; as gift economy, 212; infrastructure of, 16; MP3 on, 1, 2, 26, 52, 134, 205, 207, 208, 215, 228; musical production and distribution and, 29, 176, 204, 216–17
Internet Underground Music Archive (IUMA), 202
iPod, 10, 214, 236–37, 253n38. *See also* MP3 player
IRCAM (Institut de Recherche et Coordination Acoustique/Musique), 95, 115–16, 164
IRT (Institut für Rundfunktechnik), 142, 150
ISDN (Integriertes Sprach- und Datennetz; Integrated Services Digital Network or Isolated Subscriber Digital Network), 15, 48, 140, 143, 200, 201, 277n21
ISO (International Organization for Standardization), 136–37, 143, 145, 146, 147; standard of, 16, 207
iTunes, 14, 185, 212, 214, 231, 238

Jastrow, Joseph, 36–37
Johns, Adrian, 187

Johnston, JJ, 235; on listening tests, 167, 169–71, 173; perceptual coding and, 29, 92–93, 112, 114, 116, 154, 162
judgment: aesthetic, 149; of test subjects, 59, 152, 153, 157, 160, 164

Keightley, Keir, 238
Keil, Charlie, 190–91, 284n12, 284n13
Kelty, Chris, 213
Kircher, Athanasius, 68–69
kitten, 77–78. *See also* cat
Kittler, Friedrich, 52, 131
Krasner, Michael A., 110–13, 115, 124, 129, 153, 161, 169
Krauss, Rosalind, 235–36

Lane, C. E., 59–60, 71, 97–99, 105, 110
Larkin, Brian, 188
"layer 1," 145, 196
"layer 2," 27, 142, 145, 146, 198, 199, 202
"layer 3," 23, 27, 142, 145, 146, 175, 196, 198, 199, 200, 201, 202
Licklider, J. C. R., 118–19
listener, expert, 163–66, 169, 180–82
listening: distracted, 5; gendered, 180, 238; modes of, 10–11, 225, 236; to music, 186, 193, 217, 221, 225, 236–37, 239; normalism and, 55; positionality of, 90; praxaeology of, 25, 28, 165, 221, 237; process of, 104; public, 25, 210; subjectivity of, 94, 242; subjects, 3, 25, 26, 105, 124, 149, 155, 160, 163, 165–68, 169, 182, 193, 241; technology of, 233; ubiquity of, 225
listening tests, 24–26, 145, 147, 148–83, 189; ABX, 177; aesthetics of, 152–53, 157–59, 166, 171–76, 181; gender and, 165; subjectivity of, 180–83;

training for, 161, 162; as worst-case scenario, 151, 177, 182
living, experiments on, 35, 55, 56–57, 68
long-playing record, 12, 14; consumer demand for, 142, 185, 192, 238. *See also* LP; records
LP, 12, 14; consumer demand for, 142, 185, 192, 238. *See also* records
L3Dec, 201
L3Enc, 201, 205, 213. *See also* "thank you Fraunhofer"

Marquand, Allan, 79
marketing, 4; of digital audio technologies, 129; of MP3, 27, 201–2, 231; online, 203
markets, 24, 27, 28, 43, 45, 81, 89, 129, 132; anti-, 188; broadband, 205; capitalist, 212, 215–16; consumer electronic, 140; control of, 130, 131, 133, 135, 136, 138, 147, 211; free, 210; monopoly of, 42; of music, 184–85, 217–18; nature as, 77
Marshall, John, 87
Marx, Karl, 50, 90–91, 124, 218, 260n32, 261n44
mashups, 29, 126, 216, 229–30, 235
masking, 18, 21–22, 114, 233; behavior of ear, 52, 59–60, 94, 118; domestication of noise and, 117–24, 232; environmental, 121–23; MP3 and, 112; perceptual coding and, 115–17, 153–54, 165; research into, 94, 97–102, 109–11, 153–54; theories of, 95, 96–97; therapeutic, 118–19; threshold of, 105–6, 110. *See also* critical bands
The Mathematical Theory of Communication, 20, 64, 76, 79–80, 85–87, 92, 242. *See also* Shannon, Claude; Weaver, Warren

Mauss, Marcel, 212
Mayer, Alfred M., 97–98
McLuhan, Marshall, 9
MDCT (Modified Discrete Cosine Transform), 277n27
media, 11; digital, 2, 6, 9, 75; electronic, 5, 10; globalization of, 24, 27; history of, 2, 17; industries of, 27, 50, 85, 129, 138, 140, 141, 188, 192, 225; mass, 74; mechanical, 10; new, 2, 5, 29, 30, 33; regulation of, 136; relations among, 14–17, 141–44, 145–46, 148, 182, 188, 203, 211, 218–19; sound, 22, 33, 35, 40, 72, 94, 95; systems of, 128; technologies of, 6
mediality, 9–11, 14, 16, 243, 251n29, 252n32; development of CD and, 14, 16; of music and sound, 183
media theory, 7, 10, 131
mediation, 9
Merton, Robert K., 92, 93
Merzbach, Uta, 34
metadata, 143, 229
mice, 66
Miller, Mitch, 237–38
Mills, Mara, 30, 49, 54, 71, 108, 244, 261n49
mind's ear, 2, 20, 63, 99, 116, 122, 157, 179
mobile phone, 10, 11, 48, 54, 106, 132, 231, 232. *See also* smartphone
modeling: of cochlea, 67, 109–10, 116; devices of, 111, 116; of ear, 52, 232; of human subject, 242–43; of speech, 106–7, 232
Modified Discrete Cosine Transform (MDCT), 232, 277n27
monetization, 19, 52, 215–16
monkeys, 64
Morgan, J. P., 42–44
Morse code, 6, 81–83

Moving Picture Experts Group. *See* MPEG

MPEG, 23, 48, 95, 131–32, 148–83; audio, 26, 129, 130–31, 196, 197, 199, 200, 235; audio group, 138–40; competition in, 142–43; evolution of, 136–37; layers of, 23, 27, 145–46, 186, 196, 198–202; listening tests and, 24–25, 148–50, 153, 154–60, 164, 171–76, 179–80, 183; standard, 23, 27, 30, 134–35, 146–47, 186, 195–98; weighting scheme, 143–45

MP3 player, 10, 11, 29, 140, 175, 182, 195, 201, 203, 207, 215, 231, 239. *See also* iPod

Mumford, Lewis, 53, 194

music, 3, 7, 9; Adorno on reproduction of, 12, 192–93; aesthetics of, 152, 157, 160, 185–86; affordance of, 189, 193; Attali on political economy of, 122–25; avant-garde, 95, 115–16, 234, 244; collection of, 29, 214, 229–31; as commodity, 184, 190–92, 197–98, 213–21, 224; computer, 95, 116, 244; consumption of, 28; copyright of, 196–97, 217, 220–21, 224; effect of format on, 12, 13, 14, 127; gendered, 238; as human faculty, 68; in listening tests, 59, 150, 153, 157, 162, 169–76, 178–80; masking with, 118–22; materiality of, 31, 185–86, 214; on MP3, 25, 26, 29, 182–83, 196–98, 235–39; perceptual coding of, 94, 111–18, 153–54, 164–65; social organization of, 188–89, 204, 217, 221, 224; sound reproduction technologies' effect on, 125–27, 161, 167, 178, 181–83, 190, 235–39; speech privileged over, 244; technology of, 72, 189; ubiquity of, 10–11, 225, 236–37, 245

MUSICAM (Masking Pattern University Sub-band Integrated Coding And Multiplexing), 142, 145, 146, 150, 277n27

music distribution, 29, 228; celestial jukebox, 219–20, 223; online, 1, 185, 199–200, 202–7, 212–15; piracy, 28, 125, 186–88, 203–11, 216–20, 223–24; right to, 221; unauthorized file-sharing/copying, 27–28, 125, 187–89, 192, 203, 205, 208, 216–24, 229. *See also* Gnutella; iTunes; Napster; piracy

music industry, 185, 187, 188, 203–4, 211, 217

Musmann, Hans-Georg, 139–40, 142

Napster, 27, 89–90, 203, 206–7, 213, 214, 288n52, 289n52

nature, 77

New York League for the Hard of Hearing, 54, 56, 59

noise, 17; abatement of, 40, 108, 244; in adaptive predictive coding, 108–9, 112; Attali on, 124–26, 127; as chaos, 118, 124; cultural status of, 95, 97, 108; domestication of, 18, 21, 22, 94, 232; as extraneous disturbance, 108; as information, 87–88; levels of, 99; masking, 109–11, 117–19, 233; perceptual coding and, 21, 94; sexuality and, 168; therapeutic use of, 118–19

normalism, 55, 56, 57–59. *See also* universalism

Nyquist, Harry, 80–83, 85, 87

objectivism, 180–81, 267n67

OCF (Optimum Coding in the Frequency Domain), 153–54, 175

.ogg, 8, 227

Ogg Vorbis, 30

Ohga, Norio, 12–13
Operation Acoustic Kitty, 75

particularism, 149, 169; in listening tests, 177–83
patent, 135, 147, 191, 258n81; of FM radio, 132–33; of MP3, 26, 27, 112, 134, 186, 188, 208; perceptual coding and, 93; of Portable People Meter, 233
path dependency, 199–208, 224
peer-to-peer, 26, 28, 187, 209
perceptual capital, 48–49, 51, 161, 181, 183
perceptual coding, 2, 17, 18, 21–22, 25, 29, 33, 92–127, 245; with Huffman coding, 20, 64; industrial practice of, 128–31, 133–34; masking and, 232, 234; MPEG standards tests and, 148–68; MP3 and, 52, 234; regulation of, by MPEG, 23, 24; sound quality of, 177–79; ubiquity of, 53. *See also* perceptual technics
perceptual technics, 18, 19, 21, 22, 30, 32–60, 88, 112, 182, 183, 232, 243; decompositionism and, 126
Perlman, Marc, 177, 179
Peters, John Durham, 20, 33, 73
Philips, 12–14, 28, 134, 139, 204, 211; layer 2 and, 199; MUSICAM and, 142, 145, 150
phonograph, 3, 10, 11, 20, 62, 113, 238; industry, 21; record, 35. *See also* LP; record
Pinch, Trevor, 150, 163, 257
piracy, 28, 125, 186–88, 203–11, 216–20, 223–24
Pirate Bay, 207, 214
pirate radio, 28, 209–210
Plomp, Reinier, 241–43
plurality, auditory ontology of, 241–44
Popp, Harald: on marketing MP3, 198–99, 201, 207; on perceptual coding, 115
Portable People Meter (PPM), 233–34
Psychoacoustics, 153
psychoacoustics, 17–19, 29, 30, 32, 34, 51; aesthetic decisions in, 157; audiometer use in, 36, 38; critique of research on, 241–42; cybernetics and, 75–76, 78; *The Ear as a Communication Receiver*, 102–6; information theory and, 21, 75, 88, 90, 143, 240, 243–44; IRCAM and, 116; listening subject of, 58, 241–42; masking and, 118–21, 153–54; perceptual coding and, 94, 97, 103, 118, 153; telephone research and, 20, 33, 40, 49, 71, 77, 112
psychology, 35, 57, 72, 74, 87, 90, 159; auditory, 19, 34, 56; applied, 37, 52, 53, 55–56, 58; of hearing, 36, 38, 75; influence of, on psychoacoustics, 71

Quetelet, L. A., 57, 59, 60

race, 58, 165, 169
radio: digital, 129–31, 139, 140, 198; FM, 132–33, 141; pirate, 28, 209–10, 219
ready-to-hand, 193, 195, 221, 224, 225, 286n29
RealAudio, 200, 228
realism, 174, 250n16
recomposition, 128, 143, 224
record (phonograph), 4, 35, 195, 214, 218, 219, 224, 230; 78 rpm, 14; LP (33⅓ rpm, vinyl), 12, 14, 142, 184, 194, 199, 238; prestige, 190, 204, 221
recording industry, 1, 28, 29, 138, 184–85, 188, 190, 191–92, 215, 224, 225; response of, to unauthorized file-sharing, 203–5, 207–11, 216–17, 219–21. *See also* music industry

INDEX 339

reproducibility, 36, 162
RIAA (Recording Industry Association of America), 11–12; anti-piracy efforts of, 203, 205, 207, 214
rights, 184; of artists, 206, 224; of circulation, 191; fair-use, 220; intellectual property, 214; of listener, 221; "natural," 187; privacy, 220; sampling, 216; song as "bundle of," 191; of user, 224. See also digital rights management
ringtones, 231

sampling: of music, 125, 216–17; rate of, 13, 196
Schmundt, Hilmar, 175–76
Schroeder, Manfred, 95, 107–15, 120, 124
Seashore, Carl, 37–40
Seitzer, Dieter, 112, 113, 122, 124
sexuality, 168–69
Shannon, Claude, 76; information theory and, 20–21, 64, 83–84, 108, 241–43; *The Mathematical Theory of Communication*, 78–80, 82–88, 92
signal, 19, 79, 117, 124; adaptive predictive coding and, 107, 108–11, bandwidth and, 45–46, 80, 106; distortion and, 81; electrical, 38, 62; electromagnetic, 7; flow of, diagrammed in listening tests, 155–60, 163; mimetic, 63; in Portable People Meter, 233; redundancy of, 20, 232; system or, 167; of telephony, 45, 82–88, 114
signal processing, 4, 7, 8, 24, 79, 95, 110, 113–14, 115, 117, 127, 143, 144, 160, 232; of MP3, 1–2, 16, 26, 30, 31, 145, 231; telephony and, 3, 19, 20, 38, 45–48, 76, 80–82, 84–88
Simmel, Georg, 222
Small, Christopher, 190

smartphone, 10, 140, 231. See also mobile phone
Smythe, Dallas, 50–51
Sofia, Zoe, 194–95
software, 6–7, 11, 26, 30, 135, 136, 139, 196; copy protection of, 192, 197, 200, 221; free, 212–14; L3Enc, 186, 202; market for, 215–16, 218, 219; Napster, 206–7; Traktor, 194. See also "thank you Fraunhofer"
Sony Corporation, 12, 134, 139, 142, 208, 211
soundproof rooms, 58, 61, 63, 70–72, 99, 115
Speech and Hearing, 29, 40, 41, 60
squirrel, 65, 219, 264
standard object, 22–23, 134, 144, 146
standards, 131–36; open source, 26, 30, 213, 227; proprietary, 26, 30, 133, 134, 207. See also MPEG: standard
Stevens, S. S., 95, 103, 121
Szendy, Peter, 221

taste, 125, 150, 151, 158–60, 170–73, 176
telegraph, 6, 9, 20, 44, 45, 62, 108
telegraphy, 33, 43, 62, 80–88
telephony, 3, 10, 17, 19–20, 38–39, 46–48, 53, 90, 232; Bell System's monopoly in, 43–44; Deaf influence on, 54–55; ears and, 60, 62–63, 76, 109; hearing research and, 2, 44–45, 58–59, 64, 70–72, 73, 109, 249n11; information theory and, 79–80, 84–85, 86 (fig. 17), 249n11; national medium of, 44; perceptual coding and, 93, 112–13, 114, 117; psychoacoustics and, 40–42, 71–72, 76, 77; sound recording and, 113; speech research and, 106, 110, 113
Terranova, Tiziana, 51, 217
"thank you Fraunhofer," 202, 208

Thibault, Louis, 151, 162, 166, 167
Thomson, 23, 26, 143, 150, 175, 198
"Tom's Diner," 174–76, 278
transectorial innovation, 138, 203

universalism, 57, 149; in listening tests, 151–53, 164–69, 179–82

Vail, Theodore N., 43–44
VCD (Video Compact Disc) 129, 134, 140, 143
Vega, Suzanne, 170, 171, 173, 174–76, 178–79
verisimilitude, 4–6, 31
Video Compact Disc (VCD), 129, 134, 140, 143
vocoder, 106–7, 109, 113, 117, 126, 232
voice: as clinical tool, 35, 71; recorded, 174–176, telephony and, 41, 48, 107–8, 113, 154; Zephyr, 200. *See also* vocoder

Walkman, 10, 11, 140
.wav, 1, 4, 8, 235, 236
Weaver, Warren, 20, 79, 83, 87–88. See also *The Mathematical Theory of Communication*
Weber, Ernst Heinrich, 102–3
Weber, Max, 69
Wegel, R. L., 34, 38, 40, 59, 60, 71, 72, 97–99, 105, 110
Western Electric, 14, 38, 40, 43, 56
Western Union, 34, 44, 80, 81, 259n24
Wever, Ernest Glen, 61–75, 90, 97
Wiener, Norbert, 74–78, 108
Wundt, Wilhelm, 37, 56–57

Zwicker, Eberhard, 99, 102–6, 108, 109, 110, 117, 212, 153. See also *The Ear as a Communication Receiver*; *Psychoacoustics*

JONATHAN STERNE TEACHES IN THE DEPARTMENT OF ART HISTORY AND COMMUNICATION STUDIES AT MCGILL UNIVERSITY.

Library of Congress Cataloging-in-Publication Data
Sterne, Jonathan, 1970–
MP3 : the meaning of a format / Jonathan Sterne.
p. cm.
Includes bibliographical references and index.
ISBN 978-0-8223-5283-9 (cloth : alk. paper)
ISBN 978-0-8223-5287-7 (pbk. : alk. paper)
1. MP3 (Audio coding standard) 2. Digital media. I. Title.
ML74.4.M6S74 2012
621.382—dc23
2011053340